D1109123

WIND LOADS
ON
STRUCTURES

Proceedings of International Symposium

EXPERIMENTAL DETERMINATION OF WIND LOADS ON CIVIL ENGINEERING STRUCTURES

December 5-7, 1990, New Delhi

DEPARTMENT OF CIVIL ENGINEERING
UNIVERSITY OF ROORKEE
ROORKEE, INDIA

OXFORD & IBH PUBLISHING CO. PVT. LTD.
New Delhi Bombay Calcutta

ORGANISED BY

Department of Civil Engineering
University of Roorkee
Roorkee, U.P. India

CO-SPONSORED BY

University Grants Commission
Unitech Limited
Indian Institute of Technology, Delhi
Nuclear Power Corporation
National Thermal Power Corporation
Structural Engineering Research Centre, Ghaziabad
Bharat Heavy Electricals Limited
Central Building Research Institute
Rail India Technical and Economic Services
Oil & Natural Gas Commission

ISBN 81-204-0562-5

Published by Mohan Primlani for Oxford & IBH Publishing Co. Pvt. Ltd., 66 Janpath, New Delhi 110 001, printed at Rekha Printers A 102/1 Okhla Industrial Area, Phase II, New Delhi

1-D0-6

PREFACE

Several parts of the globe are regularly affected by moderate to severe storms causing considerable direct as well as indirect damage, including loss of life. A study of the pattern of wind disasters that have struck during the current century, show that major damages due to storms have occurred in the Asia-Pacific region. The primary reason for this recurrent distress is the lack of resources, inadequate knowhow, and, lack of application of good engineering practices.

During the last decade or so, there has been greater awareness towards the need for wind engineering research in order that the above situation can be alleviated. The state-of-the-art in many of the developed countries is a notch better than in the others. The Department of Civil Engineering at the University of Roorkee, amongst several other centres of research, has made efforts towards the development of facilities for wind engineering research and I am happy to say that a good deal of very useful work has already been produced. The symposium on Experimental Determination of Wind Loads on Civil Engineering Structures has been organised by the Department with two primary objectives. The first objective is to familiarise interested engineers and scientists in India with the state-of-the-art on the subject by inviting well-known experts, from outside India and within, to contribute papers on relevant themes. The second objective is to share the experience being gained on the subject through the efforts being made at various teaching and research institutions as well as by engineers in industry in India. It is my earnest hope that this volume containing papers contributed to the symposium will serve to achieve the above objectives.

D.N. TRIKHA

Head, Department of Civil Engg.
University of Roorkee
Roorkee, India

December, 1990

CONTENTS

TECHNICAL NOTES

ORGANISING COMMITTEE

Patron	Dr. H.C. Visvesvaraya Vice Chancellor
Chairman	Dr. D.N. Trikha Head, Civil Engineering Deptt.
Co-Chairman	Dr. Prem Krishna Co-ordinator, Centre of Wind Engg.
Members	Dr. P.K. Pande Dr. R.P. Mathur Dr. S.K. Kaushik Dr. G.L. Asawa Dr. P.D. Porey
Secretary	Dr. P.N. Godbole
Joint Secretary	Dr. A.K. Ahuja

SUB-COMMITTEES

TECHNICAL

Dr. P.K. Pande (Convenor)
Dr. P.N. Godbole
Dr. Krishen Kumar

REGISTRATION, RECEPTION AND TRANSPORT

Dr. S.K. Kaushik (Convenor)
Mr. V.K. Gupta
Mr. R.P. Gupta
Mr. M. Yahyai
Mr. Naresh Kumar

INVITATION

Dr. P.D. Porey (Convenor)
Dr. A.K. Ahuja

ACCOMMODATION

Dr. R.P. Mathur (Convenor)
Prof. M.K. Mittal
Mr. Ajai Gairola

SESSION MANAGEMENT

Dr. P.N. Godbole (Convenor)
Dr. A.K. Ahuja
Mr. Karisidappa
Mr. J. Noorjaei

PUBLICATION

Dr. G.L. Asawa (Convenor)
Dr. P.D. Porey

THEME PAPERS

ATMOSPHERIC BOUNDARY LAYER MODELLING IN WIND TUNNELS

Jack. E. Cermak

President
Colorado State University
Fort Collins
Colorado 80524, USA

SYNOPSIS

Proper scaling of the atmospheric boundary layer (ABL) is required for experimental determination of wind effects on structures. Basic similarity criteria are presented for modelling ABLs with and without thermal stratification. Features of boundary-layer wind tunnels (BLWTs) necessary to satisfy the similarity criteria and boundary conditions are discussed. Requirements for scaling the ABL for studies of high-rise and low-rise buildings are different. Modelling of the entire ABL is generally required for experimental studies of wind effects on high-rise buildings and structures. Low-rise buildings are usually subjected to flow in the lowest 100 m of the atmosphere, the atmospheric surface layer (ASL). Therefore, thickness of the modelled ASL should be magnified with the objective of increasing the size of low-rise building models used for experimental studies. This is accomplished by a technique that preserves the property of constant shear stress throughout the layer.

INTRODUCTION

Efforts to study wind effects in the laboratory were reported as early as 1759 by Smeaton [1]. His expression of a need for wind simulation was as follows: "In trying experiments on windmill-sails, the wind itself is too uncertain to answer the purpose; we must therefore have recourse to an artificial wind. This may be done in two ways; either by causing the air to move against the machine, or the machine to move against the air." Smeaton chose the latter approach and attached the sails to a long horizontal support which he rotated in his laboratory. Studies using the former approach were not reported until the 1890's--Kernot [2], Irminger [3], and Tsiolkovsky [4]. During the next decade significant wind-tunnel development was reported by Stanton [5,6], Riabouchinsky [7], Prandtl [8], and Eiffel [9]. In the period 1920-1950 efforts were made to reproduce various features of atmospheric motion in wind tunnels. Abe [10] studied lee-wave patterns downstream from a model of Mt. Fuji in laminar flow stably stratified by dry ice whereas Prandtl and

Rejchardt [11] were concerned with the effects of thermal stratification on transition of laminar to turbulent flow. Sherlock and Stalker [12] initiated studies of plumes emitted from smoke stacks in uniform flow. Studies by Nøkkentved [13] and Bailey and Vincent [14] revealed that pressure distributions on small-scale buildings are sensitive to variation of air speed with height. This significant finding, elaborated on by Jensen [15] demonstrated a need for modelling of atmospheric boundary layers (ABLs) -- natural wind up to about 500 m above ground.

The earliest efforts specifically devoted to physical modelling of the ABL in wind tunnels were by Rouse [16], Strom [17], and Cermak and Koloseus [18]. Studies of momentum and heat transfer over an 8.84-m long heated test-section floor by Cermak and Spengos [19] demonstrated that the boundary layer did not develop to a state of equilibrium with the boundary conditions. This finding motivated design and construction of a "long-test-section," boundary-layer wind tunnel (BLWT) which has been reported by Cermak [20], Plate and Cermak [21], and Cermak [22]. Flow data measured in this "meteorological wind tunnel" (MWT) and its description are referred to subsequently.

Scale modelling of the ABL is addressed by first discussing similarity criteria obtained by scaling the conservation equations for mass, momentum and energy. Requirements on initial and boundary conditions to model flows for experimental studies of wind effects on high-rise and low-rise buildings are discussed. The former requires modelling of the entire ABL while the later requires modelling of thick atmospheric surface layers (ASLs). A new technique for achievement of thick ASLs has been developed by Cermak and Cochran [23]. Flow characteristics realized by this technique are compared with field data reported by Chok [24] for the Texas Tech University research-building site.

Many types of experimental studies are made possible by proper modelling of the ABL and the ASL. This capability provides an indispensable source of wind-effect data for design purposes as well as research investigations.

THE ATMOSPHERIC BOUNDARY LAYER

Most of the time, atmospheric motion up to about 1,000 m (3,300 ft) above the surface is of the boundary-layer type and is usually designated as the planetary boundary layer. Separation in the lee of hills or mountains, severe thunderstorms, downbursts, hurricane eyewalls, and tornadoes are responsible for the greatest deviations from boundary-layer structure. A detailed description of flow characteristics in the planetary boundary layers which can account for all the perturbing effects of nonuniform and unsteady surface-flow conditions, topography and thermal stratification has not been possible. However, descriptions for idealized conditions have been formulated and serve as frames of reference. Hanna [25] reviewed fourteen different analytical and numerical models and concluded that the models developed by Lettau [26] and Blackadar [27] represent wind observations fairly well. A typical distribution of mean wind speed with height for a rough surface, as determined from the model of Lettau [26], is shown in Fig. 1. For this model the atmosphere is taken to be steady, horizontally homogeneous, dry, and adiabatic, with no vertical motion, invariant variances of velocity fluctuations, and a negligible effect of the turbulent energy dissipation rate on temperature.

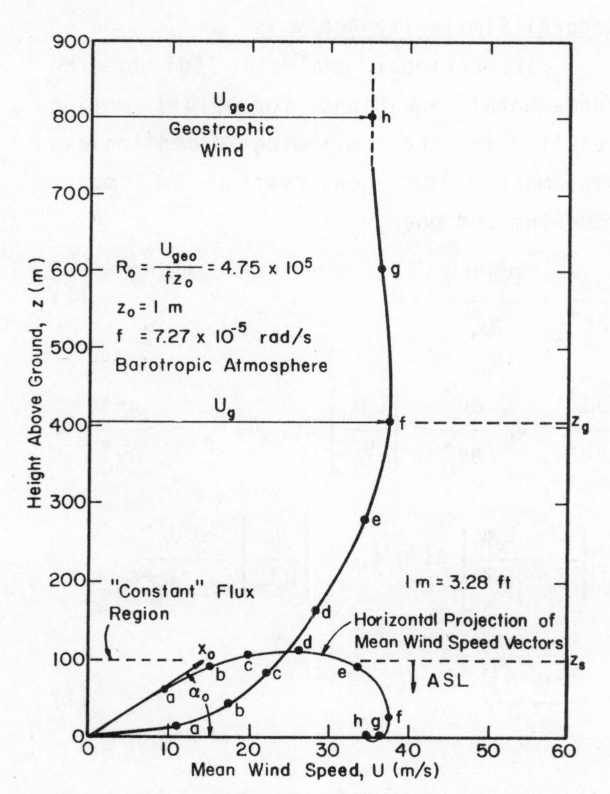

Fig. 1. Planetary boundary layer according to model of Lettau [26].

Flow above the gradient-wind level z_g is strongly influenced by thermal winds caused by horizontal temperature (density) gradients. Coriolis and pressure gradient forces are no longer in equilibrium and the wind either "veers" or "backs" from the directions shown in Fig. 1 as described by Haurwitz [28].

For wind-engineering applications the gradient wind speed is usually taken to be the maximum wind speed of the planetary boundary layer as shown in Fig. 1. Most high-rise building heights are greater than z_s and less than z_g, on the other hand, low-rise building heights are less than z_s.

ABL Approximation for Building Studies

Experimental studies of wind effects on buildings are of primary interest for strong winds (U greater than 10 m/s at a height of 10 m) with neutral thermal stratification. The boundary-layer thickness is usually the height for maximum wind speed (point f in Fig. 1) and which is designated by z_g. A power-law variation of U with height z is commonly used for convenience. Variation of the exponent α and z_g with surface roughness z_o has been determined by Davenport [29]. This dependence on z_o is shown in Fig. 2. Information of this type is particularly useful for selection of ABL properties to be modelled for a particular tall-building site.

The Atmospheric Surface Layer (ASL)

For studies of wind effects on low-rise buildings special attention must be given to modelling flow characteristics of the ASL. A distinctive feature of this layer according to Wyngaard [30] and others is constancy of vertical fluxes of heat,

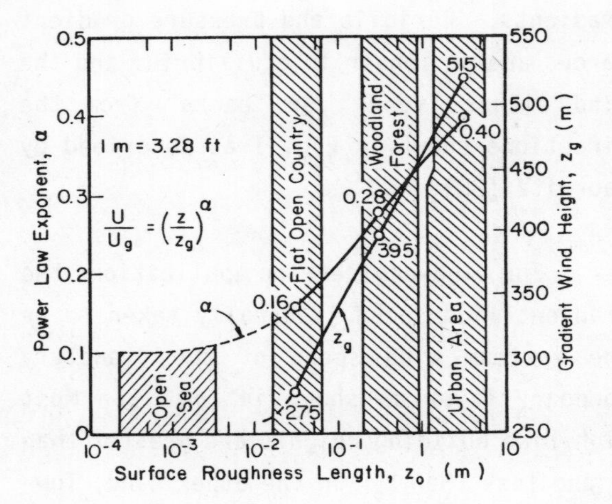

Fig. 2. Dependence of α and z_g on z_o for strong winds -- Davenport [29].

mass and momentum within 10% of surface values. This property is exhibited in Fig. 3 for shear stress (unit mass density) measured at the Kansas field program site as reported by Izumi [31].

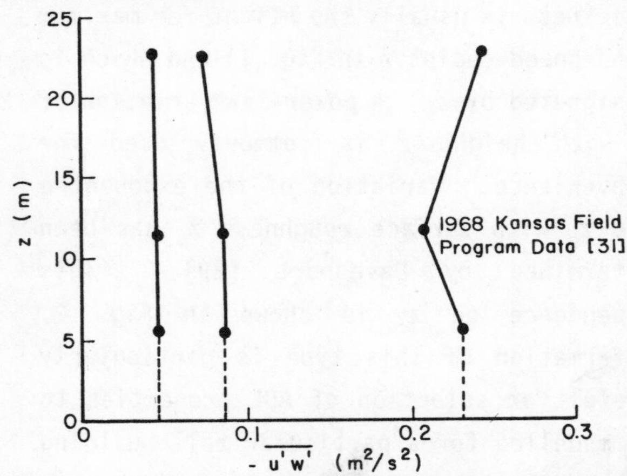

Fig. 3. Shear stress (unit mass density) distribution for three cases of near neutral stability.

A wealth of data on statistical properties and spectra of turbulence in the ASL is available in references such as those authored by Haugen [32] and Panofsky and Dutton [33].

FUNDAMENTAL SIMILARITY CRITERIA

Physical modelling for wind-engineering applications and research requires consideration of two types of similarity criteria: (1) criteria for simulation of natural winds--the ABL or regions thereof, and (2) similarity criteria for wind-effects such as pressures on buildings and structures or diffusion and transport of mass and heat. Both types of similarity criteria are discussed in the literature [22,34-39]. A brief review of ABL modelling (Type 1) follows.

General Similarity Analysis

Inspectional analysis [40] of the fundamental equations for fluid motion results in the following dimensionless statements for conservation of mass, momentum and energy:

$$\frac{\partial \rho^*}{\partial t^*} + \frac{\partial(\rho^* U_i^*)}{\partial x_i^*} = 0 \tag{1}$$

$$\frac{\partial U_i^*}{\partial t^*} + U_j^* \frac{\partial U_i^*}{\partial x_j^*} + \left(\frac{L_o \Omega_o}{U_o}\right) 2\epsilon_{ijk}\Omega_j^* U_k^* = -\frac{\partial P^*}{\partial x_i^*} - \left(\frac{\Delta T_o}{T_o}\frac{L_o g_o}{U_o^2}\right)\Delta T^* g^* \delta_{i3} + \left(\frac{\nu_o}{U_o L_o}\right)\frac{\partial^2 U_i^*}{\partial x_k^* \partial x_k^*} + \frac{\partial\left(-\overline{u_i' u_j'}\right)^*}{\partial x_j^*} \tag{2}$$

and
$$\frac{\partial T^*}{\partial t^*} + U_i^* \frac{\partial T^*}{\partial x_i^*} = \left(\frac{k_o}{\rho_o C_{po} \nu_o}\right)\left(\frac{\nu_o}{L_o U_o}\right)\frac{\partial^2 T^*}{\partial x_k^* \partial x_k^*} + \frac{\partial\left(-\overline{\theta' u_i'}\right)^*}{\partial x_i^*} + \left(\frac{\nu_o}{U_o L_o}\right)\left(\frac{U_o^2}{C_{po} T_o}\right)\phi^* \tag{3}$$

Variables in Eqs. 1-3 are scaled as follows:

$$U_i^* = \frac{U_i}{U_o},\ u_i^* = \frac{u_i}{U_o},\ \left(u_i'\right)^* = \frac{u_i'}{U_o},\ x_i^* = \frac{x_i}{L_o},$$

$$t^* = \frac{tU_o}{L_o},\ \Omega_j^* = \frac{\Omega_j}{\Omega_o},\ P^* = \frac{P}{\left(\rho_o U_o^2\right)},\ T^* = \frac{T}{T_o},$$

$$\Delta T^* = \frac{\Delta T}{\Delta T_o},\ \theta'^* = \frac{\theta'}{\Delta T_o},\ \text{and } g^* = \frac{g}{g_o}. \quad (4)$$

Although the foregoing equations are very general, some restrictions are imposed by their form. Because the effect of temperature on density has been linearized in Eq. 2 (Boussinesq approximation), flows are limited to those for which $\Delta T << T_o$. This restriction does not impose serious limitations on the use of Eq. 2 for ABL flow. Equation 3 does not allow for an exchange of energy by radiation or phase changes of water. Therefore, a strict interpretation of Eq. 3 would limit consideration to ABL without cloud content or precipitation.

Examination of Eqs. 1-3 gives the requirements for kinematic, dynamic, and thermic similarity provided that proper boundary conditions are satisfied. Equation 1 is invariant for this transformation in which all lengths are scaled equally; i.e., $x_i^* = x_i/L_o$. Therefore, *geometric similarity* will result in kinematic similarity. Equation 2 states that *dynamic* similarity can be achieved if each of the following parameters is equal for two systems: (1) $(U_o/(L_o\Omega_o)$, Rossby number (Ro); (2) $(\Delta T_o/T_o)(L_o g_o/U_o^2)$, Richardson number (Ri); and (3) $U_o L_o/\nu_o$ Reynolds number (Re.) *Thermic* similarity will be obtained if, according to Eq. 3, the following dimensionless groups are equal for two systems: (1) $\nu_o \rho_o C_{po}/k_o$, Prandtl number (Pr); (2) $U_o^2/(C_{po}T_o)$, Eckert number (Ec); and (3) Re.

Similarity Criteria for Building Studies

Primary wind-effects on buildings are associated with strong winds for which thermal effects become negligible as a result of strong mechanical mixing. Furthermore, Ec is proportional to the Mach number squared and can be neglected for atmospheric flows. Therefore, significant similarity requirements are the following:

$$(Re)_m = (Re)_p \text{ and } (Ro)_m = (Ro)_p.$$

The length scale $(Lo)_m/(Lo)_p$ for building models is in the range 10^{-1}-10^{-3}; therefore, equality of Rossby numbers cannot be achieved unless the flow facility can be given an appropriate angular velocity. This capability is not feasible for wind tunnels designed for building studies. Accordingly, changes in wind direction with height resulting from rotation of earth shown in Fig. 1 are not modelled.

Equality of Reynolds numbers $(Re)_m$ and $(Re)_p$ cannot be made equal for the model building scales commonly used unless an air-flow facility is pressurized, or a fluid such as Freon is used and/or high flow speeds are used. Fortunately, such extreme measures are not necessary because of Reynolds-number independence for flow characteristics of boundary-layers over rough surfaces. This property for local drag coefficients is illustrated in Fig. 4. Therefore, ABLs for strong winds can be modelled (except for Coriolis effects) in low-speed wind tunnels using air at atmospheric pressure.

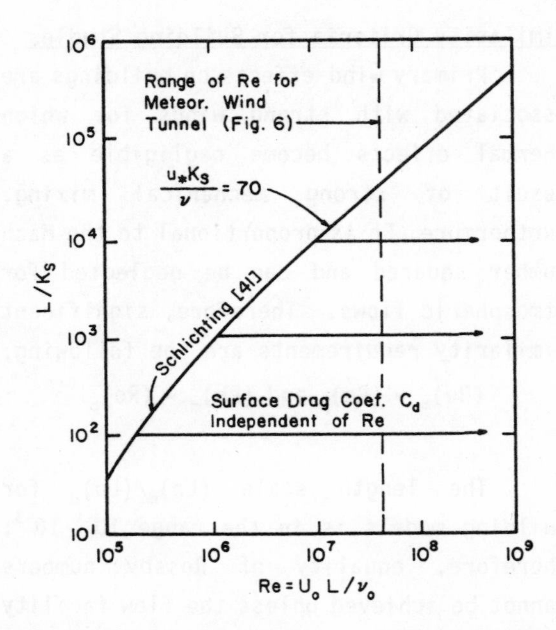

Fig. 4. Reynolds-number independence for C_d in neutral flow [34].

The requirements for geometric, dynamic, and thermic similarities obtained by inspectional analysis must be augmented by additional specifications if requirements for "exact" similarity are to be complete. These external conditions are: (1) similarity of surface roughness and temperature distributions at ground level; (2) similarity of flow structure above the ABL; (3) similarity of the horizontal pressure gradient; and (4) sufficient upwind fetch to establish equilibrium of ABL consistent with external conditions [34,35]. These conditions govern design of a wind tunnel for modelling ABLs appropriate for experimental studies of wind effects on buildings and other structures.

WIND-TUNNEL CHARACTERISTICS FOR ABL MODELLING

Basic features of a BLWT test section are shown schematically in Fig. 5 for flow without thermal stratification.

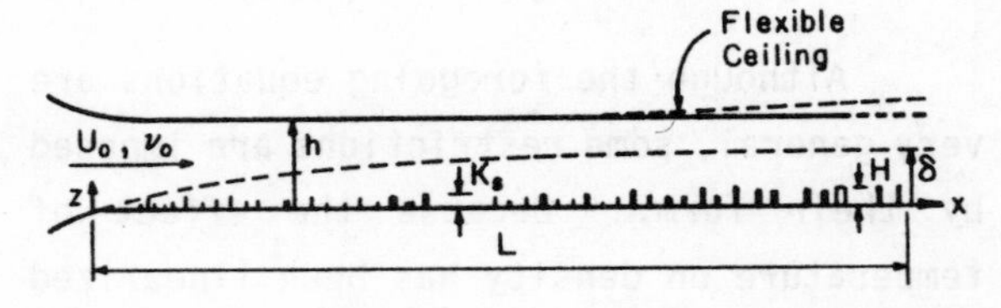

Fig. 5. Definition of BLWT test section.

The ABL depth varies in the range 300-500 m; therefore, the length of a BLWT is commonly 25-30 m, and the test-section height is 2-3 m in order to develop boundary-layer thicknesses of 0.5-1.5 m [36]. Thermally stratified ABLs can be simulated by heating (cooling) the test-section floor and cooling (heating) air in the return duct of a recirculating BLWT to develop unstable (stable) stratification. This capability is not required for experimental studies of wind effects on buildings and structures.

Wind Tunnel Types

Two basic types of BLWTs are the closed-circuit and the open-circuit. The closed-circuit BLWT shown in Fig. 6 can establish stratified boundary layers with Ri ranging from -1 to +1. Both ground-based and elevated inversions can be generated for dispersion studies of air pollutants. The range of Reynolds numbers for this facility is shown in Fig. 4. A typical open-circuit BLWT without thermal stratification capabilities is shown in Fig. 7. A list of BLWTs throughout the world is given in a recently published book by Rae and Pope [42].

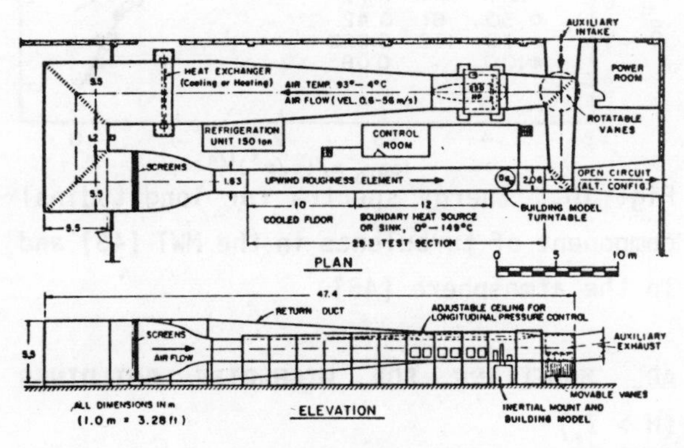

Fig. 6. Closed-circuit BLWT, Colorado State University [36].

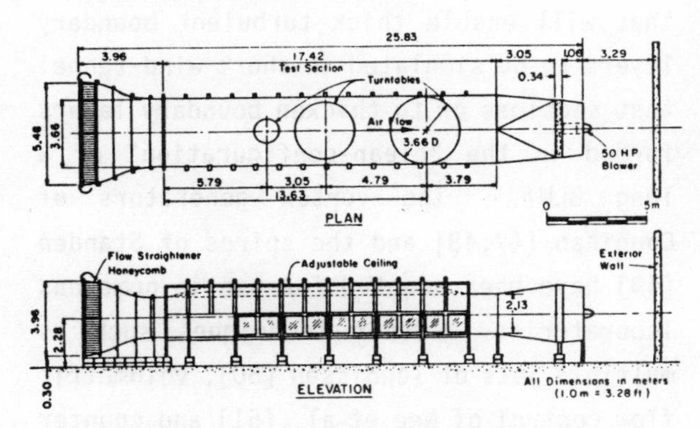

Fig. 7. Open-circuit BLWT, Colorado State University [36].

Pressure variation in the direction of ABL flow over flat terrain is essentially zero. Therefore, provisions are necessary to achieve a similar condition in a closed BLWT test section. This can be accomplished as shown in Figs. 6 and 7 through use of an adjustable flexible ceiling. Vertical displacements up to 0.6 m can accommodate a wide range of aerodynamic surface roughnesses.

Special treatment of the test-section entrance is required to fix the turbulent boundary-layer origin uniformly around the test-section periphery. A satisfactory "trip" system for the MWT (Fig. 6) is shown in Fig. 8. The "trip" system is in place for all the boundary-layer data given subsequently.

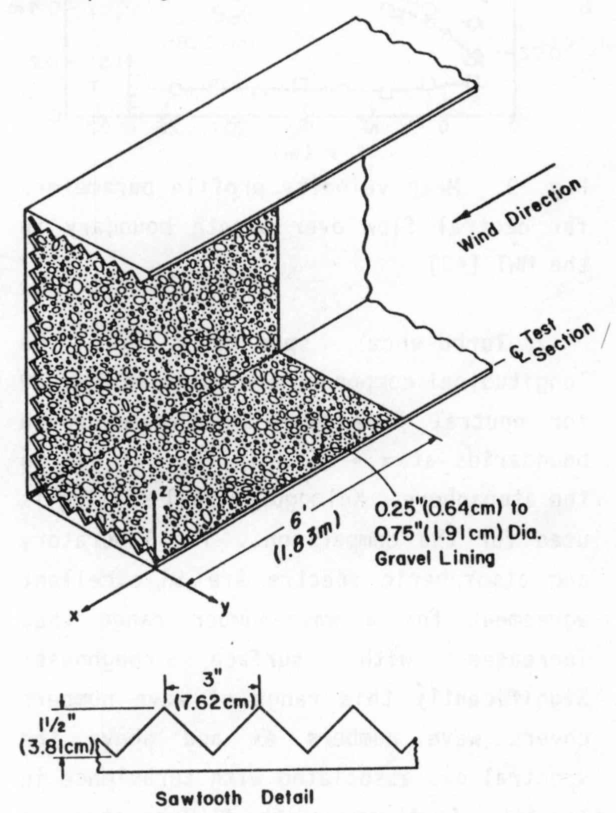

Fig. 8. Sawtooth-roughness boundary-layer trip at entrance to MWT test-section [35].

MWT Boundary-layer Characteristics

Variations of integral properties of a turbulent boundary layer formed over the smooth floor of the MWT with distance along the test section are given in Fig. 9. The shape factor H and growth rate of the boundary-layer thickness become constant at a distance of about 10 m. This indicates that the boundary layer is fully developed and is in equilibrium with boundary conditions for the following 15 m.

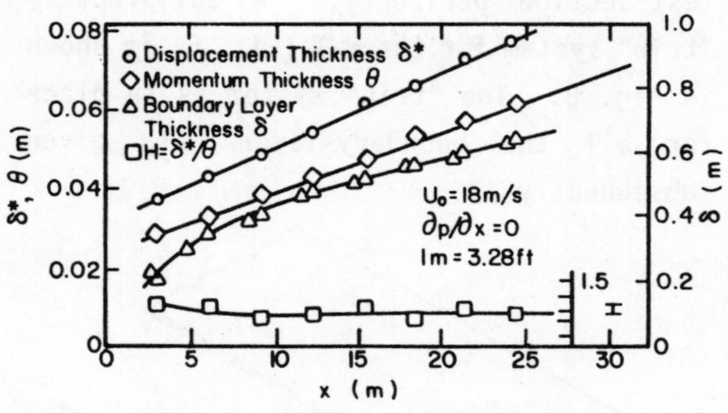

Fig. 9. Mean velocity profile parameters for neutral flow over smooth boundary in the MWT [43].

Turbulence spectra for the longitudinal component are shown in Fig. 10 for neutral flow over smooth and rough boundaries at x = 25 m in the MWT and in the atmosphere. Kolmogorov [44] scaling is used for the comparison. The laboratory and atmospheric spectra are in excellent agreement for a wave-number range that increases with surface roughness. Significantly this range of wave numbers covers wave numbers at and above the spectral gap associated with turbulence in the ASL. Furthermore, the BLWT spectra are characteristic of high Reynolds number turbulence for which a large set of wave numbers is in the inertial subrange where turbulence energy varies with wave number raised to the -5/3 power.

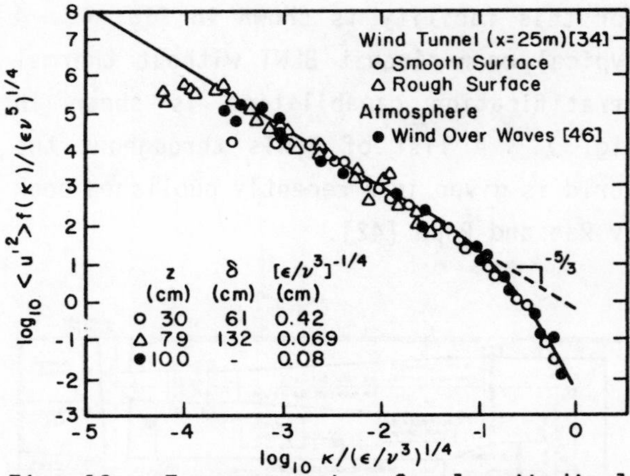

Fig. 10. Energy spectra for longitudinal component of turbulence in the MWT [45] and in the atmosphere [46].

ABL MODELLING FOR HIGH-RISE BUILDINGS ($H > Z_s$)

Much effort has been devoted in recent years to development of techniques that will enable thick turbulent boundary layers to be simulated in short wind-tunnel test sections or to thicken boundary layers formed in the "clean-configuration" of a long BLWT. The vortex generators of Counihan [47,48] and the spires of Standen [49] have been adapted for use in numerous laboratories. Other techniques such as multiple jets of Teunissen [50], volumetric flow control of Nee et al. [51] and counter jets of Morkovin et al. [52] have not found wide acceptance. In an exploratory study Peterka and Cermak [53] reviewed the various techniques and utilized combinations of spires and roughness arrays to obtain additional measurements for determination of boundary-layer development. Gartshore [54] investigated the effect of surface roughness on shape of the mean velocity profile and Cook [55] employed a combination of surface roughness, barrier and vortex generators to simulate the ABL in whole and in part.

Many questions arise regarding the characteristics of augmented boundary layers and how well they simulate characteristics of the ABL. The question of primary importance is: at what distance downstream of an augmentation device do profiles of mean velocity, turbulence intensity, and Reynolds shear stress become similar to those that form over the same boundary without augmentation?

Effects of Augmentation Devices

A systematic study [35] to determine the effect of augmentation devices on boundary layer development was made for thermally neutral flow in the MWT shown in Fig. 6. Three entrance configurations were used: (1) the 4-cm sawtooth trip for "natural" boundary-layer development described in Fig. 8, (2) 61-cm high vortex generators designed by Counihan shown in Fig. 11, and (3) 1.8-m high spires designed by Standen shown in Fig. 12. Profiles of mean and turbulent velocities were measured over three uniform surface roughness (z_o = 0.02 cm, 0.15 cm and 0.35 cm) for each entrance configuration at x = 3.3 m, 6.1 m, 12.2 m, 18.3 m, and 25.9 m downwind of the entrance. Typical profiles of mean velocity, longitudinal turbulence intensity and Reynolds stress with z_o = 0.35 cm at x = 6.10 m and 18.29 m are shown in Figs. 13-15.

The data of Figs. 13-15 reveal strong differences between the augmented profiles and the reference profiles (sawtooth trip) at a distance x = 6.10 m downstream from the entrance. As x increases to 18.29 m the profiles merge and are in equilibrium with the surface boundary condition (z_o = 0.35 cm). At this location the augmentation devices increased the boundary-layer thickness by 33%. Accordingly, boundary-layer thickness can be increased by augmentation devices at the test-section entrance provided the test-section length is sufficient for establishment of statistical equilibrium.

Figures 13-15 and other data obtained in the same study indicated that a length of about 15 m will be adequate to dissipate augmentation-device disturbances. Without augmentation devices, statistical equilibrium is achieved in a shorter distance -- about 10 m, as indicated in Fig. 9.

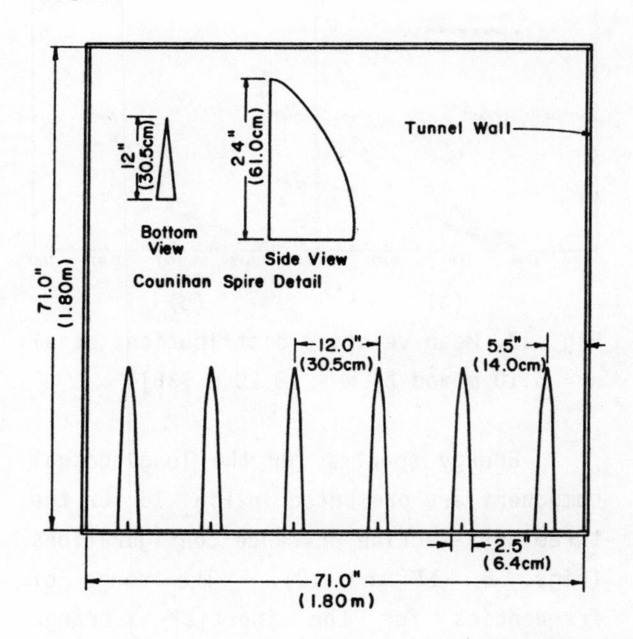

Fig. 11. Vortex generator array for boundary-layer thickness augmentation -- Counihan [47].

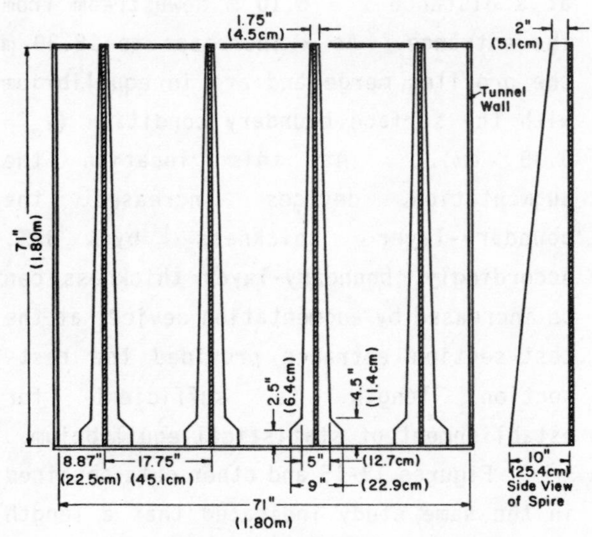

Fig. 12. Spire array for boundary-layer thickness augmentation -- Standen [49].

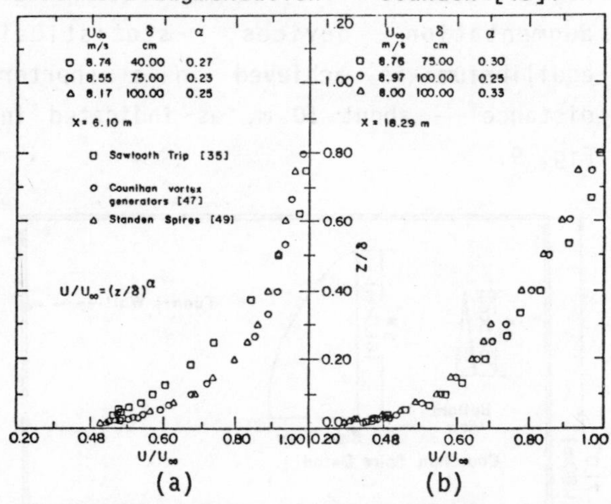

Fig. 13. Mean velocity distributions at a) x = 6.10 m and b) x = 18.29 m [35].

Energy spectra for the longitudinal component are presented in Fig. 16 for the three test-section entrance configurations (Figs. 8, 11 and 12). The range of frequencies for the inertial subrange extends over about two decades -- a common finding for atmospheric turbulence. In this frequency range the turbulence is fully developed and approaches an isotropic state independent of the initial and boundary conditions.

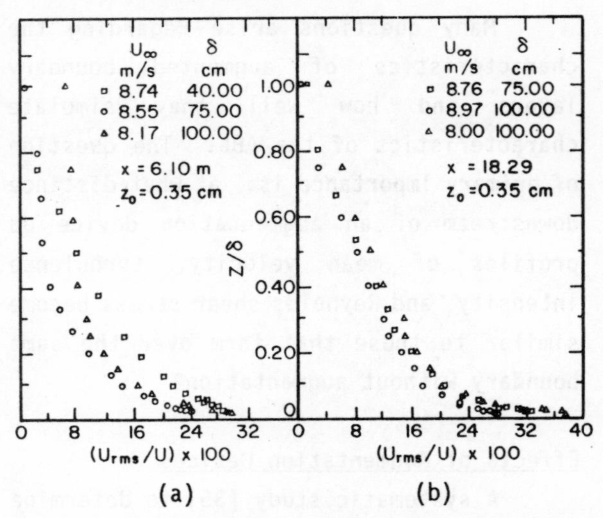

Fig. 14. Longitudinal turbulence intensity profiles for a) x = 6.10 m and b) x = 18.29 m [35].

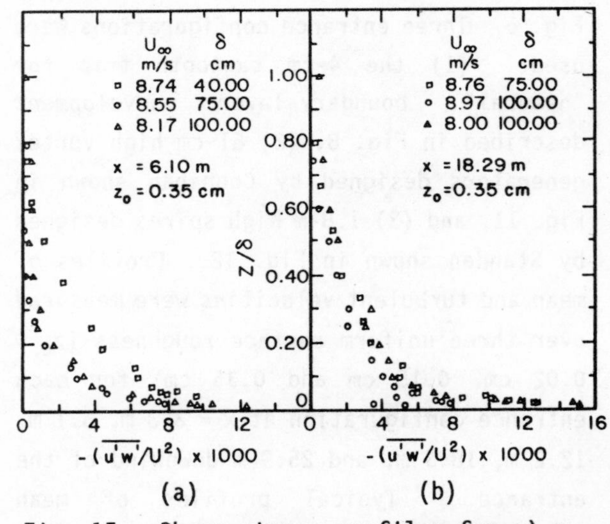

Fig. 15. Shear stress profiles for a) x = 6.10 m and b) x = 18.29 m [35].

ASL MODELLING FOR LOW-RISE BUILDINGS ($H < z_s$)

Low-rise buildings are subjected to wind in the ASL where, as indicated in Fig. 1, the vertical fluxes are sensibly con-

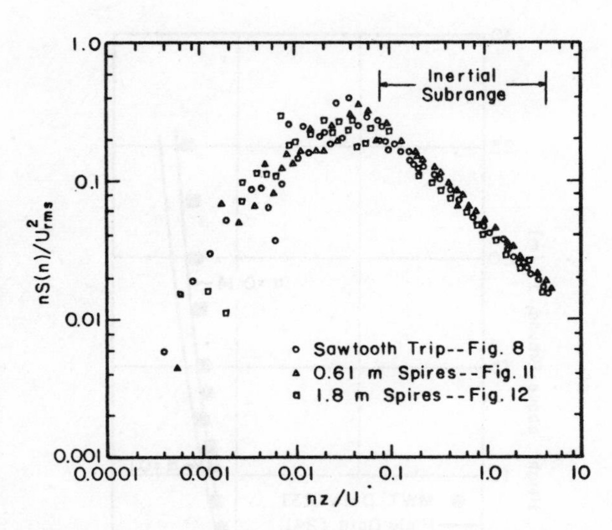

Fig. 16. Spectra of longitudinal turbulence -- x = 18.29 m; z_o = 0.35 cm; $U_\infty \cong$ 18 m/s; z/δ = 0.05 [35].

stant. The constancy of shear stress (for unit mass density) with height is illustrated in Fig. 3. Accordingly, if low-rise buildings are modelled to scales sufficiently large to avoid Reynolds number dependence of pressure distributions, the modelled ASL depth should be substantially greater than is achieved by test-section entrance configurations shown in Figs. 8, 11 and 12.

A practical method for modelling thick ASLs developed by Cermak and Cochran [23] is shown in Fig. 17. In this system the horizontal vanes generate strong u′-w′ correlations that promote transport of turbulence energy toward the lower boundary. Data obtained for flow developed by this system are compared with full-scale meteorological data measured at the Texas Tech University research building site as reported by Chok [24]. The wind-tunnel data are scaled to full-scale values using a scale ratio of 1:50. At this scale a building of 5-m height would be represented by a model 10 cm in height.

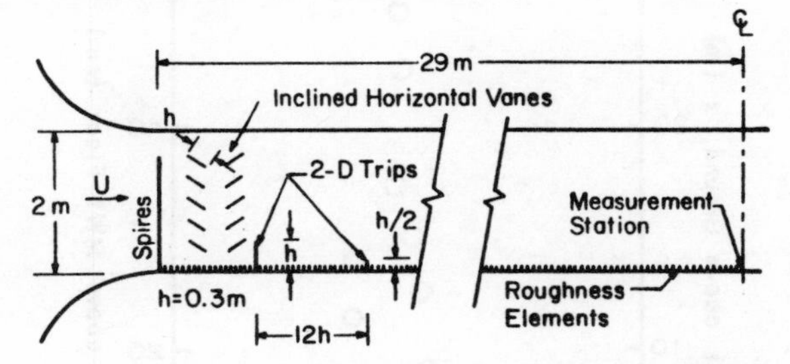

Fig. 17. Test section of MWT with flow conditioning system for ASL simulation [23].

Properties of the Simulated ASL

A vertical profile of nondimensional shear stress for flow generated by the system of Fig. 17 is shown in Fig. 18. Heights are scaled up by a factor of 50. Effectiveness of the horizontal vanes in amplifying the "constant" stress-layer depth can be evaluated by comparing with data given in Fig. 19. Data in Fig. 19 were obtained in the MWT over a smooth boundary with test-section entrance configuration shown in Fig. 8 [43].

Flow characteristics other than constancy of shear stress must be simulated if the ASL is to be properly modelled. The primary ASL features that must be reproduced are vertical distributions of mean velocity, turbulence intensity, and integral scales of turbulence.

Mean velocity distributions for wind-tunnel and field data are presented in Fig. 20. Agreement between the data sets is very good in the lower 10 m of the profiles. Roughness of the upwind fetch in

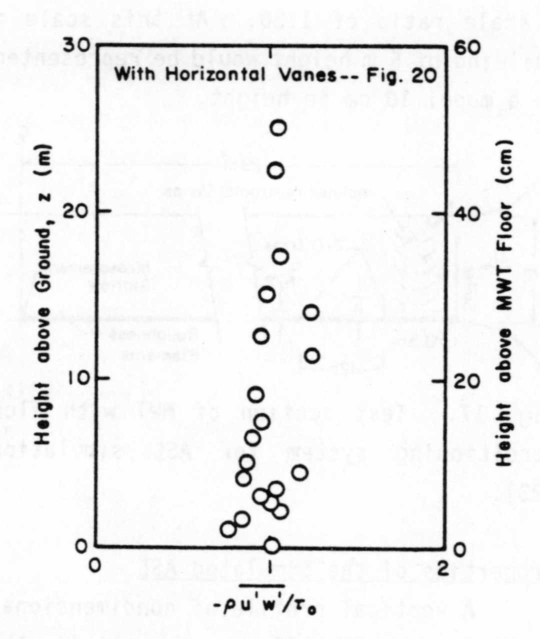

Fig. 18. Vertical distribution of shear stress with horizontal vanes [23].

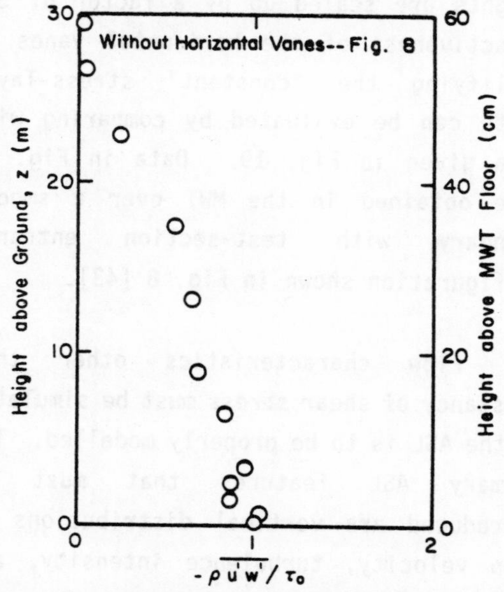

Fig. 19. Vertical distribution of shear stress without horizontal vanes [43].

the wind tunnel simulated the field conditions by chains (15 mm high links) placed on the floor transverse to the flow at a spacing of 150 mm.

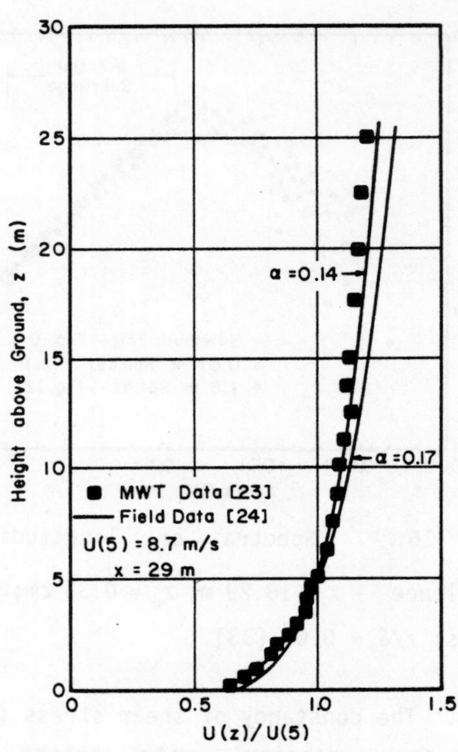

Fig. 20. Mean wind-tunnel and full-scale wind speed profiles.

Profiles of turbulence intensity for the longitudinal component at the TTU site and in the MWT for a 1:50 scale are compared in Fig. 21. Data from the two sources are in good agreement. Because peak pressures and reattachment locations for separated flow over a model building are sensitive to turbulence intensity, it is important that turbulence intensity of the modelled ASL match full-scale values.

Integral scales for the longitudinal component of turbulence are shown for the simulated ASL and the field site in Fig. 22. Integral scales measured in the simulated ASL (when scaled up by a factor 50) are approximately 50% smaller than the field values. Further research is required to develop a technique for generation of layer scales of turbulence that will not compromise the excellent correspondence of other properties of the ASL obtained with the horizontal vane system.

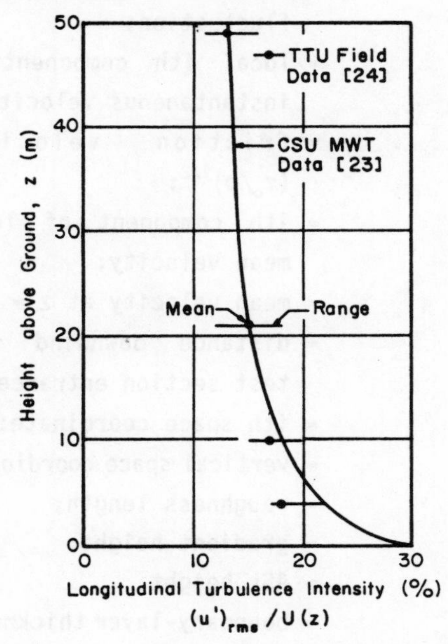

Fig. 21. Intensity of longitudinal turbulence component for simulated ASL [23] and at the TTU site [24].

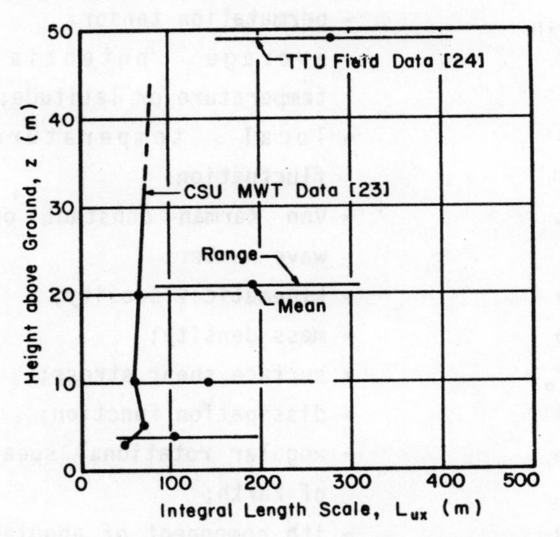

Fig. 22. Integral scale of longitudinal component for simulated ASL [23] and of the TTU site [24].

CONCLUDING REMARKS

Properly designed low-speed wind tunnels can model features of the ABL essential for experimental studies of wind effects on buildings and structures. Test sections should be sufficiently long to achieve fully-developed turbulent boundary layers approximately one meter in depth for a wide range of surface roughnesses. A test-section length of 15 m or greater with a height of 2-3 m and a width of 2-5 m satisfies this requirement.

For studies of high-rise buildings ($H > z_s$) the entire ABL should be modelled such that the boundary-layer thickness δ is scaled to z_g as H_m is scaled to H_p. Low-rise buildings ($H < z_s$) are submerged in the ASL; therefore, wind-effect studies on this class of buildings requires modelling properties of the ASL over a depth of approximately 0.5 m. This constant stress layer can be modelled (partial modelling of entire ABL) by placement of a horizontal vane system in the upper part of the test-section entrance. Shear stress, mean velocity and turbulence intensity profiles for the modelled ASL are in good agreement with full-scale field data. However, integral scales of turbulence are less than full-scale values. Further research is required to overcome this deficiency.

ACKNOWLEDGEMENTS

Support for development of a system to model atmospheric surface layers was provided by the National Science Foundation Cooperative Agreement BCS-88-21542 (Colorado State University/Texas Tech University Cooperative Research Program).

Flow measurements for this development were made by Leighton S. Cochran, CSU graduate research assistant. Wordprocessing of the manuscript for this paper was performed by Gloria Burns.

NOMENCLATURE

C_d = surface drag coefficient;

C_p = specific heat at constant pressure;

Ec = Eckert number, $U_o^2/(C_{po}T_o)$;

f = Coriolis parameter, $2\omega \sin\theta$;

g = gravitational acceleration;

H = building height;

H_o = surface heat flux;

k = thermal conductivity;

K_s = equivalent sand roughness length;

L = test-section length;

L_x = distance downwind from virtual origin of boundary layer;

L_{ux} = integral scale of longitudinal turbulence component;

n = frequency;

P = mean pressure;

Pr = Prandtl number, $\nu_o\rho_oC_{po}/k_o$;

Re = Reynolds number, U_oL_o/ν_o;

Ri = bulk Richardson number, $(\Delta T_o/T_o)/(L_og/U_o^2)$;

Ro = Rossby number, $U_o/(L_o\Omega_o)$;

R = universal gas constant;

S(n) = spectrum of longitudinal turbulence component;

T = local mean temperature;

ΔT = mean temperature difference (departure from adiabatic lapse rate);

t = time;

$u_i'(u', v', w')$ = ith component of velocity fluctuation;

u_i = local ith component of instantaneous velocity;

u_* = friction velocity, $(\tau_o/\rho)^{1/2}$;

U_i = ith component of local mean velocity;

U_∞ = mean velocity at $z = \delta$;

x = distance downwind from test section entrance;

x_i = ith space coordinate;

z = vertical space coordinate;

z_o = roughness length;

z_g = gradient height;

z_s = ASL height;

δ = boundary-layer thickness;

δ_{ij} = Kronecker delta;

ϵ = energy dissipation rate per unit of mass;

ϵ_{ijk} = permutation tensor;

θ = average potential temperature or latitude;

θ' = local temperature fluctuation;

κ = von Kármán constant or wave number;

ν = kinematic viscosity;

ρ = mass density;

τ_o = surface shear stress;

ϕ = dissipation function;

ω = angular rotational speed of Earth;

Ω_i = ith component of angular velocity;

$\overline{(\ )}$ = time average;

$(\ )_o$ = reference quantity;

$( )_{rms}$ = root-mean-square value of quantity;

$( )_s$ = depth of quantity at top of atmospheric surface layer; and

$( )^*$ = nondimensional quantity.

REFERENCES

1. Smeaton, J., An experimental investigation concerning the natural powers of water and wind, *Philosophical Transactions of the Royal Society of London*, 51:100-174, 1759-1760.
2. Kernot, W. C., Wind pressure, *Australasian Builder and Contractors' News*, 13:194, Oct. 21, 1883.
3. Irminger, I., Experiments on wind pressure, *Proceedings of the Institute of Civil Engineers*, 118:468-472, 1893-1894.
4. Tsiolkovsky, K. E., Collected works of K. E. Tsiolkovsky, *The Academy of Sciences of the USSR*, Moscow, USSR, (English Translation; TTF-236, National Aeronautics and Space Administration), 1951.
5. Stanton, T. E., On the resistance of plane surfaces in a uniform current of air, *Proceedings of the Institute of Civil Engineering*, 156, 1903-1904.
6. Stanton, T. E., Report on the experimental equipment of the Aeronautical Department of the National Physical Laboratory, Report of the Advisory Committee of Aeronautics, London, 1909-1910.
7. Riabouchinsky, D., *Bulletin of the Institute Aerodynamique de Koutchino*, Moscow, USSR, Vols. I, II, and III, 1906-1909.
8. Prandtl, L., The importance of model experiments for aeronautics and the apparatus for such tests in Göttingen, *ZVDI*, 53: 1909.
9. Eiffel, G., The resistance of air and aviation, Paris, France, 1910.
10. Abe, M., Mountain clouds, their forms and connected air currents, *Bulletin of the Central Meteorological Observatory*, Tokyo, Japan, 7(3), 1929.
11. Prandtl, L. and Reichardt, H., Einfluss von Wärmeschichtung auf de Eigenschaften Einer Turbulenten Strömung, Deutsche Forschung, Berlin, Germany, No. 21, pp. 110-121, 1934.
12. Sherlock, R. H. and Stalker, E. A., A study of flow phenomena in the wake of smokestacks, Engineering Research Bulletin No. 29, University of Michigan, Ann Arbor, MI, 49 pp., 1941.
13. Nøkkentved, C., Variation of the wind-pressure distribution on sharp-edged bodies, Report No. 7 of the Structural Research Laboratory, Royal Technical College, Copenhagen, Denmark, 1936.
14. Bailey, A. and Vincent, N. D. G., Wind-pressures on buildings including effects of adjacent buildings, *Journal of the Institution of Civil Engineers*, 20(8):243-275, Oct. 1943.
15. Jensen, M., The model-law for phenomena in natural wind, *Ingeniøren*, International Edition, 2(4), 1958.

16. Rouse, H., Model techniques in meteorological research, *Compendium of Meteorology*, American Meteorological Society, Boston, MA, pp. 1249-1254, 1951.
17. Strom, G. H., Wind tunnel techniques used to study influence of building configuration on stack gas dispersal, *American Industrial Hygiene Association Quarterly*, 13:76, 1952.
18. Cermak, J. E. and Koloseus, H. J., Lake Hefner model studies of wind structure and evaporation, Tech. Repts. CER54-JEC20 and CER54-JEC22, (AD 1054930), Fluid Dynamics and Diffusion Laboratory, Colorado State University, Fort Collins, CO, 275 pp., 1953-1954.
19. Cermak, J. E. and Spengos, A. C., Turbulent diffusion of momentum and heat from a smooth, plane boundary with zero pressure gradient, Tech. Rept. CER56-JEC22, Fluid Dynamics and Diffusion Laboratory, Colorado State University, Fort Collins, CO, 77 pp., 1956.
20. Cermak, J. E., Wind tunnel for the study of turbulence in the atmospheric surface layer, Tech. Rept. CER58JEC42, Fluid Dynamics and Diffusion Laboratory, Colorado State University, Fort Collins, CO, 31 pp., 1958.
21. Plate, E. J. and Cermak, J. E., Micro-meteorological wind tunnel facility, Tech. Rept. CER63EJP-JEC9, Fluid Dynamics and Diffusion Laboratory, Colorado State University, Fort Collins, CO, 65 pp., 1963.
22. Cermak, J. E., Wind-simulation criteria for wind-effect tests, *Journal of Structural Engineering*, ASCE, 110(2):328-339, Feb. 1984.
23. Cermak, J. E. and Cochran, L. S., Wind pressures on low-rise buildings: physical modelling, Progress Report No. 2, CSU/TTU Cooperative Wind Engineering Research Program, NSF Cooperative Agreement BCS-88-21542, Colorado State University, Fort Collins, CO, 32 pp., September 1990.
24. Chok, C. V., Wind parameters of Texas Tech University field site, Master of Science Thesis, Civil Engineering Department, Texas Tech University, 87 pp., August 1988.
25. Hanna, S. R., Characteristics of winds and turbulence in the planetary boundary layer, ESSA Research Laboratories, Air Resources Laboratory, Oak Ridge, TN, Technical Memorandum ERLTM-ARL 8, 61 pp., 1969.
26. Lettau, H. H., Theoretical wind spirals in the boundary layer of a barotropic atmosphere, *Beitr, Phys. Atmos.*, 35:195-212, 1962.
27. Blackadar, A. K., The vertical distribution of wind and turbulent exchange in a neutral atmosphere, *Journal of Geophysical Research*, 67:3095-3102, 1962.
28. Haurwitz, B., *Dynamic Meteorology*, McGraw-Hill Book Company, Inc. New York, 365 p., 1941.
29. Davenport, A. G., A rationale for determination of design wind velocities, *Proceedings ASCE, Journal of the Structural Division*, 86:39-66, 1960.

30. Wyngaard, J. C., On surface-layer turbulence, Workshop on Micrometeorology, American Meteorological Society, Boston, MA, pp. 101-149, 1973.

31. Izumi, Y. (ed.), Kansas 1968 field progress report, Environmental Research Paper No. 79, AFCRL-72-0041, Air Force Cambridge Research Laboratories, L. G. Hanscom Field, Bedford, MA, 75 pp. December 1971.

32. Haugen, D. A. (ed.), Workshop on Micrometeorology, American Meteorological Society, Boston, MA, 392 pp., 1973.

33. Panofsky, H. A. and Dutton, J. A., *Atmospheric Turbulence: Models and Methods for Engineering Applications*, John Wiley & Sons, 397 pp. 1984.

34. Cermak, J. E., Applications of fluid mechanics to wind engineering--a Freeman Scholar Lecture, Transactions of the ASME, *Journal of Fluid Engineering*, Ser. 1, 97(1):9-38, Mar. 1975.

35. Cermak, J. E., Physical modeling of the atmospheric boundary layer (ABL) in long boundary-layer wind tunnels (BLWT), *Proceedings of International Workshop on Wind Tunnel Modeling Criteria and Techniques in Civil Engineering Application*, National Bureau of Standards, Gaithersburg, MD, Cambridge University Press, Cambridge, U.K., pp. 97-137, Apr. 1982.

36. Cermak, J. E., Wind tunnel design for physical modeling of atmospheric boundary layers, *Journal of the Engineering Mechanics Division*, ASCE, 107(3):523-642, June 1981.

37. Cermak, J. E., Wind-tunnel testing of structures, *Journal of the Engineering Mechanics Division*, ASCE, 103(6):1125-1140, Dec. 1977.

38. Cermak, J. E. and Arya, S. P. S., Problems of atmospheric shear flows and their laboratory simulation, *Boundary-Layer Meteorology*, 1(1):40-60, Mar. 1970.

39. Whitbread, R. E., Model simulation of wind effects on structures, *Proceedings of Conference on Wind Effects on Buildings and Structures*, National Physical Laboratory, Teddington, England, 1:284-301, June 1963.

40. Ruark, A. E., Inspectional analysis: a method which supplements dimensional analysis, *Journal Elisha Mitchell Science Society*, 51(1):127-133, Aug. 1935.

41. Schlichting, H., *Boundary Layer Theory*, McGraw-Hill Publishing Company, Inc., New York, NY, 1980.

42. Rae, W. H., Jr. and Pope, A., *Low-speed Wind Tunnel Testing*, John Wiley and Sons, New York, NY, 1984.

43. Zoric, D. and Sandborn, V. A., Similarity of large reynolds number boundary layers, *Boundary-Layer Meteorology*, 2(3):326-333, Mar. 1972.

44. Kolmogorov, A. N., The local structure of turbulence in incompressible viscous fluid for very large Reynolds numbers, *Doklady of the Academy of Sciences of the USSR*, 30(4):299-303, 1941.

45. Cermak, J. E., Sandborn, V. A., Plate, E. J., Binder, G. H., Chuang, H., Meroney, R. N. and Ito, S., Simulation of atmospheric motion by wind-tunnel flows, Tech. Rept. CER66-JEC-VAS-EJP-GHB-HC-RNM-SI-17, Fluid Dynamics and Diffusion Laboratory, Colorado State University, Fort Collins, CO, 101 p., May 1966.
46. Pond, S., Stewart, R. W. and Burling, R. W., Turbulence spectra in wind over waves, *Journal of the Atmospheric Sciences*, 20;319-324, 1963.
47. Counihan, J., An improved method of simulating an atmospheric boundary layer in a wind tunnel, *Atmospheric Environment*, 3:197-214, 1969.
48. Counihan, J., Simulation of an adiabatic urban boundary, layer in a wind tunnel, *Atmospheric Environment*, 7:673-689, 1973.
49. Standen, N. M., A spire array for generating thick turbulent shear layers for natural wind simulation in wind tunnels, Tech. Rept. LTR-LA-94, National Aeronautical Establishement, Ottawa, Canada, 1972.
50. Teunissen, H. W., Simulation of the planetary boundary layer in a multiple-jet wind tunnel, Report 182, Institute for Aerospace Studies, University of Toronto, Toronto, Canada, 162 p., 1972.
51. Nee, V. W., Dietrick, C., Betchov, R. and Szewczyk, A. A., The simulation of the atmospheric surface layer with volumetric flow control, *Proceedings of the Nineteenth Annual Technical Meeting*, pp. 483-487, 1973.
52. Morkovin, M. V., Nagib, H. M. and Yung, J. T., On modeling of atmospheric surface layers by the counter-jet technique--preliminary results, Tech. Rept. AFSOR-TR-73-0592, Illinois Institute of Technology, Chicago, IL, 1972.
53. Peterka, J. A. and Cermak, J. E., Simulation of atmospheric flows in short wind tunnel test sections, Tech. Rept. CER73-74JAP-JEC32, Fluid Dynamics and Diffusion Laboratory, Colorado State University, Fort Collins, CO, 52 p., 1974.
54. Gartshore, I. S., A relationship between roughness geometry and velocity profile shape for turbulent boundary layers, Report LTR-LA-140, National Aeronautical Establishement, National Research Council, Ottawa, Canada, 36 p., Oct. 1973.
55. Cook, N. J., On simulating the lower third of the urban adiabatic boundary layer in a wind tunnel, *Atmospheric Environment*, 7:691-705, 1973.

ATMOSPHERIC BOUNDARY LAYER AND ITS SIMULATION IN WIND TUNNELS

R.J. Garde

Pro-Vice Chancellor

Indira Gandhi National
Open University
New Delhi 110 016, INDIA

G.L. Asawa

Professor of Civil Engineering

University of Roorkee
Roorkee 247 667, INDIA

ABSTRACT

The first part of the paper describes the formation of atmospheric boundary layer and discusses the methods for predicting mean velocity distribution and turbulence characteristics. The second part of the paper deals with similarity conditions for obtaining corresponding conditions in the laboratory wind tunnel for conducting studies on behaviour of structures kept in atmospheric boundary layer and describes experimental set up.

ATMOSPHERIC BOUNDARY LAYER

Introduction

Engineers are often interested in wind effects on structures such as tall buildings, chimneys, bridges, and T.V. towers; in the prediction and abatement of atmospheric pollution; in wind-induced discomforts in and around buildings; in the generation of wind waves and in the bumping of aeroplanes; and in the lifting of rockets and similar other phenomena. These phenomena involve at most a few Kilometers of distance and heights of the order of 300-400 m. This layer in the vicinity of earth's surface is commonly known as the atmospheric boundary layer (ABL). In this layer, atmospheric motions corresponding to micro-scale (10^{-3} m to 10^{1} m) and small scale (10^{1}m to 10^{4}m) are of primary significance in relation to the above-mentioned problems involving mass transport and wind forces. However, these motions are affected by the meso-scale motions (10^{4}m to 10^{5}m) caused by topography, nonuniform heating of earth's surface and earth's rotation.

All types of winds are caused by the variable heating of earth's surface due to sun. Solar radiation which is far more intense at the equator than at the poles tends to cause differential heating of the earth's surface which, in turn, gives rise to pressure gradients. Other factors such as earth's rotation, cloud cover, precipitation, large topographical reliefs, surface roughness and their spatial variation also affect winds. Elaborate numerical models based on fundamental fluid mechanics principles are being developed and tested to predict the future state of atmosphere for known initial state [1]. However, they have not been fully developed and tested, and hence cannot be used at present by the engineers; instead, engineers have to depend on the analysis of wind data to get design information. Winds, in general, can be classified into boundary layer winds, cyclones, anticyclones and tornadoes. We will deal only with the boundary layer winds near the earth's surface where velocity changes from zero at the surface of the earth to a near constant value at higher elevation. This is often known as planetary boundary layer or atmospheric boundary layer (ABL). Tornadoes, cyclones, severe thunder storms, and separation on the downstream side of hills, mountains etc. are responsible for greatest perturbations in the structure of atmospheric boundary layers. Typical wind speed and direction profile in the planetary boundary layer measured at Digha, India in July 1979 under MONEX experiment [2] is shown in Fig. 1. Characteristics of atmospheric layers have been described in detail by Plate [3] and present state of knowledge is summarised by Monin [4] and Plate [5].

The atmospheric motion is controlled by the following forces: (i) gravity (ii) force due to pressure gradient, and (iii) Coriolis force. The pressure gradient in the vertical direction is balanced by the gravity force. The force due to the pressure gradient in the horizontal direction is responsible for generation of wind. It may be mentioned that former is over 10,000 times greater than the latter. Coriolis force (named after the French scientist) arises when the co-ordinate system of interest is rotating and the motion of the object (here air) occurs relative to the moving co-ordinate system.

Far away from the earth's surface the wind velocity is governed by horizontal pressure gradient and Coriolis force. The resulting velocity is known as the geostrophic wind velocity and is denoted by U_g; if the isobars are curved, the centrifugal force will act in addition to pressure gradient and Coriolis forces. The resulting

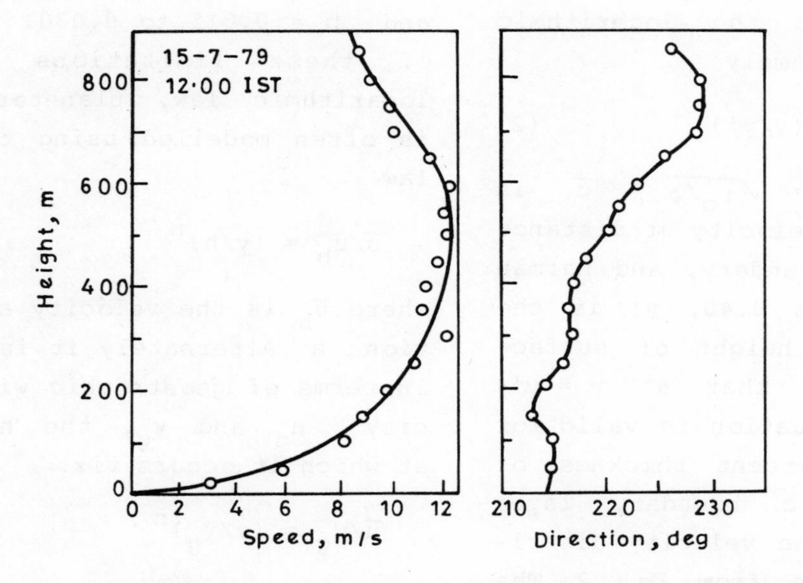

FIG. 1 - PLANETARY BOUNDARY LAYER AT DIGHA, WEST BENGAL [2]

wind velocity is then called gradient wind velocity. The geostrophic velocity U_g is given by

$$U_g = \frac{\partial\bar{p}/\partial n}{2\rho\omega.\sin\lambda_1} = \frac{\partial\bar{p}/\partial n}{\rho f} \quad (1)$$

where ω is the angular velocity of the earth, λ_1 is the latitude of the place, ρ mass density of air, $\bar{p}$ mean pressure and $f = 2\omega \sin\lambda_1 = 1.458 \times 10^{-4} \sin\lambda_1$ $[s^{-1}]$ is known as Coriolis parameter. This velocity occurs at an elevation y_g above the earth's surface: y_g is of the order of 300-800 m. Within the vertical distance y_g velocity will change from 0 to u_g due to the boundary friction. Deaves and Harris [6] give the estimate of y_g as $y_g = u_*/6f$ where u_* is the shear velocity.

Velocity Distribution in ABL

Thermal stratification in the planetary boundary layer is important in low velocity flows. Thermally stratified boundary layer flow is of great concern to environmental engineers concerned with dispersion and diffusion in the atmosphere. However, in connection with forces on buildings and other structures, since one is interested in maximum force which occurs at high velocities, the thermal stratification can be assumed to be destroyed. Such a boundary layer without thermal stratification is known as neutral boundary layer which is discussed here onwards.

Near the boundary, one can assume that the shear is constant and equal to the boundary shear τ_o.

In this layer the logarithmic law prevails, namely

$$\overline{u}/u_* = \frac{1}{k} \ln(y/y') \qquad (2)$$

in which $u_* = \sqrt{\tau_o/\rho}$, $\overline{u}$ is time-averaged velocity at distance y from the boundary, and Karman constant k is 0.40, y' is the characteristic height of surface roughness such that at $y = y'$, $\overline{u} = 0$. This equation is valid for about 15-20 percent thickness of the atmospheric boundary layer beyond which the velocity distribution deviates from Eq. 2. The length scale y' depends on the geometry, concentration and height of roughness on the earth's surface. For rough boundary made up of closely packed uniform sand particles of size d, $y' = d/30$. When y is not much larger than the height of roughness, Eq. 2 is modified to

$$\overline{u}/u_* = \frac{1}{k} \ln\left(\frac{y - y_1}{y'}\right) \qquad (3)$$

where y_1 is called the zeroplane displacement. The zeroplane displacement y_1 in the above equation is introduced to account for origin shift that is expected to occur for rough surfaces. Values of y' and y_1 are related to surface roughness such as crop height. The logarithmic law is valid upto elevation $y = y_m$ where y_m is given by

$$y_m = bu_*/f$$

and $b = 0.015$ to 0.030. Because of these limitations of the logarithmic law, planetary layer is often modelled using the power law

$$\overline{u}/\overline{u}_h = (y/h)^n \qquad (4)$$

where $\overline{u}_h$ is the velocity at elevation h. Alternately it is written in terms of geostrophic wind velocity u_g and y_g, the height y at which it occurs, viz.

$$\overline{u}/u_g = (y/y_g)^n \qquad (4a)$$

It is found [7] that y_g and n are primarily functions of ground roughness. Values of y_g, y' and n as a function of terrain, as given by Davenpart [7] are shown in Fig. 2. These values can be used for a given terrain so that for known $\overline{u}_h$ and h, one can determine velocity distribution in the vertical. Other velocity distribution law, which is valid within wall region and outside upto about 300 m is proposed by Deaves and Harris [6]. According to this law

$$\overline{u}/u_* = 5.75\left[\log \frac{y-y_1}{y'} + \frac{y-y_1}{y_g}\right] \qquad (4b)$$

Turbulent Characteristics of Atmospheric Boundary Layer

From the consideration of dynamics of the system, the atmospheric boundary layer can be divided into two parts. This is a layer near

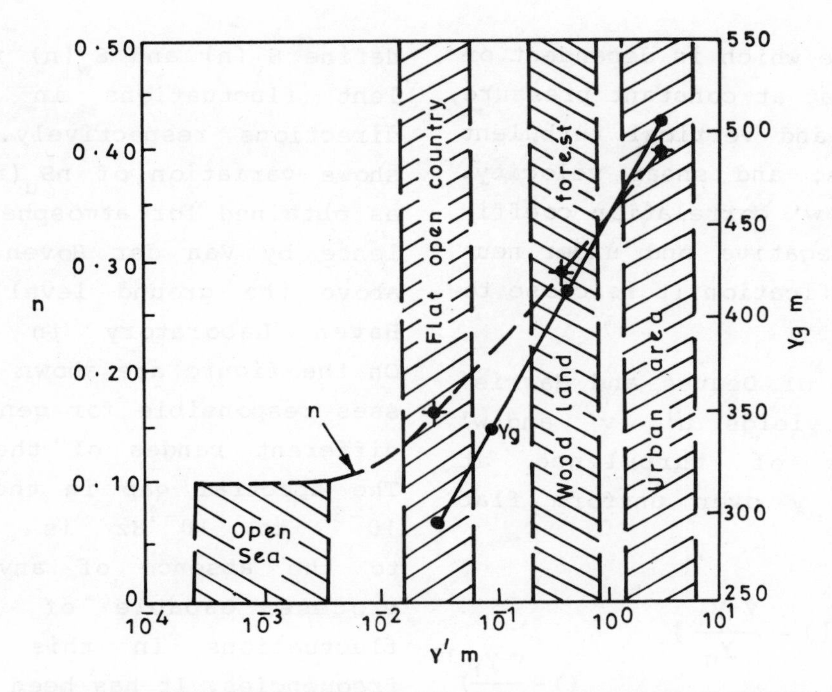

FIG. 2 - WIND PROFILE PARAMETERS FOR STRONG WINDS

the earth's surface known as the dynamic sublayer in which the effect of stratification due to humidity and temperature gradients can be neglected. This resembles the wall region of boundary layer in homogeneous fluid. All the dynamic parameters in the dynamic sublayer are well determined by u_*, ν the kinematic viscosity and mean height of surface roughness.

If the boundary acts as hydrodynamically smooth, adjacent to the boundary will be a thin sublayer within the dynamic sublayer in which the flow is laminar. This is known as laminar sublayer. The sublayer is destroyed if the boundary acts as rough which is the usual case in ABL due to the presence of a variety of large roughness elements on the earth's surface.

Above the dynamic sublayer is the surface layer in which the action of Coriolis force can be neglected. Above the dynamic sublayer heat and humidity cannot be regarded as passive substances. In the surface layer with $y >> (\nu/u_*)$ the laws governing the statistical parameters of hydrodynamic fields are determined by the components of turbulence of not too small scale. In turbulent boundary layer portion in which logarithmic velocity distribution law prevails, $\sqrt{\overline{u'^2}}/u_*$, $\sqrt{\overline{v'^2}}/u_*$ and $\sqrt{\overline{w'^2}}/u_*$ assume (4) approximately constant values of 2.3, 1.7 and 0.90 respectively. Outside this region they are functions of y/L_1 where L_1 is known as the Monin-Obukhov length scale which is dependent on specific heat at constant pressure, horizontal and vertical turbulent heat fluxes, and shear velocity. The u' and w' correlation coefficient is negative and under neutral stratification it is close to -0.50.

Development of Deaves and Harris' model [6] yields u', v' and w' components of turbulence at any height y over uniform flat terrain as

$$\frac{\sqrt{\overline{u'^2}}}{u_*} = 7.5\left[1 - \frac{y-y_1}{y_g}\right]\left[0.538 + 0.208\log\left(\frac{y-y_1}{y'}\right)\right]^{\left(1-\frac{y-y_1}{y_g}\right)^{16}}$$

$$\frac{\sqrt{\overline{v'^2}}}{u_*} = 1 - 0.32\left[1 - \frac{y-y_1}{y_g}\right]^2 \qquad (5)$$

$$\frac{\sqrt{\overline{w'^2}}}{u_*} = 1 - 0.55\left[1 - \frac{y-y_1}{y_g}\right]^2$$

It needs to be stressed that in the absence of adequate field data, it is difficult to comment on the accuracy of prediction of turbulent quantities using Eqs. 5. They need further verification.

One can also consider the contribution of kinetic energy of eddies of given frequency n by use of spectrum function. $S_u(n)$ can represent the kinetic energy contribution due to turbulent fluctuations in x direction with eddies of frequency n. Similarly, one can define $S_v(n)$ and $S_w(n)$ for turbulent fluctuations in y and z directions respectively. Figure 3 shows variation of $nS_u(n)$ with n as obtained for atmospheric turbulence by Van der Hoven at 100 m above the ground level at Brook Haven Laboratory in USA [8]. On the figure are shown the processes responsible for generation of different ranges of the spectra. The spectral gap in the range of 10^{-3} to 10^{-4} Hz is attributed to the absence of any physical process capable of generating fluctuations in this range of frequencies. It has been found that

$$\frac{n\,S_u(y,n)}{u_*^2} = \frac{4.0\,x_1^2}{(1+x_1^2)^{5/3}} \qquad (6)$$

where $x_1 = 1200\ n/\bar{u}_{10}$, $\bar{u}_{10}$ being the wind speed in m/s at 10 m above the ground. This equation gives spectral function for frequencies greater than spectral gap for the microscale fluctuations in the wind developed by thermal and mechanical effects. Spectra of vertical fluctuations upto 50 m are given by

$$\frac{n\,S_v(y,n)}{u_*^2} = \frac{3.36\,f_1}{(1+10\,f_1)^{5/3}} \qquad (7)$$

while spectra for lateral fluctuations are given by

$$\frac{n\,s_w(y,n)}{u_*^2} = \frac{15\,f_1}{(1+9.5\,f_1)^{5/3}} \qquad (8)$$

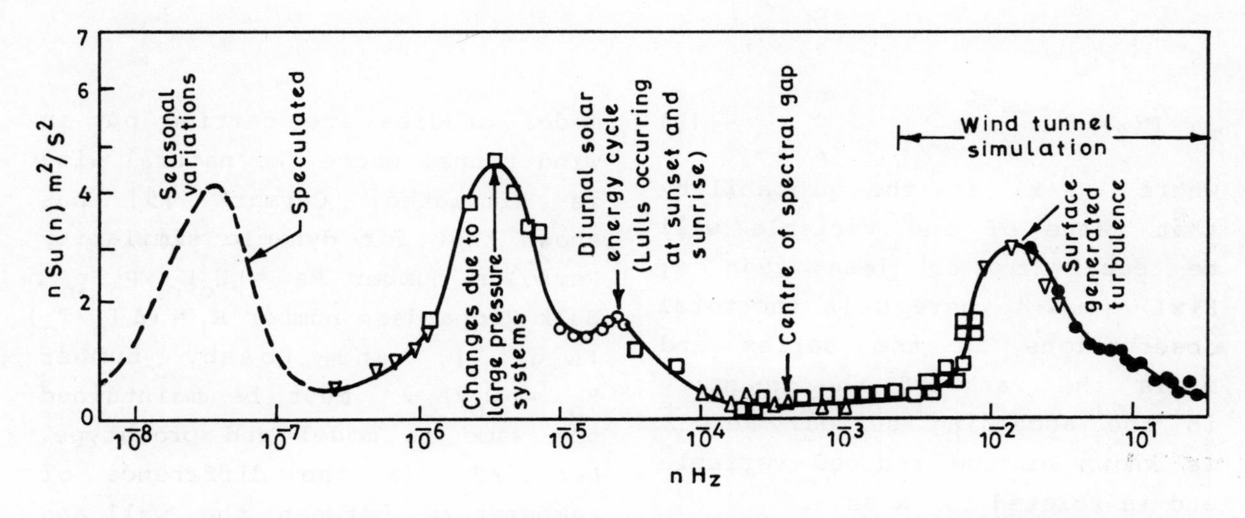

FIG. 3 - ENERGY SPECTRA OF LONGITUDINAL VELOCITY BY VANDER HOVEN [8]

where $f_1 = \frac{ny}{\bar{u}(y)}$ is sometimes known as Monin's coordinate.

Averaging of Wind Speed

Since at a given elevation the wind speed varies with time, in almost all the cases averaging in time is used. Since the turbulence spectrum is continuous over a large time period, see Fig. 3, the danger of a statistical instability of the obtained wind speed and the dependence of the average on the period of averaging arises. It has been found that average taken over a sampling period of 1 hour is fairly stable and for smaller sampling times T, $\bar{u}(T)$ is inversely proportional to T. In other words

$$\bar{u}(T) = \bar{u}_{3600} + C(T)\sqrt{\overline{u'^2}} \qquad (9)$$

where $\bar{u}_{3600}$ is average speed over 3600 s, and C(T) is found to depend on T as follows:

T,sec.	1.0	10	20	30	50	100
C(T)	3.00	2.32	2.00	1.73	1.35	1.02
T,sec.	200	300	600		1000	3600
C(T)	0.70	0.54	0.36		0.16	0

Wind Speed of Given Return Period

From the point of view of safety of a structure subjected to wind force, the wind velocity to be used in the analysis should not be maximum observed so far, but one which may occur once in T years (say 10 years or 50 years) depending on the importance of the structure. This is determined by first finding the maximum u_g or $\bar{u}$ observed each year for a number of years and forming a series arranged in an ascending order of magnitude. Such a series of extremal values is found to follow Gumbel's distribution

$$F(x) = e^{-e^{-X}} \quad (10)$$

where $F(x)$ is the probability that value of the variable will be equal to or less than x; $F(x) = m/N+1$ where N is the total observations in the series and m is the rank of the quantity in the ascending series. Here X is known as the reduced variable and is related to x as

$$X = a\,(x - x_{mo})$$

where x_{mo} and a are given as

$$x_{mo} = x - 0.454\,\sigma$$

$$a = 1/0.78\,\sigma$$

σ being the standard deviation of x and $\bar{x}$ is mean of x values. The return period will be $T = 1/(1-F(x))$. Figure 4 shows the probability distribution of maximum yearly wind speeds at Cardington during 1932-1954. The data seem to follow Gumbel's distribution remarkably well.

SIMULATION OF ABL IN WIND TUNNELS

The nature of wind to which a structure is exposed is strongly conditioned by the geometry of upwind and surrounding structures such as hillocks, large trees, cluster of buildings in the cities, etc. Since the wind conditions and its characteristics cannot be predicted correctly, model studies are carried out in wind tunnel where the natural wind is simulated. Cermark [9] has shown that for dynamic similarity Reynolds number $Re = U_o L_o \rho_o / u_o$, Bulk Richardson number $R_i = (\Delta T_o / T_o)(L_o g / U^2)$, and Rossby number $R_b = U_o / L_o \omega$ must be maintained the same in model and prototype. Hence ΔT_o is the difference of temperature between the wall and the fluid and T_o is the wall temperature : U_o and L_o are the characteristic velocity and length respectively. The conditions of thermal similarity obtained from energy equation give equality of

$$\text{Prandtl number } Pr = \frac{\nu_o}{(K_o / C_{po} \rho_o)}$$

$$\text{Eckert number } Ec = \rho_o U_o^2 / C_{po} \Delta T_o$$

Here K_o is the thermal conductivity of fluid of mass density ρ_o and C_{po} is specific heat at constant pressure. When the same fluid is used in model and prototype, Prandtl number Pr will automatically remain the same in both cases. Eckert number equality is essential at speeds approaching sonic speed. Hence, for wind engineering problems, where wind speed is not so high, equality of Ec is not essential. Since effect of earth's rotation, which is Rossby number cannot be

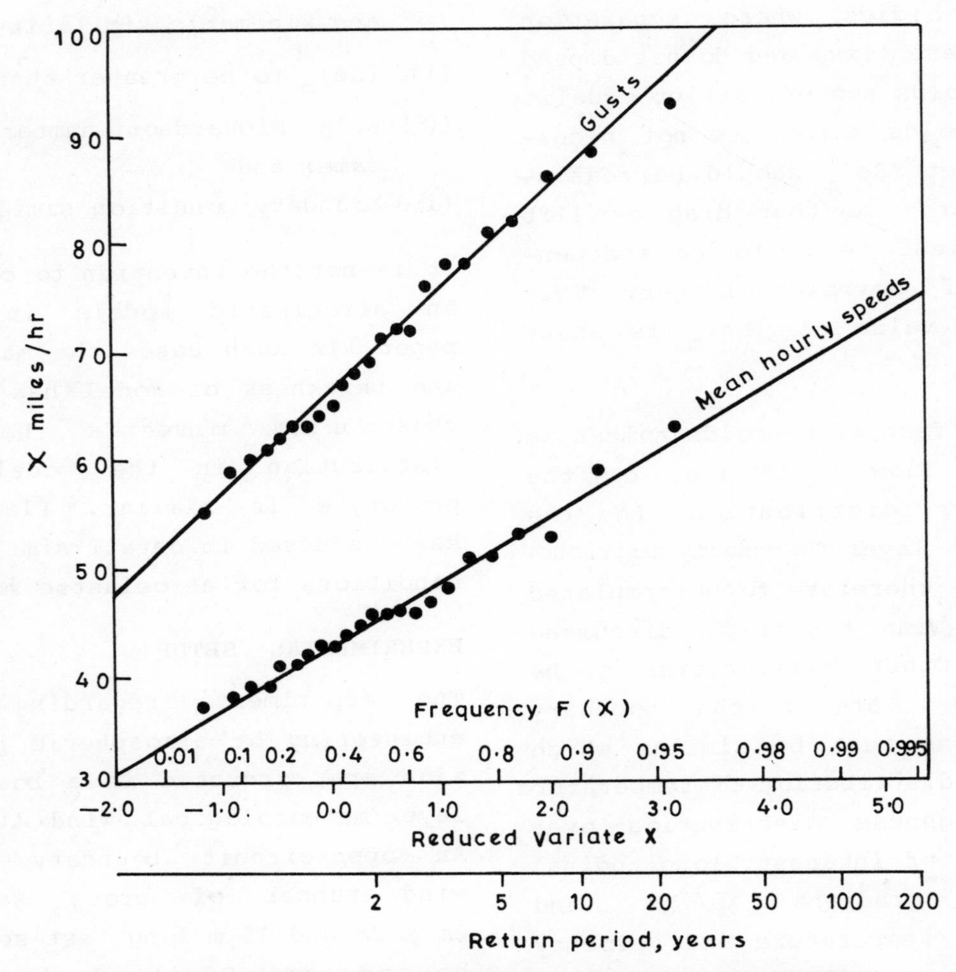

FIG. 4 - PROBABILITY DISTRIBUTION OF WIND SPEEDS

achieved. However, it may be mentioned that the earth's rotation causes the mean wind speed to change direction by 5° over a height of 300 m of the boundary layer. The tendency, therefore, is to relax the condition of equality of Rossby number also. Hence, one comes to the general conclusion that for geometrically similar models, Reynolds number be maintained the same, and in heat transfer problems Richardson number should be same.

In a wind tunnel where small scale (1:300 to 1:500) models are tested, $(Re)_m$ will be less than $(Re)_p$ if air under normal pressure is used: Otherwise compressed air will have to be used to produce high velocities so that $(Re)_m = (Re)_p$. Here subscripts m and p refer to model and prototype respectively. However, if one is dealing with sharp cornered or bluff bodies where separation points are fixed and do not depend on Reynolds number, strict equality of Reynolds number is not necessary. Yet $(Re)_m$ should be reasonably large so that drag or lift coefficient tends to be independent of Reynolds number. This minimum value of $(Re)_m$ is about $5x10^5$.

Other effect of Reynolds number is on the flow field i.e. on the velocity distribution in the boundary layer. The velocity distribu - tion has therefore to be simulated in the wind tunnel as discussed below. Other similarities to be maintained are : the boundary conditions must be similar which include distribution of temperature and roughness distribution over the area of interest, longitudinal pressure gradient $\partial \bar{p} / \partial x$ and vertical temperature and velocity distribution of the approaching flow. Surface roughness conditions are usually met if adequate care is taken. Surface roughness is scaled in accordance with the prototype roughness. However, when this results in an aerodynamically smooth surface, the roughness height in the model needs to be increased by increment equal to viscous zone thickness $10\, \nu / u_*$. In summary, the conditions to be satisfied are

(i) geometrically similar model and kinematic similarity;

(ii) $(Re)_m$ to be greater than $5x10^5$;

(iii) Bulk Richardson number to be same: and

(iv) boundary condition similarity.

It is not the intention to comment on aeroelastic models in this paper. In such cases the material and thickness of model has to be chosen in such a manner that the mass distribution in the model and prototype is similar. Plate [5] has discussed in detail similarity conditions for aeroelastic models.

EXPERIMENTAL SETUP

The experiments regarding wind engineering or atmospheric pollution are conducted in a boundary layer/meteorological wind tunnel. An open-circuit boundary layer wind tunnel of cross section 2m x 2m and 15 m long test section housed in Civil Engineering Department of University of Roorkee is shown in Fig. 5. The fan of the wind tunnel is of 125 HP capable of producing velocities of the order of 20 m/s in the wind tunnel. The details of this tunnel have been described in detail elsewhere [10]. With the use of a barrier wall, vortex generators and suitable roughening devices, it is possible to simulate the desired wind profile in the tunnel. One such typical wind

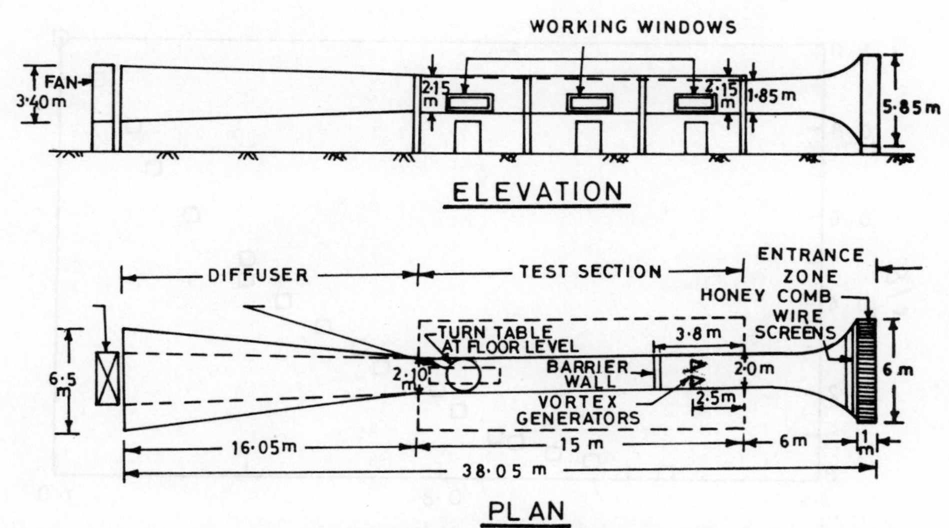

FIG. 5 – LINE DIAGRAM OF THE INDUSTRIAL WIND TUNNEL AT UNIVERSITY OF ROOEKEE [10]

profile with power-law index of 1/7 has been shown in Fig. 6. The UOR tunnel has been used for obtaining wind forces on different structures such as cooling towers, antennas, curved roofs etc. The variations of force and bending moment coefficients, viz. C_x, C_y, C_{my} on a rigid model of a typical cooling tower [11] measured in UOR tunnel are shown in Fig. 7. These coefficients are defined as follows:

$$C_x = \frac{F_x}{\frac{1}{2}\rho V^2 S} \; ; \quad C_y = \frac{F_y}{\frac{1}{2}\rho V^2 S}$$

$$\text{and} \quad C_{my} = \frac{M_y}{\frac{1}{2}\rho V^2 S D_o}$$

Here, F_x and F_y are the measured values of drag forces in x- and y-directions respectively, M_y, the base bending moment, ρ, the mass density of air, V, the reference velocity, S, the cross-sectional area of the cooling tower at throat i.e. $(\pi/4).D_o^2$, and D_o is external diameter of the throat.

The values of maximum and minimum pressure coefficients on a model of typical hyperbolic-paraboloid circular roof with side openings ranging from 0 to 100% [12] obtained in UOR tunnel were found to be in the ranges of (-0.2 to +0.90) and (-1.75 to -0.15) respectively. The boundary layer wind tunnels can also be of closed-circuit type such as shown in Fig. 8. The tunnel and method of testing are described by Cermak et al. [9,13,14].

The lower boundary of the wind tunnel is heated if necessary and ambient air temperature can

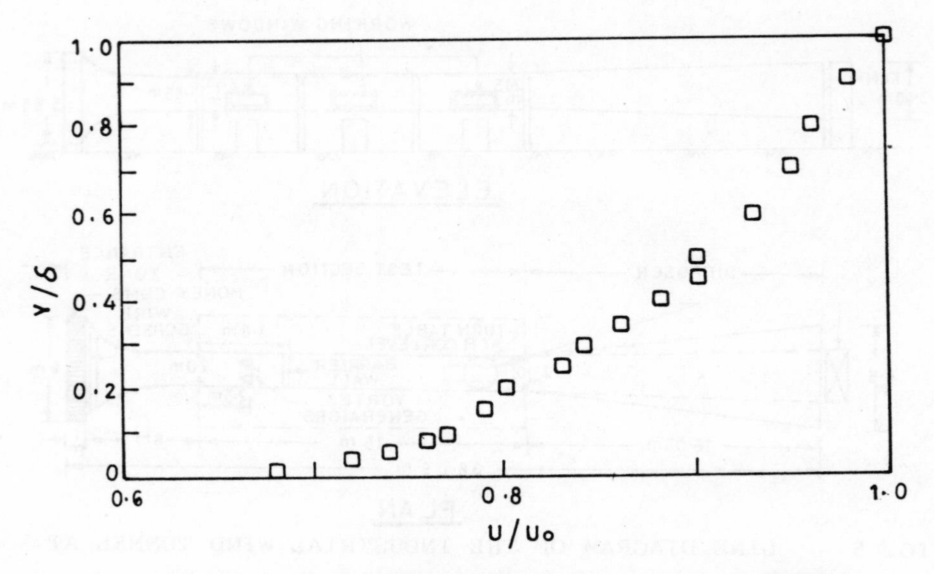

FIG. 6 - TYPICAL WIND PROFILE IN U.O.R. BLWT

also be modified. This is done by providing heating and cooling arrangements. For studies related to forces on structures such heating or cooling is generally not required. The top ceilling of the tunnel is adjustable so that $\partial\bar{p}/\partial x$ can be adjusted to desired value. The boundary layer thickness of the order of 1.0 m at wind speeds of 40 m/s or so needs to be obtained. To get such a thick boundary layer, one would require large length of wind tunnel since boundary layer thickness increases as $L^{0.80}$. Hence length requirement is reduced by using certain devices at the entrance to the tunnel. These include vortex generators or spires, curved screen or a screen with longitudinal bars whose spacing increases with increasing values of y, or sparesely placed gravel or other roughness at the entrance to the wind tunnel. The surface roughness downstream from the stimulator must be equivalent to the roughness which without the stimulator would have produced an equilibrium boundary layer with the same characteristics as those generated by stimulator. With these roughness and stimulator, velocity distribution should be plotted according to power law and if the exponent n is the same for field data and wind tunnel data, one can assume that similarity of mean flow conditions is achieved. Normally even tall buildings are submerged in atmospheric boundary layer. Hence boundary layer in wind tunnel has to be sufficiently thick so that model of a tall building is submerged in wind

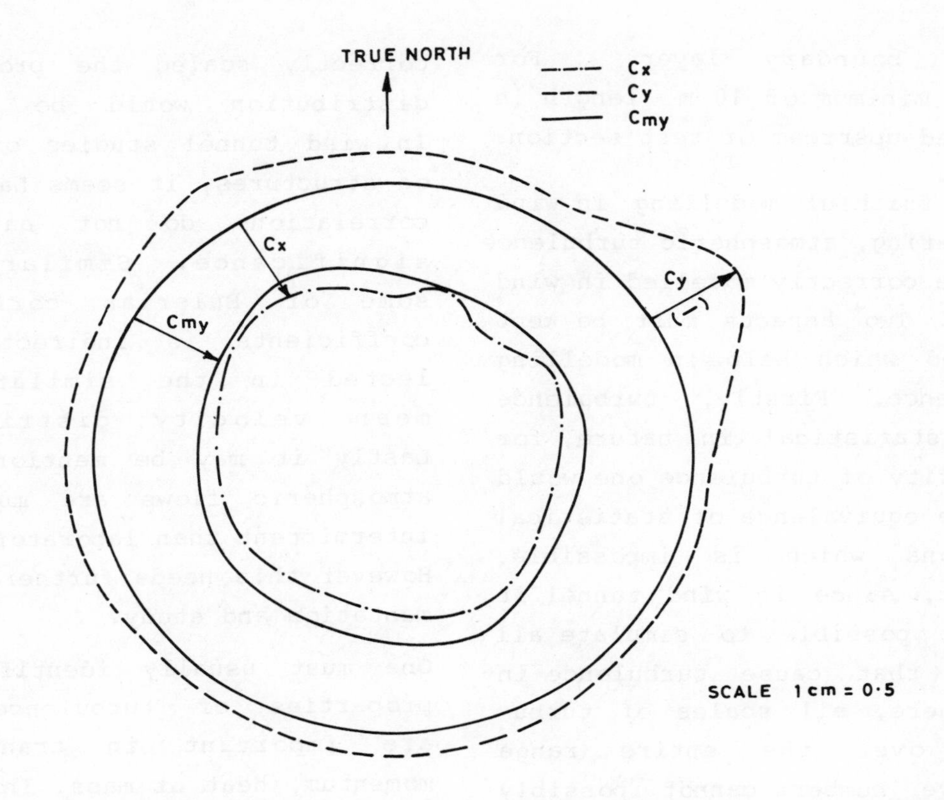

FIG. 7 – FORCE AND BENDING MOMENT COEFFICIENTS FOR A TYPICAL COOLING TOWER CORRESPONDING TO DIFFERENT DIRECTIONS OF WIND

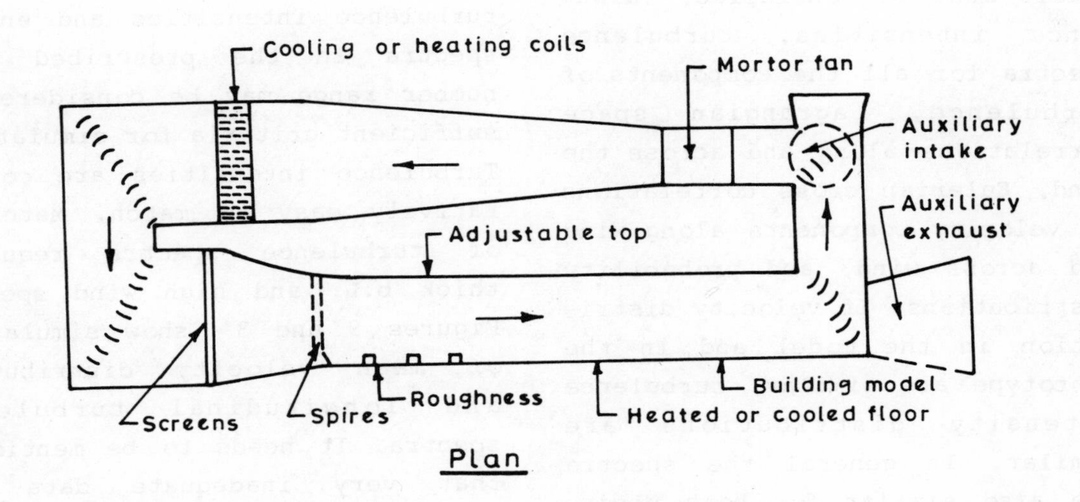

FIG. 8 – TYPICAL WIND TUNNEL FOR WIND ENGINEERING AND POLLUTION STUDIES [13]

tunnel boundary layer. For this a minimum of 10 m length is required upstream of test section.

For a faithful modelling in wind engineering, atmospheric turbulence must be correctly modelled in wind tunnel. Two aspects must be kept in mind which help in modelling turbulence. Firstly, turbulence being statistical in nature, for similarity of turbulence one would require equivalence of statistical functions which is impossible. Further, since in wind tunnel it is not possible to simulate all forces that cause turbulence in atmosphere, all scales of turbulence over the entire range of wave number cannot possibly be simulated.

When one deals with similarity of turbulence, the following parameters must be considered: turbulence intensities, turbulence spectra for all the components of turbulence, Lagrangian space correlations along and across the wind, Eulerian cross correlations of velocity components along wind and across wind, and probability distributions. If velocity distribution in the model and in the prototype are similar, turbulence intensity distributions are similar. In general the spectra are also similar for both undisturbed as well as disturbed flows. Futher if $\overline{u}$ and $\overline{u'^2}$ are correctly scaled the probability distribution would be similar. In wind tunnel studies on forces on structures, it seems Lagrangian correlations do not have much significance. Similarity of some of Eulerian correlation coefficients is indirectly reflected in the similarity of mean velocity distribution. Lastly it may be mentioned that atmospheric flows are much more intermittent than laboratory flows. However this needs further experimentation and study.

One must usually identify those properties of turbulence which are important in transporting momentum, heat or mass. In certain cases simulation of turbulent energy in a particular narrow eddy range would be required. For this, equivalence of normalised turbulence intensities and energy spectra in the prescribed wave number range may be considered as sufficient criteria for simulation. Turbulence intensities are comparatively easy to match. Matching of turbulence spectra requires thick b.L. and high wind speeds. Figures 9 and 3 show simulation of mean velocity distribution and longitudinal turbulence spectra. It needs to be mentioned that very inadequate data are available about the turbulence intensities and spectra in winds over large cities. Figure 3

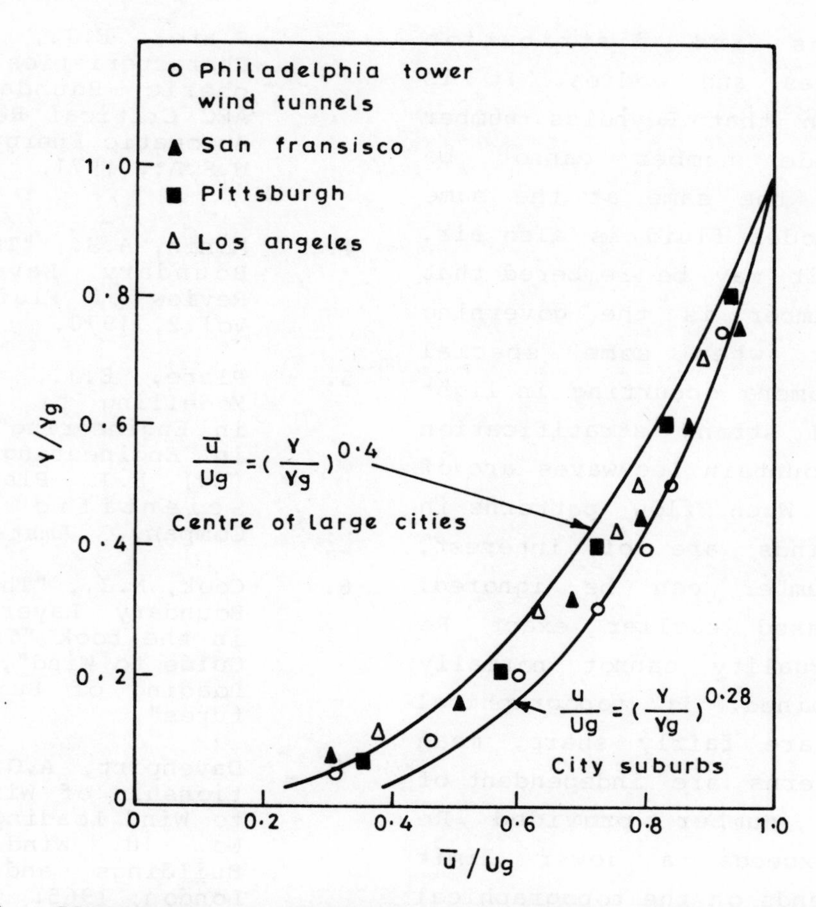

FIG. 9 - COMPARISON OF SIMULATED AND NATURAL WIND PROFILES

indicates that within model scale range of 1:100 to 1:1000 simulation is achieved down to scales associated from maximum turbulent energy to end of spectral gap. This simulation is considered adequate. In such simulation it is extremely important that all major upwind buildings be represented in the model since the wakes produced by them will add energy to the spectrum at wave lengths comparable to the dimensions of the spectrum.

SIMILARITY OF MOTION OVER TOPOGRAPHICAL FEATURES

This aspect has been examined in detail by Cermak et al. [14] and Nemoto [15]. They have found that to obtain similarity of flow patterns and turbulence structure in topographic modelling, Euler number, Froude or Richardson number, and Reynolds number need to be matched, in addition to geometric similarity. Matching of Euler number only requires geometric similarity of gross wind directions and distribution of vortices and eddies. It is well known that Reynolds number and Froude number cannot be maintained the same at the same time if model fluid is also air. However, it may be rembered that Froude number is the governing parameter when some special flow phenomena occurring in light winds and strong stratification such as mountain lee-waves are of interest. When flow patterns in strong winds are of interest, Froude number can be ignored. As discussed earlier exact Re number equality cannot normally be maintained. If topographical features are fairly sharp, mean flow patterns are independent of Reynolds number provided Re number exceeds a lower limit which depends on the topographical features. Smooth surfaces of topographical features should be roughened. In a few cases one may find it necessary to exaggerate the vertical scale.

REFERENCES

1. Smagorinsky, J. General Circulation Experiments with the Primitive Equations. Monthly Weather Reviews, Vol. 91, No.3, 1963.

2. Sethuraman, S., P. Michael, W.A. Tuthill and J. McNiell, "Atmospheric Boundary Layer Measurement During Summer Monex 79 at Digha India", Brook Haven National Laboratory, U.S.A., 1979.

3. Plate, E.J., "Aerodynamic Characteristics of Atmospheric Boundary Layers", AEC Critical Review Series, Automatic Energy Commission, U.S.A., 1971.

4. Monin, A.S., "The Atmospheric Boundary Layer", Annual Review of Fluid Mechanics, Vol.2, 1970.

5. Plate, E.J., "Wind Tunnel Modelling of Wind Effects in Engineering", Chapter 13 in Engineering Meteorology (Ed) E.J. Plate, Elsevier Scientific Publishing Company, Amsterdam, 1982.

6. Cook, N.J., "The Atmospheric Boundary Layer", Chapter 7 in the book "The Designer's Guide to Wind", by N.J.Cook, Loading of Building Structures".

7. Davenport, A.G., "The Relationship of Wind Structures to Wind Loading", Symposium No. 10. Wind Effects on Buildings and Structures, London, 1965.

8. Van der Hoven, I., "Power Spectrum of Horizontal Wind Speed in the Frequency Range 0.0007 to 90 cycles per hour., Jour. of Meterology, Vol.14, 1957.

9. Cermak, J.E., "Wind Tunnel Testing of Structures", ASCE-EMD Specially Conference, Univ. of California (CEP 75-76 JEC 15), March 1976.

10. Asawa, G.L., S.K. Pathak and A.K. Ahuja, "Industrial Wind Tunnel at University of Roorkee", Proc. of Asia-Pacific Symposium on Wind Engineering, Roorkee, 1985.

11. Asawa, G.L., M.M. Tilak, P.K. Pande and U.S.P. Verma,

"Experimental Determination of Wind Loads on a Pair of Cooling Towers", National Seminar on Cooling Towers held at New Delhi in 1990.

12. Asawa, G.L., Krishen Kumar and S.K. Sharma, "Wind Pressures on Hyperbolic-Paraboloid Circular Roof Models", Sent for publication.

13. Cermak, J.E., "Laboratory Simulation of the Atmospheric Boundary Layer", AJAA Journal, Vol.9, No.9, Sept. 1971.

14. Cermak, J.E., V.A. Sandborn, E.J. Plate, E.H. Binder, H, Chug, R.W. Meroney and B. Itd., "Simulation of Atmospheric Motion by Wind Tunnel Flows", Fluid Mechanics Programme, CUS, CER 66-JEC-VAS-EJP-GHB-HC-RNM-SI, Colorado State University, U.S.A., 1966.

15. Nemoto, E., "Similarity Between Natural Load Wind in the Wind Tunnel", Papers, Meteorological Geophysics, Vol.19, 1968.

EXTRAPOLATION OF WIND TUNNEL MODEL RESULTS TO FULL SCALE

G. N. V. Rao

Professor of Aerospace Engineering

Indian Institute of Science
Bangalore 560 012, INDIA

SYNOPSIS:

In this paper some issues concerning the methods for extrapolation of wind tunnel model results of non-aerospace structures, to full scale are discussed. These are classified as those due to, (1) Wind tunnel flow such as blockage,wind gradient and turbulence, (2) Model dissimilarities such as different mass, sectional model etc., (3) Scaling parameters like Reynolds number, Elastic parameter etc. and (4) Miscellaneous effects like dependence of aerodynamic force on the amplitude of oscillations. It is concluded that while corrections can be made for some effects like mass dissimilarity and Reynolds number, information available do not allow one to correct for all possible sources but the excercise of correcting the model results must be undertaken in all cases to the extent possible.

1. INTRODUCTION:

Ever since the advent of wind tunnel testing of models, the issue of extrapolation of the results to full scale conditions has engaged the attention of research workers. The earliest correction methods were developed for the lift and later the drag of airfoils (1), and beginning with the requirement of small blockage, the correction techniques have slowly improved as the demands of high angle of attack, bluff body flow and large blockage tests became necessary. The state of the art upto about 1965 is probably best summarised by Garner et al (2) for aerospace work, which is valid to a substantial extent for tests involving non-aerospace structures.There have been two significant developments in wind tunnel techniques since that time, namely the development of flexible wall wind tunnels and the development of blockage corrections,based on flow properties on a control volume enclosing the body (Boundary measurement method, BMM see (3)). Since that time accelerated uses of special purpose and general purpose wind tunnels for non-aerospace work,particularly for studying wind loads on civil engineering structures is more evident. There have been sporadic attempts to develop correction rules that are specific to the wind tunnel flows characteristic of the fully developed pressure system flow in the atmosphere. Much of the work has been directed towards corrections for drag (4,5,6) and pressure distribution (4). But the developments have not been very great, possibly because many of the corrections need to be validated from full scale data and there is a great paucity of relevant full scale data. If one may judge from the remarks of the various 'summaries' in (3) users are becoming more and more pessimistic about the ability of the wind tunnel to yield 'exact' values for design. What many of these users do not realise is that the 'accuracy' of wind tunnel results on models depends on how well the full scale or prototype (an inapt word in the writer's opinion) conditions have been reproduced. Few designers are prepared to commit themselves to the magnitude of structural damping, material properties or even provide the wind tunnel engineer with the properties of the atmosphere. Instead, they expect the wind tunnel engineer to commit himself to these values and blame him for refusing to do so. The cause for this source of uncertainity is that few owners are prepared to undertake or support measurement on full scale structures.

The sources of uncertainty in wind tunnel results are many and it seems appropriate to classify them for non-aerospace usage in a manner different from that used in aerospace work. In this paper, classification under the four broad categories of, (i) Wind tunnel flow, (ii) Model dissimilarity, (iii) Scaling parameters and (iv) miscellaneous effects will be considered (Fig. 1). At this stage, it is perhaps well to keep in view that the dynamic response tests form a larger proportion of tests on earth fixed structure models like chimneys than in aerospace work. It may be observed that, emphasis has not apparently be placed on a statement or description of the scaling laws as would normally be done but the 'scale' effect has been treated as one of the, no doubt, important contributing elements.

In the discussions that follow, the effect of non-fulfilment of scaling requirements has been postponed to the last since it is more convenient to apply them after assessing the effects of the other parameters. An uninitiated person may, to repeat what has been said before, become uneasy at the large number of parameters which can vitiate model tests in so far as extrapolation to full scale is concerned and may indeed wonder whether a full scale value obtained after applying so many corrections can be relied upon at all. Such fears are not wholly unjustified but the search for making corrections taking into account all possible causes must continue (7). Such corrected values can perhaps be relied upon to within 5% of the absolute percentages if the corrections do not exceed 25% to 30% of the value obtained without carrection, unless a large correction due to one or two of the causes is more precisely known, in which case one can perhaps accept even 100% difference between 'raw data' and corrected data. For example, one knows that the Strouhal number of circular cylinders at transcritical Reynolds numbers is about 0.27 while at subcritical Reynolds numbers is about 0.2 and response amplitudes vary as the inverse square of the Strouhal number, which means a correction of 1.82 or 82% ± 2%. Likewise, with $C_D = 1.2$ at subcritical R_N and 0.72 (say) at transcritical R_N, one can confidently apply a correction of 60% to measured drag forces on two dimensional cylindrical bodies tested at subcritical Reynolds numbers. With these preliminary remarks, a discussion of the various corrections required in the light of the classification given in Fig. 1 is made below.

2. EFFECT OF WIND TUNNEL FLOW:

There are three main effects of the flow in the wind tunnel, namely, blockage effects, effects of non-simulation of mean velocity profile and effect of non-simulation of turbulence properties (intensity, spectra and scale).

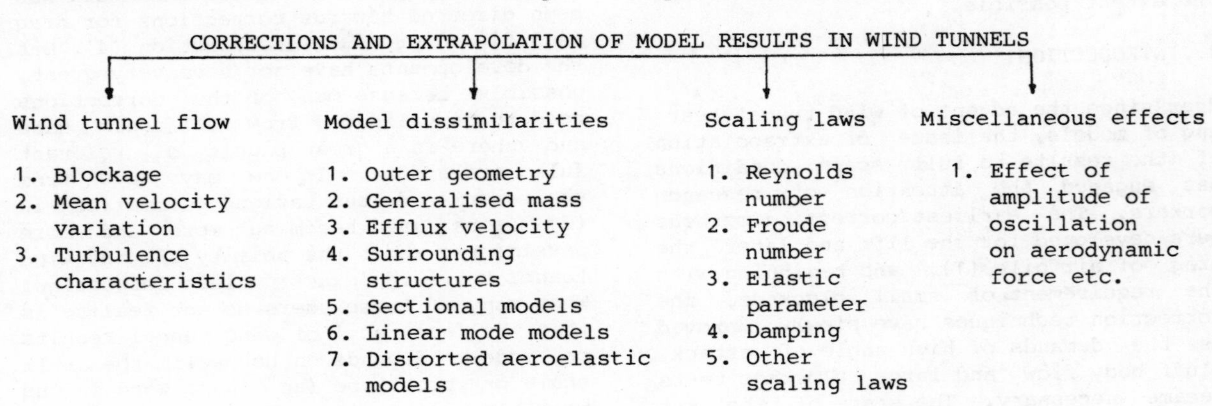

CORRECTIONS AND EXTRAPOLATION OF MODEL RESULTS IN WIND TUNNELS
Wind tunnel flow
1. Blockage
2. Mean velocity variation
3. Turbulence characteristics
Model dissimilarities
1. Outer geometry
2. Generalised mass
3. Efflux velocity
4. Surrounding structures
5. Sectional models
6. Linear mode models
7. Distorted aeroelastic models
Scaling laws
1. Reynolds number
2. Froude number
3. Elastic parameter
4. Damping
5. Other scaling laws
Miscellaneous effects
1. Effect of amplitude of oscillation on aerodynamic force etc.

Fig. 1. Factors that need to be Considered for Extrapolating Wind Tunnel Model Results to Full Scale

2.1 Blockage Effects:

Blockage effects mean the effect of the confinement of flow between the walls of the wind tunnel in contrast to the fact that the flow extends to infinite distance in the horizontal direction and to the infinite region above the surface of the earth, in the atmosphere. This effect depends on the ratio of the solid silhouette area of the bodies in the wind tunnel on a plane perpendicular to the flow to the cross-sectional area of the wind tunnel. The study of blockage or boundary restraint effects has a very vast literature and it is not possible to discuss them in detail here. The state of the art upto 1965 particularly with respect to wind tunnels with solid boundaries is probably best summarised by Garner et al (2).

The only new input since that time, so far as wind tunnels with solid boundaries are concerned has been the development of boundary measurement methods (Holst (3) for a summary of work in this direction). The most important of the correction, namely to drag, has been repeatedly investigated, since the original formulation by Maskell (8) for large bluff bodies. Limitations of Maskell's corrections is now well accepted(9). Lift effect,also covered by Maskell, has attracted less attention. It seems fairly clear that the nearness to one wall, common in civil engineering structures, requires a different formulation from that at the centre. Fig. 2 shows an example of 'corrected and uncorrected' drag of a bluff body. One of the constraints faced by the wind tunnel engineer in testing models of earth fixed structures or surface transportation vehicles is that the aeronautical engineer's normal thumb rule of 'less than 2% blockage' cannot be easily met. Even if the primary structure of interest has less than 2% blockage,the surrounding structures, which have to be modelled, add substantially to the blockage, exceeding the 'norm' of 2% on the whole, in any plane perpendicular to the flow or based on the silhouette area on a plane perpendicular to the direction of flow.

The effect of blockage on dynamic response of structures is less known. It is generally accepted that such effect may not become important for blockage upto about

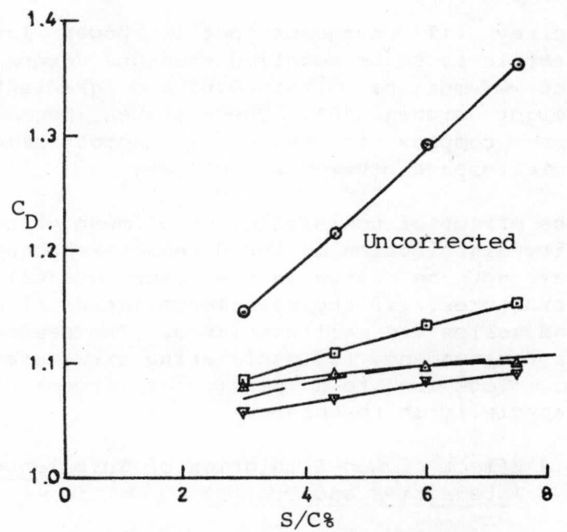

Fig. 2. Corrected Drag Coefficient according to different Criteria

8% to 10% for some parameters such as frequency of vortex shedding, but not for other parameters like oscillatory lift force which are corrected as for steady lift forces, for want of satisfactory corrections techniques. It may be mentioned here that the corrections depend also on the amplitude of the response, but such effects can be neglected if the amplitude is less than probably 0.5% of the tunnel dimension in the direction of response.

2.2 Effect of Non-Simulation of Mean Velocity Distribution:

Although this is not usually the case, it has been observed that the mean velocity variation with height, generated in aeronautical wind tunnels with insufficient 'fetch length' i.e. distance over which the 'atmospheric' boundary layer has developed, departs from the 'exact' simulation slightly when such techniques as those developed by Cowdrey (10) are used. The usual practice in such cases for direct or static force correction is to estimate the ratio of theoritical force due to exact velocity profile and that due to the actual profile and apply this ratio to the measured static forces and moments. Such corrections should not exceed probably 5%. An additional and important requirement is the matching of Jensen number, the ratio of the significant linear dimension of the model to the characteristic 'roughness' height Z_o of the logarithmic profile which is proposed to be modelled.

Surrey (11) suggests that if power law profile is to be modelled then the exponent α must be within 0.05 and gradient height within 10%. These issues become more complex if there is aerodynamic interference between structures.

The effect of non-simulation of mean velocity distribution on the dynamic response may not be large in the case of tall structures, if the turbulence intensities and scales are well simulated. The reason is the tendency for oscillating structures to 'lock on' to a particular frequency especially at resonance.

2.3 Effect of Non-Simulation of Turbulence Intensities and Scales:

This issue has already been addressed indirectly in emphasising the importance of the roughness scale Z_o. After a sufficient fetch length, this roughness will create an equilibrium turbulent boundary layer with certain distribution of fluctuations, intensity and length scale across the atmospheric boundary layer. It is sometimes presumed that a scaled down roughness will produce a scaled down thickness of the turbulent boundary layer with the same levels of fluctuations, intensities and scales. This is obviously not true for the simple fact that the roughness Reynolds numbers in the model and full scale are not the same. Assuming that appropriate higher roughness has been used, so as to create, the same intensities, fluctuations and scaled down turbulence length scales, one can be reasonably confident that the effect of intensities, fluctuations and turbulence length scales have been built into the experimental programme. But in reality, such idealisations are rarely achieved.

In general, one may state that there are no clearly defined methods for accounting for differences in the model and full scale properties. The nearest that one can come to, for along wind response is to scale the model results according to the codal methods involving response to gusts. For example, the well known expression

$$F_o = \bar{F}\left[1 + g\sqrt{\frac{K}{Ce}\left(B + \frac{SE}{\beta}\right)^{1/2}}\right] \ldots\ldots (1)$$

can be theoretically computed for the wind tunnel values of B and E and the value that should have been present. The ratio of these can be used to correct the actual measured results. A similar procedure cannot, in principle, be adopted for across wind response as Fig. 3 taken from Surrey (11) shows. The pattern of across wind response is so drastically changed due to turbulence, that one should be extremely cautions in looking for well correlated response patterns due to turbulence. Perhaps, one can attempt to evolve experimentally, effects of turbulence properties for specific body shapes at least in the first instance, before deciding whether any general rules can be formulated. At present, this issue must be considered as not quantitatively known but as one with some qualitative information. Figs. 4 and 5 show some more examples of the effect of turbulence intensity and scale.

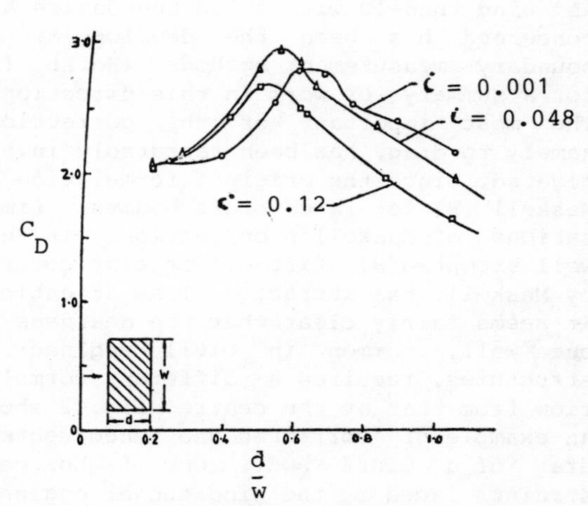

Fig. 3. Effect of Aspect Ratio and Turbulence Intensity on Drag Coefficient of Rectangular Blocks

3. EFFECTS OF MODEL DISSIMILARITIES:

3.1 General Remarks:

The range of model dissimilarities that one encounters is so vast that one can only attempt to describe the possible corrections for full scale only in some common cases. The most common dissimilarities can be classified into, (i) Distortion of outer geometry, (ii) Distorted generalised mass, (iii) Incorrect efflux velocities (where they have to be model-

over the structure of interest can be very critical. This issue becomes quite complex if some of the interfering structures have circular or even significant rounded portions. It must also be emphasised that modeling the far field need not ensure correct modeling of local interfering flow. It is obviously not possible to suggest or aim to obtain information as to how to extrapolate the results with improper or insufficient modeling of surrounding structures to full scale conditions. One can only state that care must be taken in the experiments to ensure that the interfering flow is properly modeled.

3.1.5 Sectional models:

The along wind load on sectional models, which are usually quite bluff bodies with fixed flow separation points can be taken directly for design. The same consideration can be taken to be valid for measured steady state lift forces. But oscillatory responses have to be corrected for mode shape differences, Rao (12). It has been shown, (12) that the model response amplitude of a sectional model to that of a fully aeroelastic model is

$$\frac{\eta_{aeroelastic}}{\eta_{sectional}} = \frac{2\int F_n\,(Y/L)\,d(Y/L)}{3\int F_n^2\,(Y/L)\,d(Y/L)} \quad ..\ (2)$$

where $F_n(Y/L)$ is the mode shape of vibration of full scale structure. In such models, prediction of flutter speed is also very important. To the extent that the flutter derivatives are measured in properly simulated (in terms of turbulence quantities and mean velocity variations (where necessary),the results can possibly accepted although the issue of reattachment point of the separated flow, remains to be cleared up in terms of its relation to Reynolds number.

3.1.6 Linear mode models:

The along-wind response of linear mode models are treated, for extrapolation to full scale, in a manner similar to that of any sectional model, including correction for Reynolds number, which will be discussed later. The across-wind response has to be corrected for mode shape (Rao (12)), which can be written as

$$\frac{\eta_{aeroelastic}}{\eta_{linears\ mode}} = \frac{2\int F_n\,(Z/L)\,d(Z/L)}{3\int F_n^2\,(Z/L)\,d(Z/L)} \quad ..\ (3)$$

3.1.7 Distorted aeroelastic models:

Due to difficulties in fabricating fully replicated aeroelastic models or enforced partial distortions due to limitations in tunnel speed, suitable multiplying factors for response have to be worked out. One example will be given here and another will be presented in section 4 under 'Designed Scaling Ratios'.

The example which will be taken up is that of a tall free standing R.C.C. Chimney with the following characteristics. Height 200 m; top diameter 6.0 m; bottom diameter 12.0m; thickness at the top 200 mm; thickness at the bottom 600 mm. Natural frequency in the first mode 0.40 Hz, Second mode 1.55 Hz. The issue is interference between two similar chimneys spaced at 75m apart centre to centre. Using the approximate criterion that the critical wind speed is $5\,n\,D_e$ where n is the natural frequency and D_e is the outer diameter at 5/6 height, one observes that the effective diameter is 7.0 m and the critical wind speeds are 14.0 m/sec & 54.25 m/sec. Both these wind speeds are observed to be possible over the top 1/3 height of the chimney and it is therefore decided that the wind tunnel speeds must be adequate to excite both the first and second modes. Assuming that the wind tunnel height and width are 1.8 m x 2.8 m, one finds that the model scale has to be limited to about 1:200. If the maximum speed that the wind tunnel is capable of is 100 m/sec, one has the possibility of making a 'distorted aeroelastic model' of say Aluminium (whose modulus of elasticity is about twice that of R.C.C.), which has geometrically scaled down dimensions, i.e., Height 100 cm; top diameter 3.0 cm; Bottom diameter 6.0 cm; thickness at the top 1.0 mm; thickness at the bottom 3.0 mm. We then get the following scale ratios, with N = 200.

$$\frac{\omega_m}{\omega_f} = \left(\frac{1}{N}\,\frac{E_m}{E_f}\,\frac{\rho_f}{\rho_m}\right)^{1/2} ; \quad \left(\frac{\omega_m D_{em}}{U_{em}}\right) = \frac{\omega_f D_{ef}}{U_{ef}} \text{ and}$$

$$\text{Hence } U_{em} = \frac{D_{em}}{D_{ef}}\,\frac{\omega_m}{\omega_f} ; \quad U_{ef} = \frac{1}{N}\,\frac{\omega_m}{\omega_f}\,U_{ef}$$

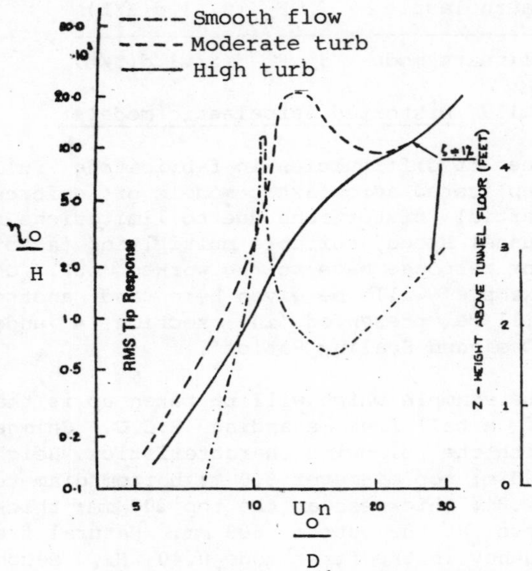

Fig. 4. Effect of Turbulence Intensity on Across-Wind Response of a Square Prism due to Vortex Shedding

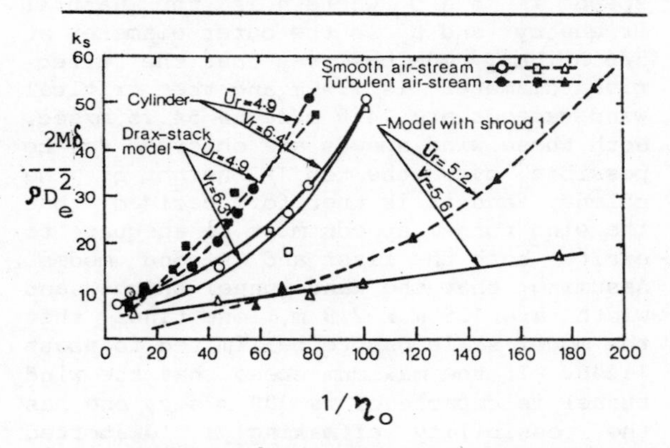

Fig. 5. Effect of Turbulent Air Stream on Across-Wind Response of a Circular Cylinder

led), (iv) Insufficient modeling of surrounding structures, (v) Sectional models, (vi) Linear mode models and (vii) Distorted aeroelastic models. Possible corrections, where they can be justified are discussed below.

3.1.1 Distortion of Outer Geometry:

In most cases, models are fabricated such that the outer geometric shape is preserved. Occationally, it becomes necessary to distort the geometric shape. Examples are mounting large sized stub cylinders on geometrically scaled down cables to achieve correct mass scale, deliberate increase of rounding radius to achieve reasonably large Reynolds numbers based on rounding radius etc. No hard and fast rule can be prescribed to correct for such situations but an approach, at least for along wind load, of correcting the measured values by the ratios of theoretically or similarly carried earlier tests with sufficient scaling, seems acceptable. It is obviously not justifiable to correct dynamic response differences due to distortions in geometric scaling.

3.1.2 Distortion due to generalised mass difference:

In some of the dynamic tests involving models of suspension bridges, chimneys and cooling towers, one finds that the generalised mass in each of the modes of vibration (calculated, not inferred from frequency measurement) achieved is different from the desired value by as much as 20%. Amplitudes due to vortex excited, galloping and such responses depend on the inverse square root of the generalised mass and corrections can be made accordingly. The along wind response according to random input analyses is not so easily estimated and one has therefore to estimate the theoretical responses due to 'exact' and actual general masses and use the ratio to correct the measured response, using Eqn (1).

3.1.3 Incorrect efflux velocity:

Dimensional analyses suggests that the ratio of the momentum of the efflux velocity from the structure (generally from the top) per unit area to the momentum of the free stream should be preserved. No work on this issue seems to have been done except that, in the case of stacks, the efflux has shown some effect, possibly by inducing an upward flow near the top and reducing the back pressure. Since, it is usually not difficult to model the stated momentum ratio, this parameter is not considered to be important for the purposes of this paper.

3.1.4 Insufficient modeling of surrounding structures:

The interference of the surrounding structures and terrain in modifying the flow

Recalling that the maximum response oscillatory amplitude depends inversely on the generalised mass, one finds that oscillatory amplitude ratio can be be written as

$$\frac{\eta_{om}}{\eta_{of}} = \frac{1}{N} \frac{\rho_f}{\rho_m}$$

(for same strouhal number and similar mode shape). The direct stress can be easily shown to be the same for model and full scale at corresponding points at the same dynamic pressure. However, since the model is stiffer, it must be tested, according to the requirement of equality of parameter $(E/\rho U^2_{max})$, upto speeds equal to $(E_m/E_f \ \rho_f/\rho_m)^{1/2}$ times the design full scale speed.

4. SCALING PARAMETERS:

The correction for non-fulfilment of the conditions for equality of scaling parameters needs to be addressed with care. As is well known, the parameters are, (i) Reynolds number, (ii) Froude number, (iii) Elastic parameter $EI/\rho U_o^2 L^4$ (iv) Damping, (v) Density ratio between that of the full scale and model, (vi) Other scaling ratios and (vii) Mean velocity and distribution of turbulence properties. The effect of the turbulence properties has already been discussed. The damping and mass parameter are sometimes combined into a single parameter $(2M\delta/\rho D_e^2)$ where M is the equivalent generalised mass in the mode of interest. The effect of each of the parameters and the correction rules which may be adopted (but not, like some others, fully justified but considered reasonable) are described below.

4.1 Reynolds Number Effect:

The Reynolds number effect applies for rounded bodies, or bodies with significant curved portions such as circular cross-section chimneys, antenna dishes etc. The approaches that are presented here are the ones followed at the Indian Institute of Science.

4.1.1 Steady drag force:

Here corrections are applied by reference to the variation of C_D with R_N (Fig. 6). Experiments in atmospheric pressure wind tunnels cannot be done at full scale Reynolds numbers and our practice has been to carry out tests at subcritical R_N making sure that there is laminar separation on the model (which can be ascertained by seeing the separation zone at about 80° to 90° on bodies of circular cross-

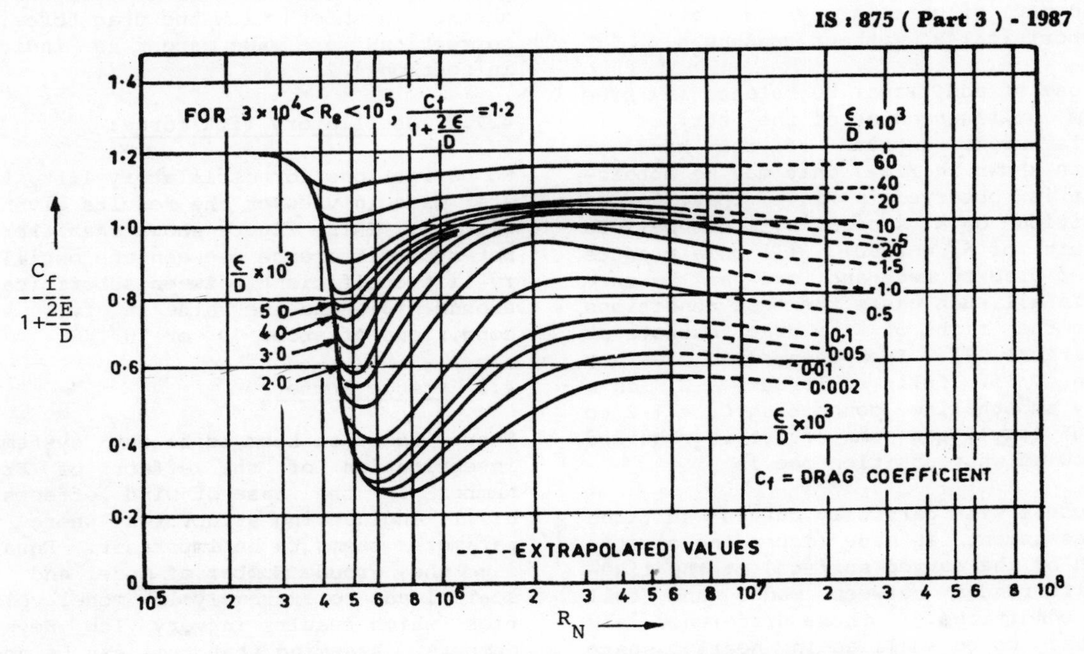

Fig. 6. Variation of C_D of Circular Cylinder with R_N and Surface Roughness

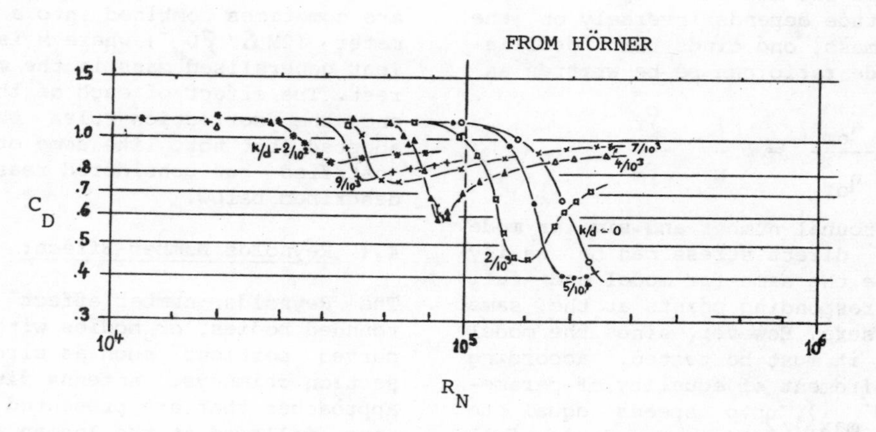

Fig. 7. Effect of Different Transition R_N on Drag Coefficient of Circular Cylinders

section but not so easily ascertained in the case of other bodies) then the full scale drag force is estimated as (13),

$$\frac{\text{Full Scale Drag Force Coefficient}}{\text{Model drag force Coefficient}} = \frac{\text{Known } C_D \text{ of circular cylinder at full scale after allowing for roughness effect}}{\text{Known } C_D \text{ of circular cylinder at subcritical } R_N \text{ without roughness}}$$

The use of additional turbulence in free stream or roughenning of the surface is not favoured, because of the kind of results shown in Fig.7 that may be obtained. It is observed in Fig. 7 that forced transition on a smooth cylinder well to the left of $R_N = 2x10^5$, will only produce smaller drag buckets and one has be sure that in all such cases,the test conditions are to the right of the drag buckets at subcritical R_N. Even then, one does not obtain C_D of full scale but a C_D on a nearly smooth line connecting C_D =1.2 to C_D=0.65 (say), and some effort is required to know at what position one is.

For models with partially curved surfaces an assessment is made (depending on the length of the curved surface) of the likely differences between model and full scale conditions. If these differences are expected to be small as in nearly sharp edged bodies or antenna dishes, no R_N corrections are made. Otherwise, the correction is reduced in proportion to the curved perimeter. Thus, if the perimeter (or the cross section of the body) has 30% curved portion, and the C_D measured is 1.6, on the model at a R_N of $7x10^4$, the full scale C_D is assessed as 1.6 x 0.7 + 0.3 (0.65/1.20) = 1.2825.

If the along wind gust response is being measured, then corrections only for the hourly mean wind component as above are made, by estimating theoretically or measuring in smooth flow the drag force. No corrections are made except as indicated in section 1.2.

4.1.2 <u>Oscillatory lift force:</u>

No corrections for oscillatory lift force are made in view of the results given in Fig. 8. This figure shows that there is not much difference between the oscillatory lift coefficient between subcritical R_N around 10^5 and the value in full scale conditions of about 10^7 or 10^8.

4.2 <u>Froude Number:</u>

There does not seem to be any systematic investigation of the effect of Froude Number in the case of wind effects on civil engineering structures where this parameter seems to be important. Equalising the Froude Number of model and full scale leads to extremely low model velocities which results in very low Reynolds numbers. Assuming that this can be accepted because of the sharpness of the body edges, then the relation between the model and full scale properties will be as follows: $U_m = U_f/N^{1/2}$; $n_m = n_f N^{1/2}$ corresponding stiffness and torsion parameters ($EI/\rho U_o^2 L^4$ and $GJ/\rho U_o^2 L^4$) as well as the mass ratio, ρ_m/ρ_f have to be modelled. In the absence of information on Froude number effect, no corrections are made due to differences in model and full scale Froude numbers; it seems that such deviations are not very critical.

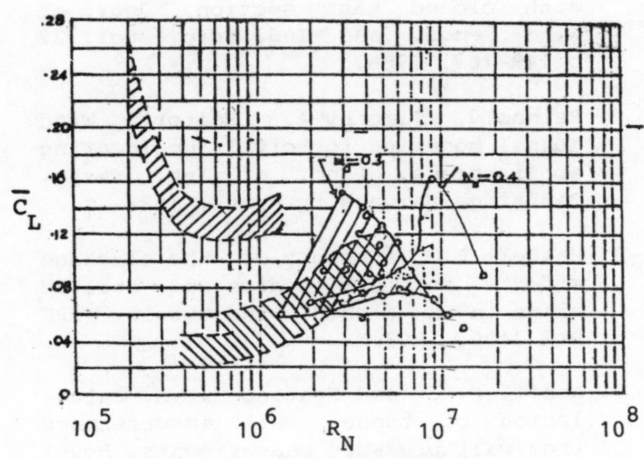

Fig. 8. Variation of Drag Coefficient with Reynolds Number Showing Flat Upper Bound Beyond $R_N = 2 \times 10^5$

4.3 Elastic Parameter and Other Ratios:

In general, except in the cases of Reynolds number, Froude number and turbulence parameters, one can arrive at suitable modeling ratios depending on the nature of the problem, wind tunnel limitations and available materials. In the case of vertical structures one starts with the need for equality of $EI/\rho U_o^2 L^4$, $GJ/\rho U_o^2 L^4$ (where necessary) and arrives at the required E, I or possible U_o. Let us assume that $U_{om} = K_1 U_{of}$. Then equality of stroubal number gives us the natural frequency to which the model has to be designed, say $n_m = K_2 n_f$. Such considerations usually lead to non-fulfilment of generalised mass for which amplitude corrections have to be made. A similar approach is made in the case of horizontal structures such as suspension bridges where one starts with equalising the Froude number and evaluates the frequency scaling, and the rest of the parameters from equality of $EI/\rho U_o^2 L^4$, $GJ/\rho U_o^2 L^4$.

4.4 Damping:

Normally, it is preferable to make the damping of the model equal to that of full scale, estimated or known. Alternately one can try to equalise the mass damping parameter $2M\delta/\rho D_e^2$ where M is the equivalent generalised mass ($\int m_f\, dh/\int F^2\, dh$), where F is the mode shape and this has to be made equal for each expected mode of oscillation. Although experiments have sometimes shown non-linear variation of the response amplitude with damping, it is usually sufficient to assume an inverse variation of response amplitude with damping for extrapolation to full scale.

5. MISCELLANEOUS EFFECTS:

There are some additional effects, not all of which will be clear initially but for which one has to be alert to make corrections. One example is the interaction of the amplitude of response with aerodynamic force (Fig. 9). In such cases if the ratio of the response amplitude of model to a characteristic length of the model is known to be likely different from that in full scale, corrections have to be made. Normally such corrections are not made if the directly scaled up amplitude (or stress) on full scale is more than the estimated value after applying other corrections to retain some conservatism. In some cases, information may not be available of the amplitude effect and one has to make a judicious decision as to whether corrections (in some arbitrary' but 'reasonable' manner) should be made. Such a situation will probably arise if the estimated full scale response is more than that of the response from model values obtained by direct scaling.

6. CONCLUDING REMARKS:

An uninitiated person who had the patience to go through all possible corrections to model results will no doubt wonder as to how far he can trust the final numbers. He must not forget that, with all the uncertainties, most model results will yield, (i) generally somewhat conservative results, which are within reasonable uncertainties accepted by designers, (ii) more importantly, reveal unknown and unsuspected and possibly a totally new phenomenon (like galloping, stall flutter etc) and

(iii) there is no substitute in the foreseable future (say 25-30 years) of quantifying the effects of aerodynamic interference effects reasonably well. It is also

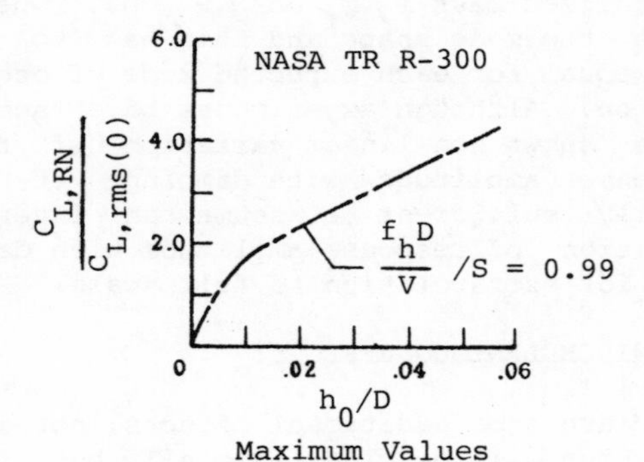

Fig. 9. Effect of Oscillatory Amplitude on Oscillatory Lift Coefficient

necessary to remind oneself that the sacred `codal values' are also wind tunnel derived values. Thus with a realistic and positive approach to wind tunnel results, one can expect to come up with a safe and economical design.

REFERENCES:

1. Glauert H. Wind tunnel interference on wings bodies and Airscrews. ARC R&M 1566, 1933.

2. Garner H.C., Rogers E.W.E., Acum W.E.A. and Maskell E.C. Subsonic wind tunnel wall corrections. Agardograph 109, 1966.

3. Holst H. (Editor). Wind tunnel wall interference in closed, ventilated and adaptive test sections. NASA C.P. 2319, 1983.

4. Melbourne, W.H. and Mckeon R.J. Wind tunnel blockage effects and drag on bluff bodies in a rough wall boundary layer. Prof. of 3rd Int. Conf. on wind effects on Str. Tokyo, Vol I, 1971.

5. Kirshnaswamy T.N., Rao G.N.V. and Reddy K.R. Blockage corrections for large bluff bodies near a wall in a closed jet wind tunnel. Jour. of Aircraft. Vol 10, No-10, P 634-636, 1973.

6. Mercker E. A blockage correction for automotive testing in a wind tunnel with closed test section. Jour. of wind engg. and Ind. aero. vol. 22 p 149-167, 1986.

7. Reinhold, Timothy A. (Editor). Wind tunnel modeling for civil engineering applications. Proc. of Int. Wkshp, Cambridge University Press, 1982.

8. Maskell E.C. A theory of the blockage effects on bluff bodies and stalled wings in a closed wind tunnel. ARC R&M 3400, 1963.

9. Ashill P.R. and Keating R.F.A. Calculation of tunnel wall interference from wall pressure measurements. Royal Aircraft Establishment, TR 85086, 1985.

10. Cowdrey C.F. A simple method for the design of wind tunnel velocity profile grids. NPL Aero. Note 1055, 1967.

11. Surrey D. Consequences of distortions in the flow including mismatching scales and intensities of turbulence. Proc. of Int. Wksp. on wind tunnel modeling criteria and techniques in Civil Engineering Applications, Gaithersburg, Maryland, USA, Edited by Timothy A. Reinhold, PP 136-185, 1982.

12. Rao, G.N.V. Extrapolation of wind tunnel tests on sectional and linear models to full scale. Jour. of Aero. Soc. of Ind. Vol 24, No.1, PP 2G2-2G5, 1972.

13. Rao G.N.V. A survey of wind engineering studies in India. Sadhana, Vol 12, parts 1 and 2, PP 201-218, 1988. Pub. by the Indian Academy of Sciences, Bangaldore 560 - 012.

Wind Loads on Structures, 1990

WIND LOADS ON LOW RISE BUILDINGS

Carl Kramer

Fluid Mechanics Laboratory
Goethestrasse-1
D-5100 Aachen, GERMANY

H.J. Gerhardt

Institute of Industrial Aerodynamics
Aachen, GERMANY

The paper deals with the wind loading mechanism on the outer surfaces of low-rise buildings which usually have a rectangular planform and flat or inclined roof surfaces. The outer surface of those buildings can be airtight or wind permeable like tiled roofs or wall claddings providing rain protection. For these cases the wind load is described as the resulting pressure difference between external pressure and internal or underelement pressure. Based on a physical description of the wind loading mechanism on low-rise buildings a comparison of the German standard DIN 1055, the Indian standard IS 875-1964 and the ISO standard 4354 (draft 1990) is given. Finally the essential rules for wind resistant design of cladding and sheeting for low-rise buildings are formulated and an example of a relevant method for wind resistance testing of such building elements is described.

1.0 INTRODUCTION

In all parts of the world, the majority of buildings are low-rise buildings. The design loads for such buildings are normally the wind loads and - in cold climates - possibly the snow loads. An enormous amount of damage is encountered every year on such buildings, in particular to their roof and facade coverings due to wind action. Thus, it appears that in spite of the significance of low-rise buildings the knowledge of either the wind loads to be expected and/or the know-how to master those loads are inadequate. In many areas of the world low-rise buildings - as opposed to the more spectacular high-rise structures - are non-engineered structures, meaning that the available engineering knowledge has not been introduced into the design. It is interesting to note that low-rise buildings are quite often "not engineered" in areas of extreme wind conditions, whereas in the moderate wind climates of central Europe <u>all</u> low-rise buildings have to be carefully engineered. To use the existing knowledge of wind loads and the wind loading mechanisms and to apply this knowledge in designing wind resistant buildings and building components would lead to a decrease in building damage and thus to an appreciable saving in public and private expenditure.

This paper gives a short description of the wind load acting on walls and roofs of low-rise buildings and discusses various methods of codification. The main body of the work is concerned with the wind resistant design of cladding and sheeting systems, both, for facades and roofs.

2.0 WIND LOAD ON BLOCK-TYPE BUILDINGS

2.1 Physical Background

The flow field around buildings is determined by the flow displacement due to the building. Locally, it may lead to large velocities and correspondingly low pressures. On block-type buildings, the flow will separate on all windward edges, thus increasing the effective displacement. For buildings of larger aspect ratios the flow may reattach at the sidewalls for wind direction parallel to the long sides. This reattachment will lead to an increase in pressure, in particular concerning the base pressure. Tall buildings will displace the windflow predominantly to the sides. In contrast, low-rise buildings will lead to large flow displacements over the building roof.

The outer, wind exposed layer of many roofing and facade systems is windpermeable. The wind load on a roofing or facade element of such systems is determined by the pressure difference across the individual element. The external pressure is not only due to the building flow field but may also be influenced by the element flow field, e.g. on roof tiles, see Figure 1. For smooth outer sur-

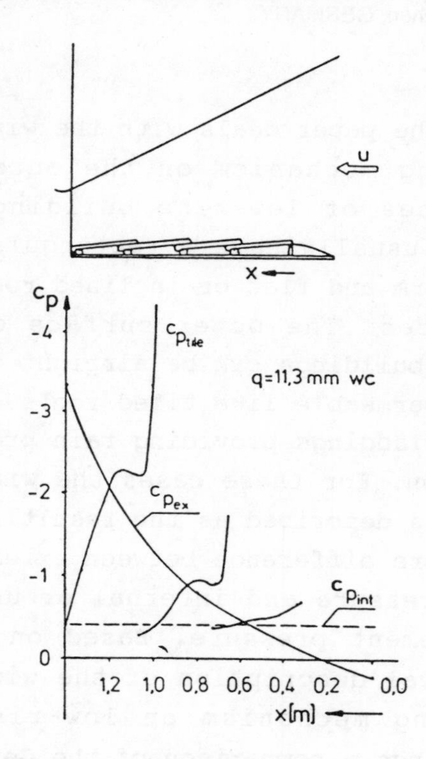

Figure 1: External and internal pressure distribution on a tiled roof area, simulated in a wind tunnel

faces, like curtain-wall facades, the external pressure acting on an element coincides with the building induced external pressure distribution. The underelement pressure is influenced by the external pressure distribution and the resistances for flow equilibration across the building surface and in the gap between external, windpermeable surface and windimpermeable building wall. Some examples of roof and wall coverings are given in Figure 2. Typical roof coverings for inclined roofs are tiles, Figure 2 a. Here, the resistance to the underelement flow - the batten space flow resistance - is much smaller than for the flow through the tiled surface.

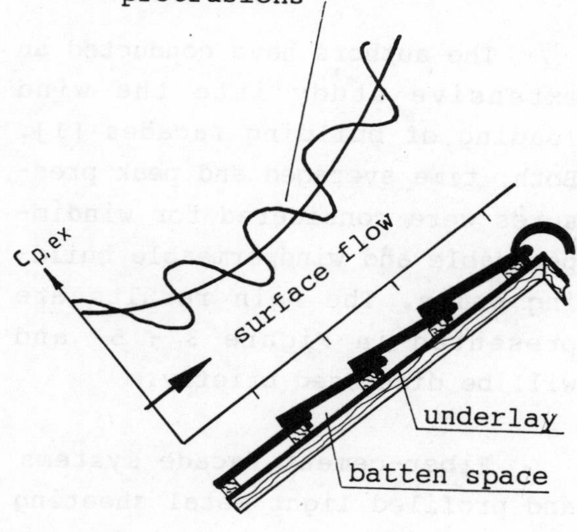

Figure 2a: Wind permeable roof coverings

Using pavers to ballast loose laid roofing membranes on flat roofs, Figure 2 b, leads, in contrast, to

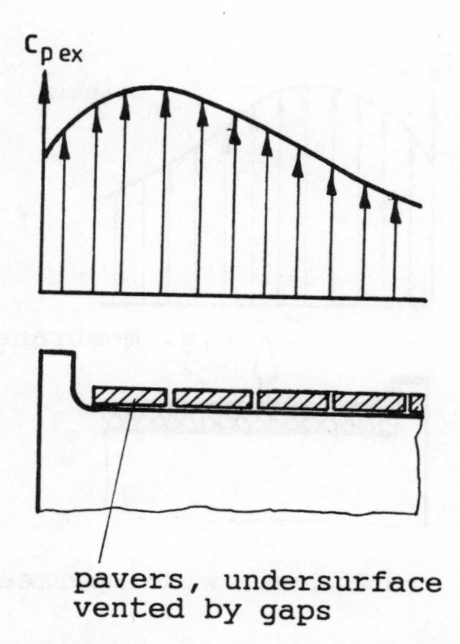

Figure 2b: Wind permeable roof coverings

high underelement flow resistance and relatively small resistance for the gap flow. Loose laid, mechanically fixed roofing membranes on profiled sheet metal decking is commonly used on industrial buildings, Figure 2 c. Here, there is no flow through the membrane, i.e. infinite flow resistance, but very little flow resistance for the pressure equilibration underneath the membrane.

Concerning facade coverings, either curtain-wall facades or rain-screen facade elements to protect the external thermal insulation are two windpermeable systems quite widely used. For both systems underelement flow resistance and through-flow resistance may be of the same order of magnitude. In fact, facade systems cov-

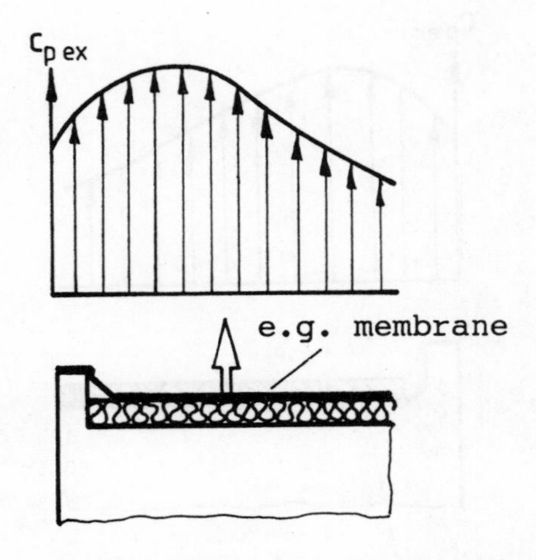

roof deck: windimpermeable

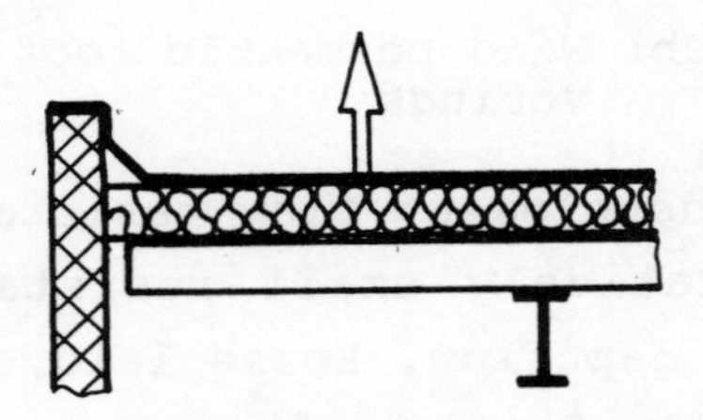

windpermeable roof deck
(e.g. steel deck)
venting through roof deck
corrugations

Figure 2c: Windimpermeable roof coverings

ver a wide range of ratios of those two flow resistances. As will be shown in Section 4.1, the clue for wind resistant and economical designs of facade coverings is a proper usage of those resistances.

2.2 Windload On Walls

The largest pressure acting on a building wall is given for flow direction perpendicular to the wall and amounts to an area and time averaged value of about $c_{p\,max}$ = 0.8. Due to the flow separation at the windward edges a suction acts on the windparallel walls. The magnitude of the corresponding negative pressure coefficient depends on the relative building dimensions h/w (height/ width) and l/w (length/width). It may vary considerably over the wall area and locally with time. The largest suction, both time averaged and short duration peaks, occurs in the vicinity of the windward edges. This may not be important for the wind safety of the basic structural elements but it certainly has to be considered when evaluating the wind safety of wall sheetings or claddings.

The authors have conducted an extensive study into the wind loading of building facades [1]. Both, time averaged and peak pressures were considered for windimpermeable and windpermeable building walls. The main results are presented in Figure 3 ÷ 5 and will be discussed briefly.

Fiber-cement facade systems and profiled light metal sheeting were investigated, the windpermeability of which varied greatly. The volume flow through a permeable wall covering will be related to the applied pressure difference by an exponential law. The proportionality constant of this relation is called the permeability parameter C_D and is a measure of the windpermeability of the system. It depends directly on the flow effective cross section area A_L of the system perforation, e.g. the overlapping gaps. Figure 3 shows the wide variation of permeability parameter of wall coverings (1 through 8).

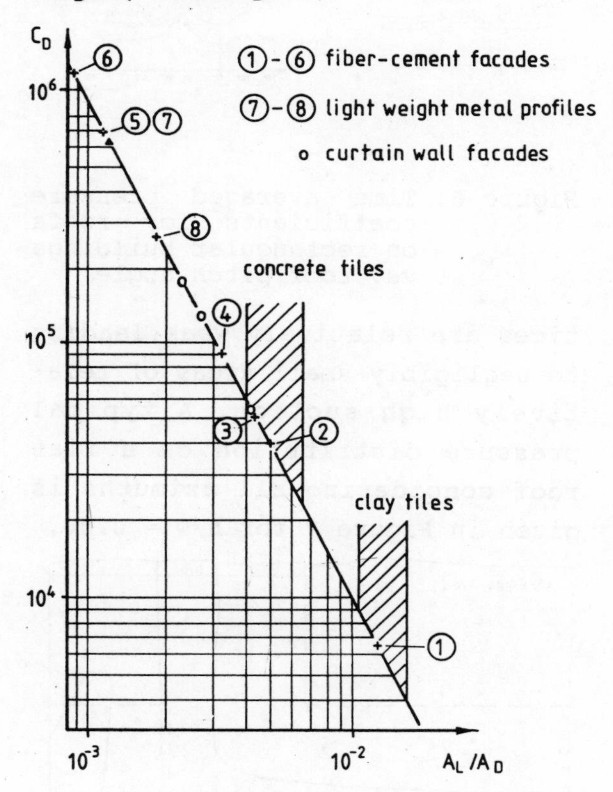

Figure 3: Windpermeability factor C_D vs. relative, effectively free cross section area

Figure 4 gives as a typical example the underelement pressure between windimpermeable building wall and permeable facade on the windparallel side of a building

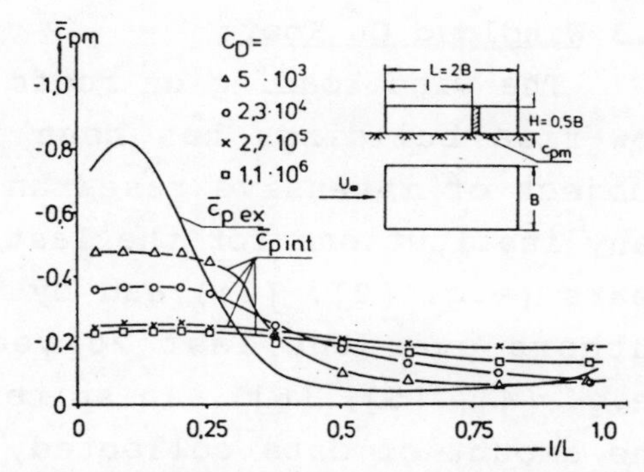

Figure 4: External and internal pressures on wind permeable model facades without substructure

model. The difference between external and internal pressures decreases with increasing facade permeability. Whereas Figure 4 is valid for very low underelement resistance, Figure 5 gives the equivalent pressure distributions for large underelement flow resistances. Now, the pressure equilibration between outer surface and underelement surface occurs locally. Thus, the net pressure difference and the resulting wind load is very small.

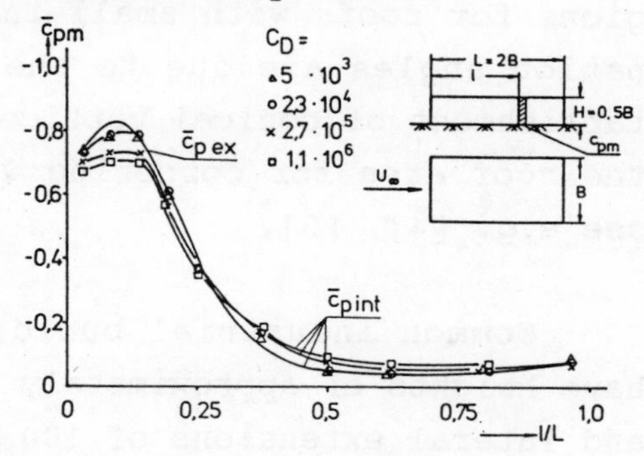

Figure 5: External and internal pressures on windpermeable model facades with simulated windimpermeable substructure

2.3 Windload On Roofs

The wind loading on roofs of low-rise buildings has been the subject of intensive research in many institutions for the last 15 years (e.g. [2], [3]) and by the authors over the last 20 years (e.g. [4], [5], [6]). In spite of the amount of data collected, in particular for time averaged pressure distributions, the influence of the relative building dimensions, and thus the flow displacement, has not been fully taken into account by most researchers and most standards. Figure 6 gives a summary of results obtained by the authors. The pressure coefficients for corner-, edge- and middle-region of the roof are plotted versus roof inclination angle for building aspect ratio $l/w < 1.5$ and $l/w \geq 1.5$. The strong influence of the roof inclination angle is apparent. The relatively high suction in the corner and edge regions for roofs with small inclination angles are due to the establishment of conical vortices on the roof area for cornering wind, see e.g. [4], [5].

Common industrial buildings have heights of approximately 10 m and lateral extensions of 100 m or more. Those buildings have recently been investigated by the authors, [7], [8]. For cornering wind condition, the conical vortices are relatively weak leading to negligibly small areas of relatively high suction. A typical pressure distribution on a flat roof considering all azimuths is given in Figure 7 for $h/w = 0.04$.

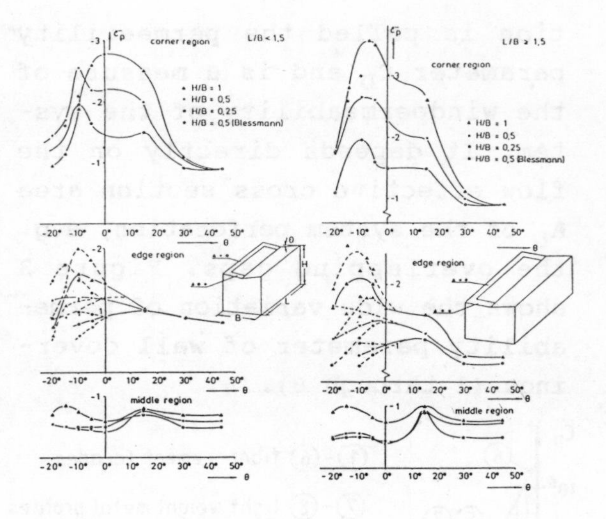

Figure 6: Time averaged pressure coefficients for roofs on rectangular buildings vs. roof pitch angle

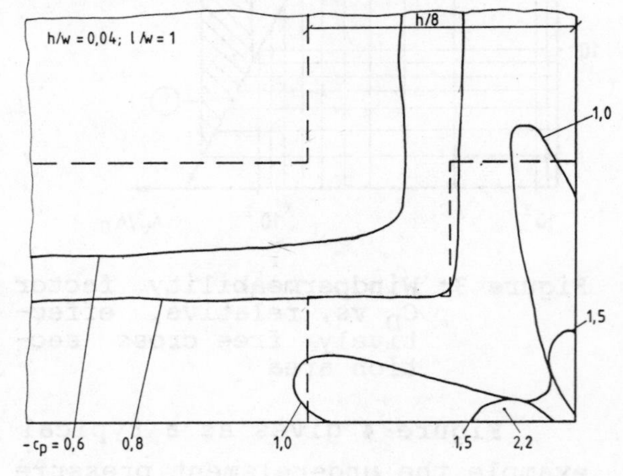

Figure 7: Typical pressure distribution on a flat roof of a very low building for all azimuths

3.0 CODIFICATION

All wind loading codes or standards give the design wind load as the product of a shape dependent load coefficient and an appropriate wind stagnation pressure. The latter one is either based on a gust windspeed or on a time averaged wind speed. Usually, the load coefficient is given as a time averaged coefficient, very rarely a peak load coefficient is taken. The following types of codes or standards may be distinguished according to the loading model used:

1. static model
2. quasi-steady model
3. extreme-value model

3.1 German Code Of Practice DIN 1055 Teil 4

The German wind loading standard is a typical example of the static load approach. The design wind load is given by

$$w_p = c_p \cdot q$$

with c_p = time averaged pressure coefficient and q = gust stagnation pressure. The gust stagnation pressure is based on the 5 sec speed and is presented as a function of building height and of building site. Typical gust stagnation pressures are of the order of q = 1000 Pa for buildings of height h > 20 m. Only time and area averaged pressure coefficients for various basic building shapes are given. The pressure coefficients for flat roofs as given in Figure 6 are e.g. approximated by the data given in the table. It should be noted that the influence of the flow displacement due to the relative building dimension is taken into account.

l/w	h/w	c_p for corner	edge
	> 0.4	- 2.8	- 1.5
≤ 1.5	0.4 - 0.1	- 2.0	- 1.5
	< 0.1	- 1.0	- 1.0
	> 0.4	- 3.0	- 1.7
> 1.5	0.4 - 0.1	- 2.5	- 1.0
	< 0.1	- 1.0	- 1.0

for field area: c_p = - 0.6

3.2 Indian Standard IS 875-1964

IS 875 is a code of practice for structural safety of buildings. Similarly to the German code of practice it may not be used readily for the evaluation of the wind safety of building components like cladding and sheeting. The Indian wind loading standard is of the static loading type, too. The basic wind stagnation pressures include winds of short duration, i.e. gusts. The Indian subcontinent is divided in 3 areas for which the gust stagnation pressure for h < 30 m is q = 1000 Pa; 1500 Pa and 2000 Pa respectively. In chapter 4.3 - of the standard - building shape factors are given and in chapter 4.3.5 values for the internal air pressure depending on the permeability of the building surfaces. For buildings with normal permeability - however, no values of permeability are given - a pressure coefficient of ± 0.2 has to be taken into account. For buildings with large openings (open area equivalent to at least 20 % of wall area) a pressure coefficient c_{pi} = ± 0.5 has to be used.

A special section (4.6) is devoted to wind pressures on roofs. For the leeward slope of the roof a pressure coefficient c_p = - 0.5 has to be applied for the whole roof area. For the windward slope the pressure coefficient depends on the roof inclination angle. For a flat roof c_p = - 1 is given. No corner or edge region is specified, neither is the influence of relative building dimensions mentioned. However, the pressure coefficients for building components have to be increased by Δc_p = ± 0.3 (section 4.6), thus leading to a maximum pressure coefficient on a flat roof of c_p = - 1.3. All sheeting and fastening components have to be designed according to this pressure coefficient. In addition, within the distance of 15 % of the building length l a pressure coefficient c_p = - 2.0 has to be taken into account. This edge area parallel to the gable will cover the critical corner area for most buildings. However, the pressure coefficient given (c_p = - 2.0) may be too small for relative building heights h/w > 0.4, see the table in section 3.1. No edge area is considered parallel to the gutter. The design wind loads for the critical edge region parallel to the long side of roofs with small inclination angle are greatly underestimated by the Indian standard.

3.3 ISO 4354 (Draft 1990)

ISO 4354 is a typical example of the quasi-steady wind loading approach. The wind load is determined as

$$w_p = q_{ref} \cdot C_{exp} \cdot C_{fig} \cdot C_{dyn}$$

with q_{ref} = reference velocity pressure; C_{exp} = exposure factor; C_{fig} = aerodynamic shape factor and C_{dyn} = dynamic response factor. The reference velocity pressure is based on the velocity averaged over approximately 10 min and with a recurrence interval (return period) of once-in-50 years.

The aerodynamic shape factors are again time averaged coefficients, e.g. pressure coefficients. The dynamic response factor for cladding is $C_{dyn} = 2.5$. The exposure factor accounts for the surface roughness surrounding the building site.

For walls and roofs of low-rise buildings the products of the aerodynamic shape factors and the dynamic response factors are given, Figure 8. Corner and edge regions are distinguished. However, the influence of relative building dimensions is not taken into account neither is there an edge region parallel to the gutter side. Dividing the coefficients given in Figure 8 by the dynamic response factor $C_{dyn} = 2.5$ gives pressure coefficients roughly equivalent to the time averaged pressure coefficients as specified e.g. in the table sectiion 3.1 for the German code of practice DIN 1055 Teil 4. Concerning flat roofed buildings, for the edge regions of buildings with relative height $h/w \leq 0.4$ the load coefficients given in the two codes are the same. However, the coefficients given differ greatly for the corner region of low buildings ($h/w < 0.4$). The values specified by the ISO code may overestimate the corner wind loads greatly. On the other hand, the ISO code will underestimate the wind loads for buildings of larger relative height ($h/w > 0.4$).

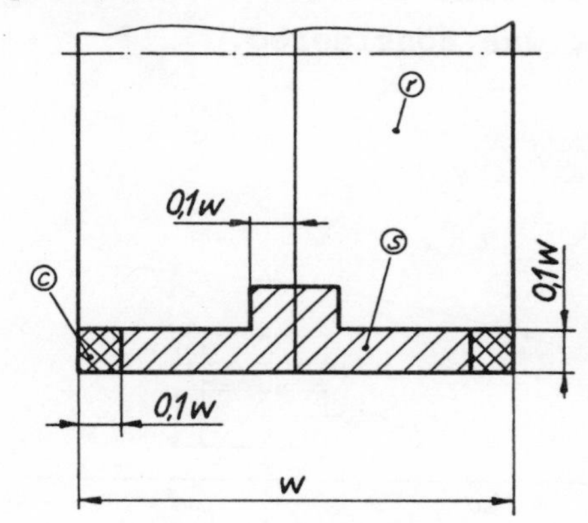

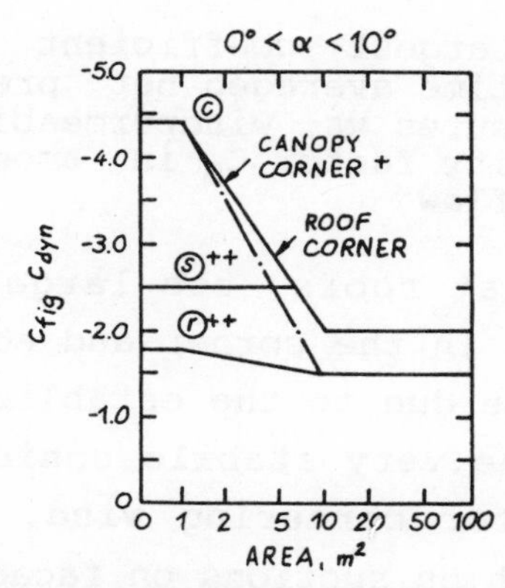

Figure 8: Values for C_{fig} C_{dyn} for low buildings

4.0 WIND RESISTANT DESIGN OF CLADDING AND SHEETING

4.1 Facades

The net pressures acting on facade elements have been determined for a wide range of relative building dimensions ($h/w = 0.25$ to 1.0 and $l/w = 1$ and 2) for facade permeabilities in the range of $C_D = 5 \cdot 10^3$ to 10^6. The results for the most critical building dimensions and the most critical flow direction are given in Figure 9.

The case of small underelement flow resistance, leading to the largest wind loads on facade elements, was considered.

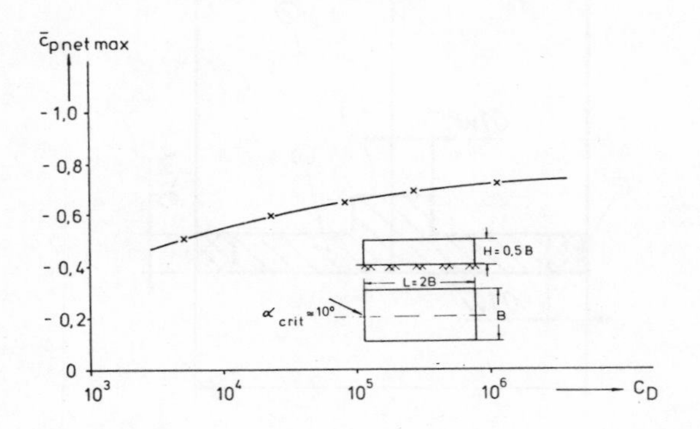

Figure 9: Largest coefficient of time averaged net pressures vs. windpermeability factor C_D in smooth flow

On flat roofs, the largest wind loads in the corner and edge regions are due to the establishment of the very stabile conical vortices for cornering wind. In contrast, high suctions on facades may occur intermittently just downstream of the leading edges. To investigate this phenomenon, the authors have measured the transient pressures in the critical facade regions. Since the pressure equilibration occurs with the speed of sound those external pressures are almost immediately transmitted to the gap between windimpermeable building wall and windpermeable facade coverings. The resulting net loads are again very small. Figure 10 gives the results equivalent to those in Figure 9, but for peak pressures. Again the critical building, the critical flow direction and low

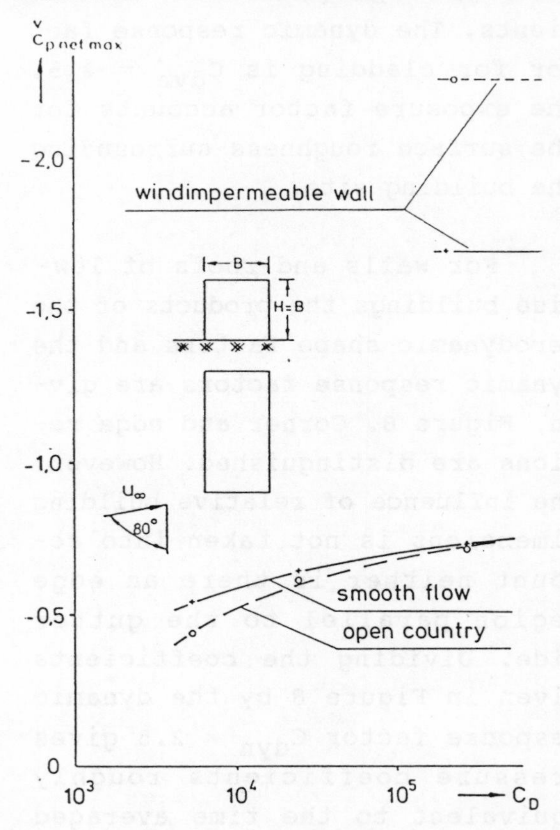

Figure 10: Largest coefficient of net peak pressures vs. wind permeability factor C_D

underelement flow resistance has been considered. Even for the high suction intermittent pressure peaks the net wind pressures are quite low.

There are some simple ways to ensure high wind stability of facade systems:

1. large facade permeability
2. large underelement flow resistance
3. airtight separation of the batten spaces of two adjacent walls.

Conditions 2. and 3. may easily be achieved by an appropriate design of the vertical battens.

4.2 ROOFS

4.2.2 Pitched roofs

At pitched roofs flow separation mainly occurs on the leeward roof part. The external pressure distribution in zones with separated flow is not affected remarkably by the roughness of the roof surface structures, e.g. the shape of the roof tiles. For the windward roof part, however, the shape of the roofing elements strongly influences the external pressure. This was investigated earlier by the authors [9]. An example was already shown in Figure 1. In order to simulate the convectively accelerated flow on the windward surface of a pitched roofs for flow direction perpendicular to the ridge, a narrowing flow duct was put into a windtunnel. The diagram, Figure 1, shows the surface pressures measured on the roof tiles with numerous pressure tabs, the area averaged external pressure which corresponds to the pressure distribution on a smooth surface and the internal pressure which was measured in the batten space between the tiles and the test section flow. This batten space pressure (measured against ambient pressure) is significantly higher (less negative) than the area averaged external pressure. The reason for this is the stagnation of the flow at the tile overlaps facing the wind.

The lifting-up mechanism of a tile may be described by a pitching moment turning the tiles upward around the pivoting point on the batten. The moment consists of a lifting force and a force couple resulting from the difference between the external and the internal pressures. The most critical external pressure distribution occurs for flow direction more or less perpendicular to the ridge and is strongly influenced by the leading edge curvature, see Figure 11. The internal pressure depends

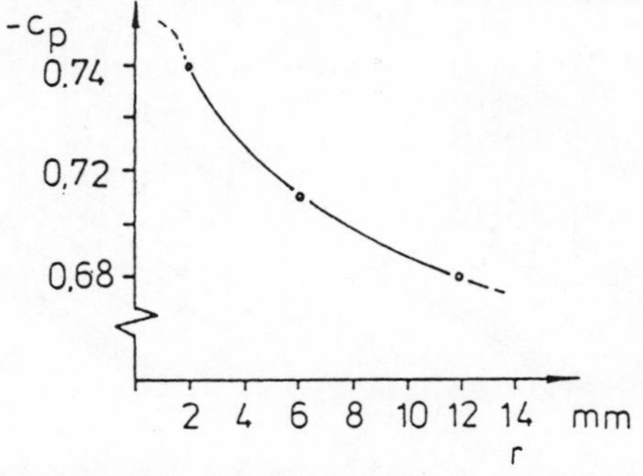

Figure 11: Influence of edge curvature on peak suction on a tile

on the permeability and the position of the gaps between the tiles. If, for example, the interlocking gaps between the tiles are positioned in a zone, where suction occurs for flow parallel to the ridge a much lower batten space pressure and, consequently, a much lower net wind load will be obtained for this wind direction than for gaps positioned in a stagnation zone. Examples of the external pressure distribution on a concrete roof tile for various flow directions are shown in Figure 12.

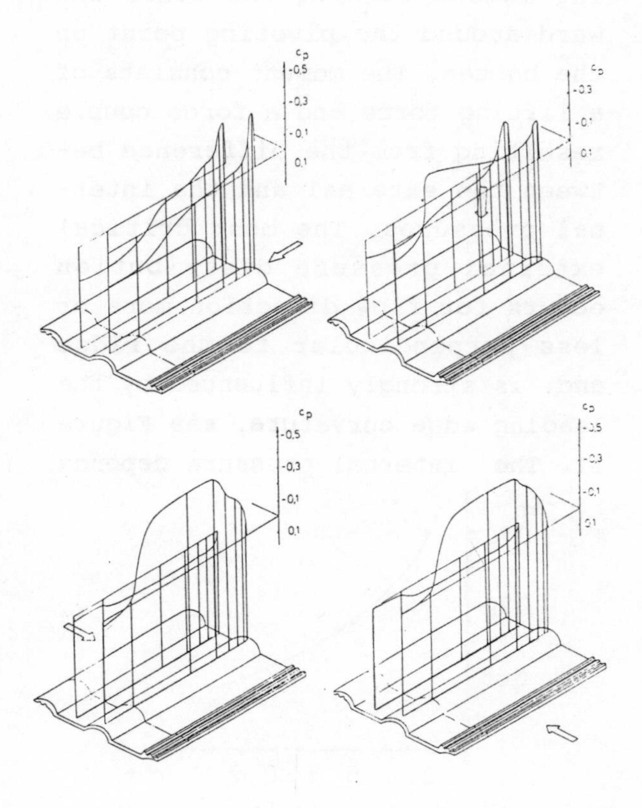

Figure 12: External pressure distribution on a tile

Of course, another important design criterion for roof tiles is the water tightness especially for driven rain, which can lead to design requirements conflicting with optimized wind uplift safety.

Taking into account the wind load due to the building flow field <u>and</u> the local flow field on the tiles, the critical areas for tile lifting-up on pitched roofs as shown in Figure 13 can be defined.

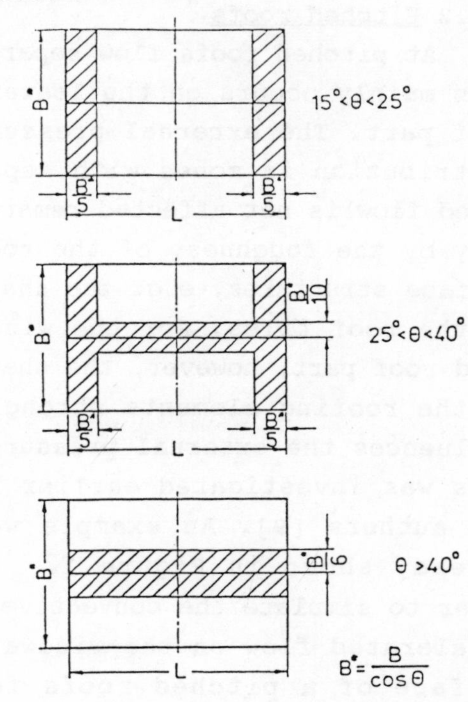

Figure 13: Critical areas for tile lifting up of pitched roofs

An underlay or boards sealing the batten space against the space underneath the roof will not prevent the cladding elements on the windward roof part from being lifted up. For a local flow direction perpendicular to the ridge the batten space pressure may even become positive due to the stagnation, and the vented net wind load may be increased by the sealing effect of the underlay. For such a roof, the windward side will be the most critical. By an underlay, however, the pressure equilibration in the gable space of the roof is prevented, leading to much lower net wind load for the leeward roof tiles. Favourable are tiles with a shape avoiding strong stagnation at the overlaps. The wind safety can be increased furthermore, if the permeability of the overlapping gaps parallel to the ridge is smaller than the permeability of the interlocking gaps, which should be positioned in a zone where suction occurs due to the tile surface flow field.

4.2.2 Flat Roofs

For flat roofs covered by sheeting with low stiffness or by roof membranes the fasteners preventing wind uplift must be distributed on the roof surface corresponding to the worst case the wind pressure distribution. This leads to a concentration of fasteners in the edge and corner regions. A typical example for the fastener frequency distribution on a flat roof is shown in Figure 14.

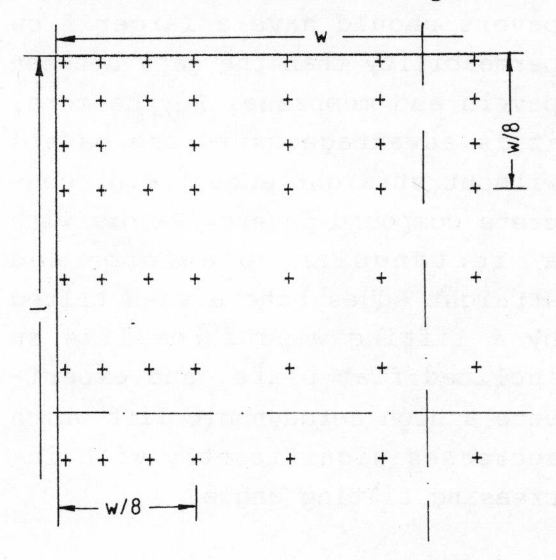

Figure 14: Fastener frequency distribution on a flat roof

The necessary number of fasteners can be reduced by making the roof deck as windimpermeable as possible and by sealing the roof edges carefully. For such a roof, by lifting up the roof membrane, the internal pressure will decrease, and the resulting pressure difference will be much smaller than the wind load calculated from the external pressure distribution only. Following this idea, large industrial roofs with loosely laid plastic roofing membranes were constructed successfully without any fasteners.

When using ballast on flat roofs in order to secure the roofing membrane against wind uplift, the gaps between the ballasting pavers should have a larger flow permeability than the gaps between pavers and membrane. Furthermore, it is advantageous to use pavers without straight edges, e.g. concrete compound pavers. Pavers with a rectangular planform and straight edges behave when tilted by a lifting wind force like an inclined flat plate and experience a high aerodynamic lift which increases significantly with increasing tilting angle.

Light weight roofs, especially for industrial buildings, have roof decks consisting of corrugated sheet metal. This sheet metal roof decks are not very stiff, and even moderate wind forces lead to remarkable deflections. If the roof membrane is bonded to the roof deck this deflections will cause cracks in the roof membrane due to lack of flexibility or lead to failure of the bonding between the membrane and the roof deck. A bubble will occur underneath the roofing membrane and the resulting failure mechanism is illustrated in Figure 15. Therefore for such types of roofs a mechanical fixing of the sealing membrane is appropriate.

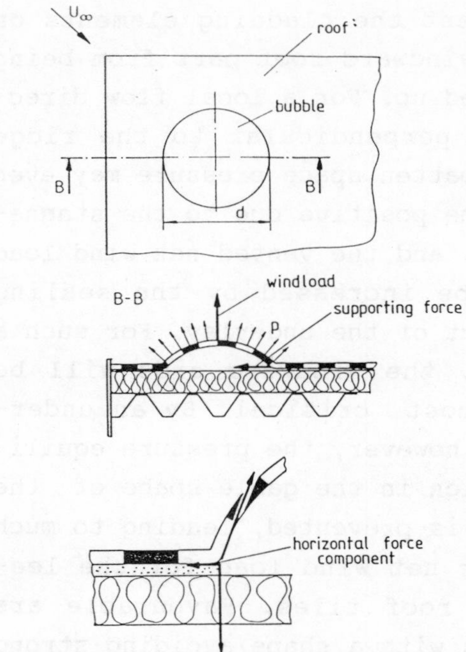

Figure 15: Failure mechanism of adhered roofing sheets

Very common fasteners are screws drilled into the roof deck and securing a load distribution plate which presses the membrane on the roof deck or the thermal insulation respectively. Figure 16 shows the typical situation for so called inlap spot fasteners and illustrates the non-symmetrical loading of the fastener itself.

Due to the wind pressure fluctuation caused by the wind gustiness and by flow instabilities in the building flow field this load is a fluctuating load which may cause fatigue failure if the fastener is only designed to resist pure static loading. Field observations on

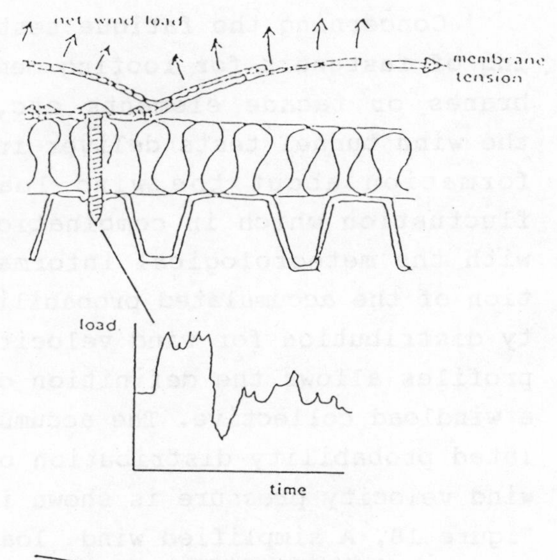

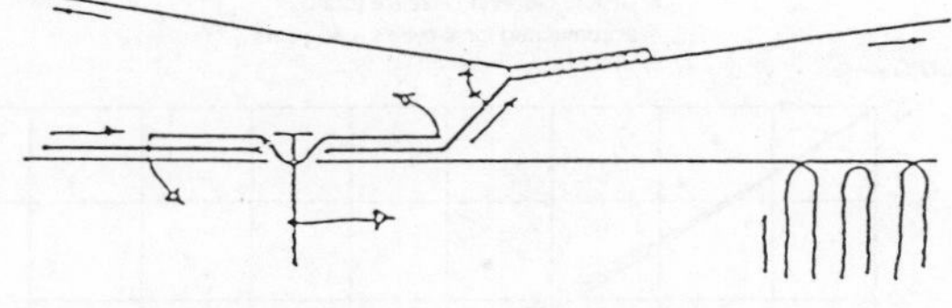

spot fastener (non symmetrical loading)

note: for determination of max allowable load for fixation system the fluctuating wind pressure has to be taken into account to allow for fatigue behaviour

Figure 16: Mechanical attachment of roof membranes on metal roof deck

various mechanically attached roof membrane systems give a clear indication for this type of failures. The roof membrane slips below the load distribution plate because the compression between roof membrane and insulation substrate will be reduced due to fastener back-out or fastener failure. Furthermore the attachment plate can be deformed by high membrane forces.

Fatigue cracks in the sheet metal deck originating from the bore of the fasteners will ultimately lead to fastener pull-out already at moderate wind velocities well below the design wind velocity. This failure mechanisms must be considered as a long-term effects. Therefore, a safe design of such roof membrane fasteners should be based on an appropriate fatigue testing in order to guarantee a sufficient life expectancy.

5.0 TESTING FOR WIND RESISTANCE

The most important tool for wind resistant testing is the windtunnel which simulates the natural wind flow. In addition to the well known practice of pressure measurements on building models a sand erosion technique may be applied, which gives a quick and reliable overview of the flow field and the wind loading to be expected on the roof of a certain building. Figure 17 shows an example. Photographs obtained by multiple exposure and constant time intervals indicate clearly the most endangered roof areas.

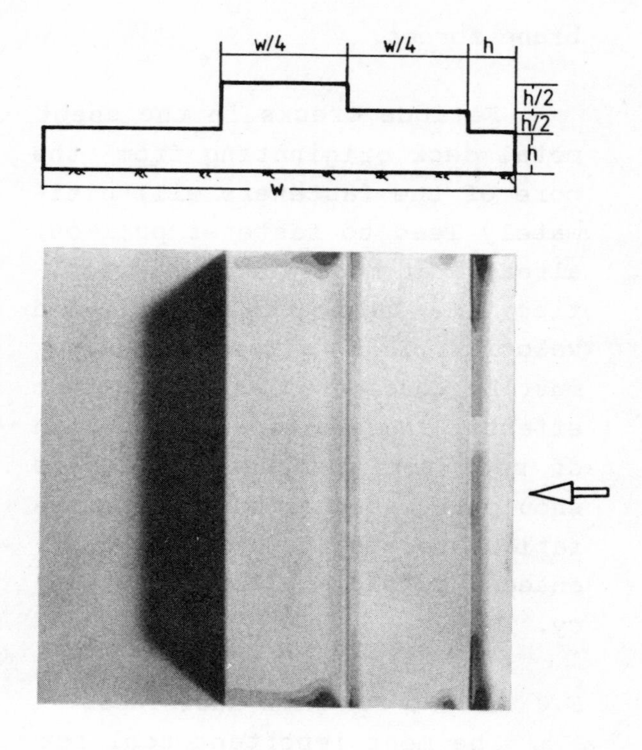

Figure 17: Visualisation of roof surface flow field using sand erosion technique

The pressure distribution on a flat roof of a certain building depends on the structure of the wind flow which is mainly defined by the wind velocity profile. For block-type, low rise buildings, however, the most critical case is a velocity profile at low surface roughness corresponding to open country exposure which can be simulated with sufficient accuracy even in a usual smooth flow wind tunnel.

Concerning the fatigue testing of fasteners for roofing membranes or facade elements etc., the wind tunnel tests deliver information about the wind load fluctuation which in combination with the meteorological information of the accumulated probability distribution for wind velocity profiles allows the definition of a windload collective. The accumulated probability distribution or wind velocity pressure is shown in Figure 18. A simplified wind load

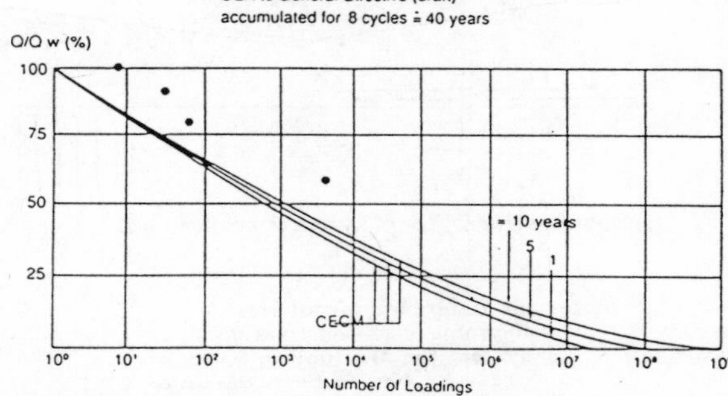

Figure 18: Accumulated probability distribution of wind velocity pressure

cycle for a five year return period based on this probability distribution is shown in Figure 19. For the tests a special appa-

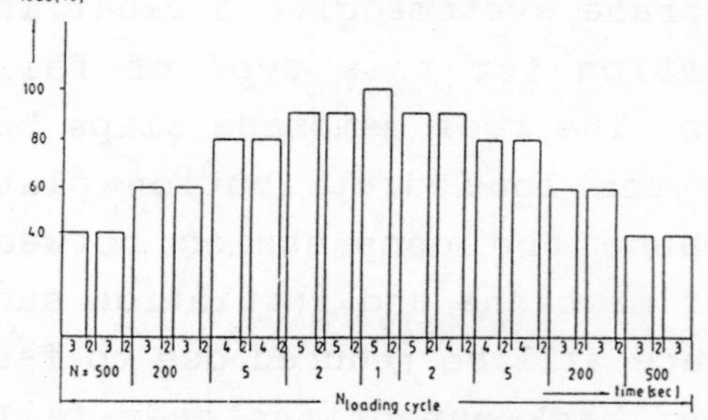

Figure 19: Wind load cycle (5 year return period) based on accumulated probability distribution

ratus was designed. It consists of a suction chamber placed on a rigid frame around the test specimen, which is an excerpt of the full size roof or wall construction. The suction acting on the test specimen is created by a centrifugal fan evacuating the test chamber. The gust action is simulated by a control valve. The fan speed and the position of the valve is controlled by a PC according to the wind load cycles. The 100 %-level for the wind load cycle is increased in steps from return period to return period until the specimen fails, Figure 20. The number of failure free return periods gives a safe indication for the expected lifetime.

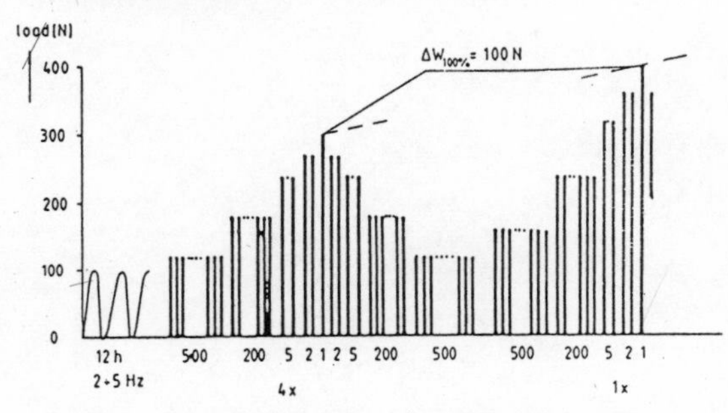

Figure 20: Wind load test cycle (draft of UEAtc General Directive)

6.0 LITERATURE

1. Gerhardt, H.J. and C. Kramer. Wind Loads on Wind-permeable Facades. Betonwerk und Fertigteil-Technik, Heft 1: 46 - 53, 1985

2. Kind, R.J. Worst suctions near edges of flat rooftops on low-rise buildings. Journal of Wind Engineering and Industrial Aerodynamics 25: 31 - 47, 1986

3. Surry, D. et al. Wind loads on low-rise structures. ASCE fall convention, Atlanta, Georgia, October 1979

4. Kramer, C. Untersuchungen zur Windbelastung von Flachdächern. Der Bauingenieur 50: 125 - 132, 1975

5. Kramer, C., H.J. Gerhardt and S. Scherer. Wind pressures on block-type buildings. Journal of Industrial Aerodynamics 4: 229 - 242, 1979

6. Gerhardt, H.J. and C. Kramer. Windlasten auf nach innen geneigten Dachflächen. Der Bauingenieur 59: 53 - 59, 1984

7. Gerhardt, H.J. Windlastannahmen für große Industriehallen mit Flachdach (accepted for publication in Bautechnik)

8. Gerhardt, H.J. and C. Kramer. Windlasten auf Dächern großer niedriger Industriebauten (accepted for publication in Journal of Flight Sciences and Space Research (ZfW))

9. Kramer, C. and H.J. Gerhardt. Windloads on permeable roofing systems. Journal of Wind Engineering and Industrial Aerodynamics 13: 347 - 358, 1983

WIND INDUCED RESPONSE OF BUILDINGS

Ahsan Kareem

Department of Civil Engineering
University of Notre Dame
Notre Dame, Indiana 46556-0767, USA

INTRODUCTION

Under the influence of dynamic wind loads, typical high-rise buildings vibrate in the alongwind, acrosswind, and torsional directions. The alongwind motion primarily results from pressure fluctuations on the windward and leeward faces, which generally follow the fluctuations in the approach flow, at least in the low frequency range. The acrosswind motion is introduced by pressure fluctuations on the side faces which are induced by the fluctuations in the separated shear layers and wake flow fields. The wind induced torsional effects result from the unbalance in the instantaneous pressure distribution on the building surface. The wind load effects are further amplified in asymmetric buildings as a result of inertial coupling. Modern trends towards unconventional shaped buildings with complex exterior geometry have led to buildings dynamically more sensitive to torsion induced loads resulting from asymmetric wind pressures, and static and/or dynamic coupling. The ratio of displacements, and accelerations near the building corners at the top floors, to that at the center can significantly increase due to torsional effects. This results in higher stresses in exterior columns and more importantly causes human discomfort.

Modern high-rise buildings designed to satisfy static lateral drift requirements still may oscillate excessively during wind storms. The level of these oscillations may not be significant enough to cause structural damage but may cause discomfort to the occupants. Human comfort needs to be emphasized due to human biodynamic sensitivity to angular motion which is accentuated by the increase in awareness of building motion provided by visual cues of moving surroundings. An accurate assessment of building motion is an essential prequisite for serviceability. Such an exercise ensures that the building remains functional under normal loads. Extreme loading conditions resulting from dynamic wind effects are needed for the strength design to ensure structural integrity. This paper focusses on the aerodynamic loads, dynamic response estimation procedures, effects of uncertainties on the response estimates and overall reliability considerations in the design of tall buildings.

AERODYNAMIC LOADS

As the wind encounters a structure, it exerts positive pressure on the windward face. The wind is then deflected around the structure and accelerated such that the velocity passing the upwind corners is greater than the velocity approaching the structure. The high-velocity fluid cannot regotiate the sharp corners and thus separates from the building, leaving a region of high negative pressure. The separated flow forms a shear layer on each side and subsequent interaction between the layers results in their rolling up into discrete vortices which are shed alternately. This region is generally known as the wake region. In this manner, the pressure fluctuations on the surface of a structure exposed to the atmospheric boundary layer result from the turbulence present in the approach flow, from flow separation and reattachment, from wake effects, and from possible impingement of vortices shed upstream objects. The structural motion may induce additional aerodynamic forces. These mechanisms do not always take place in isolation. Rather more than one may collectively contribute to the pressure fluctuations and hence loads on a structure. Therefore, the aerodynamic loading may be categorized as due to the far-field which is the undisturbed flow pattern and the near-field which results from the fluid-structure interaction and the wake fluctuations. The scales associated with the far-field, near-field and wake fluctuations are quite different. The dominant far-field scale is determined by the scale of atmospheric turbulence in the boundary layer, whereas the scales associated with the wake and near-field fluctuations may be expressed in terms of a characteristic structural dimension and shear layer thickness, respectively.

The aerodynamic loads may be expressed as

$$\text{Aerodynamic loading} = F_T(t) + F_W(t) + F_1(t) + F(x, \dot{x}, \ddot{x}) \qquad (1)$$

in which $F_T(t)$ = forces induced by incident turbulence, $F_W(t)$ = forces induced by wake fluctuations, $F_1(t)$ = aerodynamic forces due to interference of upstream and adjacent structures, and $F(x, \dot{x}, \ddot{x})$ = motion induced loading expressed directly as stiffness, damping and inertia forces which are expressed in terms of the structural displacement, velocity and acceleration and corresponding aerodynamic derivatives.

Notwithstanding the improved knowledge of wind effects on structures over the past few decades, our understanding of the mechanisms that relate the random wind field to the various wind induced effects on structures has not developed sufficiently for functional relationships to be formulated. Not only is the approach wind field very complex, the flow patterns generated around a structure are complicated by the distortion of the wind field, the flow separation, the vortex formation, and the wake development. These effects cause large pressure fluctuations on the surface of a structure which in turn impose large overall aerodynamic loads upon the structural system and lead to intense localized fluctuating forces over the envelope of the structure. Under the collective influence of these fluctuating forces, a structure may vibrate in rectilinear modes or torsional modes or both.

The alongwind motion primarily results from pressure fluctuations in the approach flow, at least in the low frequency range. The across-wind motion that is perpendicular to the direction of motion is introduced by pressure fluctuations on the side faces which are primarily induced by the fluctuations in the separated shear layers, vortex shedding and wake flow fields. Bluff bodies of all cross-sections exposed to fluid motion for a broad range of Reynolds numbers shed vortices, usually alternating from each side of the body. The vortex shedding induces alternating forces on bodies in both crossflow and in the direction of flow. The fluctuations in the crossflow direction are predominantly periodic and characterized by the Strouhal number than is equal to N_sD/U, in which N_s= frequency of vortex shedding. D= structural dimension, and U= wind speed. The phenomenon is more pronounced when flow around a body has two-dimensional features; whereas, in the case of three-dimensional situations the severity of the shedding process is vitiated due to a lack of correlation. Unlike vortex shedding from sharp cornered sections, the shedding is strongly dependent on the Reynolds number for bodies of circular cross-section. The wind induced torsional effects result from the unbalance in the instantaneous pressure distribution on the building surface. The wind load effects are further amplified on asymmetric buildings as a consequence of inertial coupling.

The aerodynamic loads in alongwind direction are quantified analytically utilizing quasi-steady and strip theories. The dynamic effects are customarily represented by a "Gust Factor Approach". Random-vibration-based gust factors translate the dynamic loading caused by unsteady wind into an equivalent static loading [1-11]. The fluctuating pressure field acting over the side and leeward faces is greatly influenced by the wake dynamics, e.g., vortex shedding which renders the applicability of the foregoing theories doubtful. Therefore, a lack of a convenient transfer function, between the velocity fluctuations in the incident turbulence and the pressure fluctuations on the side faces of a building with separated flow, has prohibited any acceptable formulation, to date, of the across-wind and torsional loads on structures. Hence, physical modelling of fluid-structure interaction provides the only viable means of obtaining information on the aforementioned loads. Coordinated efforts in the field of computational fluid dynamics are in progress to numerically generate flow fields around bluff bodies exposed to turbulent flows [e.g., 12].

EXPERIMENTAL EVALUATION OF AERODYNAMIC LOADS

The aerodynamic loads on structures may be obtained by mapping and synthesizing the random pressure fields acting on structures [10, 12-15]. The structure of random pressure fields through simultaneously monitored multiple-point realizations of pressure fluctuations, and measurement of local averages of the space-time random pressure fields by means of spatial and temporal averaging techniques can be mapped. The spatial averaging procedure may employ local averaging of the random pressure field utilizing an electronic summation circuitry, a pneumatic manifolding device, or a pressure-sensitive surface element like PVDF [13].

Details of load distribution over parts of a structure and over its entirety may be accomplished through multi-point measurements of the pressure field in time and space, or continually at specific locations. A lack of spatial and temporal coherence in the random pressure field requires simultaneous monitoring of a large number of pressure taps on a building surface which may either become prohibitively expensive or may impose a difficult data acquisition, management and reduction problem. One example of a measurement system equiped to monitor a very large number of pressure sensors is reported in reference 16. Alternatively, off-line computation of simultaneous multi-channel sampling permits statistical averaging as well as conditional sampling permits statistical averaging as well as conditional sampling to examine the spatiotemporal distribution of pressure peaks. Practical means have been introduced for overcoming the difficulties that have resulted in prohibiting a wider use of the direct pressure measurements for the space-time averaging of random pressure fields. Kareem [17] employed a number of pressure transducers in a large number of test configurations on which the transducers were moved to different locations to map the pressure field. The spatially averaged loads and their covariances were obtained. The statistical information of the local averages was assimilated through statistical integration or covariance integration procedure to obtain estimates of the integral aerodynamic loading function on the building as well as the desired mode-generalized loading [17 and 18]. The synthesis of homogeneous pressure fields may also be accomplished by utilizing frequency-dependent spatial scales of the pressure field obtained from second-order information about point-to-point variations contained in the covariance function [13 and 19]. This procedure provides a computational procedure involving simple algebraic operations such as sums and products instead of multiple integrations. This format also provides an amenable adaptation to a finite element discretization of a random field into which the local averages are expressed over a finite element and the inter-element covariance is obtained from the covariance of the local averages [20 and 21].

The aerodynamic loads may be synthesized by analog circuits from simultaneously monitored multi-point measurements [14]. Alternatively, space-time averaging may be accomplished by a pneumatic averaging technique which, through a pneumatic manifolding procedure, determines time varying local area averages of aerodynamic loads. In its simplest application, it makes use of a manifolding device (multiple-input single-output manifold), which interconnects a labyrinth of tubing leading from several pressure taps that are distributed uniformly over an area [13 and 15]. A detailed treatment of the mapping and synthesis of random pressure fields that highlights the concept of point, spatially and/or temporally averaged random pressure fluctuations in the context of local or integral aerodynamic loads is presented in Reference [13].

The generalized aerodynamic loading on a structure in the ith mode may be obtained by synthesizing the spatially averaged random pressure field by means of covariance integration

$$S_{F_i}(f) = \sum_{k=1}^{2} \sum_{l=1}^{2} \sum_{A_k} \sum_{A_l} \sqrt{S_{p_k}(\Delta A_k, f)\, S_{p_l}(\Delta A_l, f)}$$

$$XCo_{p_{kl}}(\Delta A_k, \Delta A_l; \mathrm{f})\, \phi_i(z_k)\, \phi_i(z_l)\, \Delta A_k \Delta A_l \quad (2)$$

in which subscripts k and l denote different faces and $S_{p_{kl}}(\Delta A_k, f)$ and $Co_{p_{kl}}(\Delta A_k, \Delta A_l; f)$ represent the spectrum and co-spectrum of spatially averaged pressure fluctuation areas ΔA_k and ΔA_l that represent the differential areas over which the random pressure field is spatially averaged and $\phi_i(z_k)$ is the ith mode shape evaluated at the centroid of the area ΔA_k. The preceding equation can be conveniently expressed in a discrete matrix form so that it is compatible with experimental measurements. Based on experimental data, Kareem [18] has developed closed-form expressions for auto- and co-spectra of the random pressure field responsible for the acrosswind force on isolated square cross-section buildings for any desired approach flow condition, i.e., open country, suburban, or urban. The model provides flexibility in the selection of appropriate input parameters, thus broadening the scope of its application and serving as a useful tool for tailoring the preliminary design of tall buildings. Following the foregoing procedure for synthesizing spatio-temporal pressure fluctuations, mode-generalized spectra of the torsional moment are given by

$$S_{T_i}(f) = \sum_{m=1}^{4} \sum_{n=1}^{4} \sum_{A_n} \sum_{A_n} \sqrt{S_{p_m}(\Delta A_m f)\, S_{p_n}(\Delta A_n f)}$$

$$XCo_{p_{mn}}(\Delta A_m, \Delta A_n; f)\, \Delta l_m \Delta l_n \phi_i(z_m)\, \phi_i(z_n)$$

$$\Delta A_m \Delta A_n \quad (3)$$

in which the variables are defined previously, and Δl_m and Δl_n are distance between the centroid of areas ΔA_m and ΔA_n from the building centre. More details regarding the matrix representation of the preceding equation may be found in Ref. [22]. The previous covariance integration approach used to quantify mode-generalized loads may be simplified by utilizing a weighted pneumatic averaging technique. In this approach, the building at the instrumented levels is divided into tributary areas or segments which are assigned a cluster of pressure taps to attain the appropriate weighting function proportional to the moment arm for each top location [23]. Alternatively, distributed and continuous weighted averaging may be obtained utilizing a porous polyethylene or piezopolymer triangular surface that includes appropriate weighting functions [13].

More recently, high-frequency force balance techniques for determining the dynamic wind induced structural loads from scale models of buildings and structures have been implemented at various boundary layer wind tunnel laboratories [24, 27]. These techniques have dramatically reduced both the time and cost required to obtain estimates of wind loads and structural response levels. The force balance provides dynamic load information for a specific building geometry and setting which may be used to calculate loads and response levels for a wide range of structural characteristics, damping values, and building masses. The basis of the technique is the measurement of power spectral densities of mode-generalized wind loads on the building model as a whole. This requires that the balance has sufficient sensitivity to measure small fluctuations in the modal loads while either having extremely well-defined and stable dynamic properties or a high enough natural frequency to insure that the mechanical admittance function is nearly unity throughout the frequency range of interest.

In a recent study by the author, a wide range of ultra-light models were fabricated to be used in conjunction with a sensitive base balance. A host of generic buildings with a wide range of shapes and modular combinations were utilized to cover mid-rise to super tall buildings. This study has made possible a collection of unique data set that can be used for estimating response of a wide range of buildings.

The force balance technique has some shortcomings, e.g., only approximate estimates of the mode-generalized torsional moments are obtained and the lateral loads may be inaccurate if the sway mode shapes of the structure differ significantly from a linear mode shape [23, 27]. Therefore, the mode-generalized spectra obtained from a force-balance study requires adjustments if the building mode shapes depart from those implied in the derivation of the force balance theory. This is especially true for the torsional loads. These adjustments may require invoking either the quasi-steady or strip theories, and the spatial averaging technique may facilitate a precise basis for establishing reliable mode correction factors for any arbitrary mode shape [13]. A second generation of force balances permits overcoming the aforementioned limitations [28]. A typical configuration includes several torsional flexures mounted on a still spine. The mode-generalized torsional moments may be obtained by weighting the torques measured with each torsional flexure according to the mode shape and adding the torques either digitally or by means of a simple analog summing circuit. The spine is instrumented so that lateral forces may be measured for each segment. This force may in turn be combined to produce mode-generalized lateral loads which more closely correspond to the expected mode shape of the prototype building or structure.

The covariance integration and high-frequency force balance techniques do not include motion-induced aerodynamic loads. These motion-induced loads may be conveniently expressed in terms of aerodynamic damping and be effectively implemented in the estimation of structural response through the modification of the building transfer function utilizing appropriate values of aerodynamic damping [17]. It is a general consensus that in most of the tall buildings the influence of motion-induced loading is insignificant for typical design wind speeds [18, 29]. However, for exceptionally slender, flexible and lightly damped structures, the motion-induced effects may reach a significant level [30]. An experimental system has been reported in Reference 31 which complements the base-balance technique by providing information about motion-induced forces acting on the structure. Alternatively, aeroelastic models provide the most recognized type of model for use in determining wind induced motion of a building and in determining resultant fluctuating loads acting on the building as a whole [32]. While aeroelastic models provide all the necessary information a designer may require, they are not instrumented to provide information on the space-time distributions of forces over the surface of a building.

The verification or validation of experimental procedures or theoretical methods lies in the comparison with the observed full-scale response. Notwithstanding the uncertainties associated with the full-scale response measurements, they offer at present the only possible means of demonstrating the validation of the laboratory simulation or theoretical predictions. At the same time, full-scale measurements may help to guide both numerical and experimental studies. Davenport [33] has summarized the full-scale measurements from a historical perspective and outlined many of the difficulties encountered, particularly in pressure measurements. There are a number of structures currently being monitored all over the globe for wind loads and associated response. Further details may be found in Reference 6.

DYNAMIC RESPONSE

The equtions of motion of a building represented by a discretized lumped-mass system are given by

$$[M]\ \{\ddot{Y}\} + [C]\ \{\dot{Y}\} + [K]\ \{Y\} \ = \ \{F(t)\} \tag{4}$$

in which M, C, and K are assembled mass, damping and stiffness matrices of the discretized system, respectively. In general, the assembly process involves transformation and condensation that reduces the system degree-of-freedom to the global coordinate system consisting of two translations and one rotation per story level. The preceding equations of motion may be solved by means of modal superposition, direct frequency and time domains, and recursive z-transform techniques. In the following, a brief outline of the model superposition technique is presented.

The normal mode approach may be utilized wherein the undamped eigenvectors facilitate decoupling of coupled systems of equations. Employing the standard transformation of coordinates involving undamped eigenvectors of the system offers the following uncoupled system of equations.

$$\ddot{z}_i + 2\xi_i \omega_i \dot{z}_i + \omega_i^2 z_i = P_i(t) \tag{5}$$

in which $P_i(t) = \{\phi^{(i)}\}^T \{f(t)\}$ and ξ_i is the critical damping ratio in the *ith* mode. The transformation used herein leads to a mode displacement method in which the problem coordinates are related to modal coordinate by

$$\{y\} = [\phi]\{z\} \tag{6}$$

in which $[\phi]$ is the modal matrix. In an alternate format generally known as the mode-acceleration approach, the displacement is given by

$$\{y\} = [K]^{-1}\{F\} - [\phi][\omega^2]^{-1}\{\ddot{z}\} \tag{7}$$

The first term in the above equation is the pseudo-static response, while the second term gives the method its name [34]. The presence of ω^2 term in the denomonator improves the convergence of the superposition technique. Therefore, this approach may be useful in reducing errors in calculated responses introduced by conventional modal truncation in a mode-displacement method.

By the properties of the Fourier transforms and the orthogonality of random Fourier components, the mean square value of response components due to wind loads at the *ith* node is given by

$$\sigma^2_{y_i^{(r)}} = \sum_{j=1}^{n} [\Phi] [\int G^{(r)}_{z_{ij}}(f)\, df] [\Phi]^T \tag{8}$$

$$[G_z(r)(f)] = [H^{(r)}(i2\pi f)]^* [G_F(f)] [H^{(r)}(i2\pi f)] \tag{9}$$

$$[G_F(f)] = [\Phi]^T [S_F(f)] [\Phi] \tag{10}$$

$$[H^{(r)}(i2\pi f)] = \frac{(2\pi f)^r}{(2\pi)^2 [(f^2 - f_n^2) + 2i\xi f f_n]} \tag{11}$$

in which $[H(i2\pi f)]$ = *a* diagonal matrix of frequency response function or mechanical admittance function; $[S_F(f)]$ = the cross-power spectral density matrix of the forcing function; and $[\Phi]$ = the modal matrix that is normalized with respect to the mass matrix. The superscript, *r*, represents the derivative of response, i.e., r = 1, 2, 3 denotes velocity, acceleration and jerk, respectively. The acceleration and jerk responses are required for occupancy comfort [35-38].

The integration involved in the preceding equations for lightly damped structures may be performed by separating the resonant and background response components. The resonant component is evaluated by means of the residue theorem in which the excitation is idealized as white noise. The implicit assumption in this simplification is that the forcing function is replaced by a white noise with a constant spectral density function at the structural natural frequency. The response due to background effects may be evaluated on the basis of a quasi-static assumption

$$\sigma^2_{y^{(r)}} = \sum_{n=1}^{N} \frac{\phi_i^2 \pi f_n G_{Fn}(f_n)\,(2\pi f_n)^{2r}}{4\,(2\pi f_n)^4 \xi_n m_n^2} \tag{12}$$

$$+\sum_{n=1}^{N} \frac{\phi_i^2 \left(\int_o^{f_n} G_{Fn}(f)\,(df)\,(2\pi f_n)^{2r}\right)}{(2\pi f_n)^4} \tag{13}$$

In the time domain the equations of motion may be integrated directly using a numerical step-by-step procedure. Like the previous section, the term 'direct' implies that the equations of motion are not transformed into a different coordinate system prior to the numerical integration. The Step-by-step integration is based on the equilibrium of forces acting on a dynamic system at discrete time points, separated by small time intervals, within the interval of the solution. The variation of the response components, e.g., displacement, velocity and acceleration within each small time interval is assumed to follow a given form which determines the accuracy, stability and cost of the solution procedure. Some of the few commonly used effective direct integration methods are the central difference method (explicit integration scheme), the Houbolt method, Wilson θ method, the Newmark method (implicit integration schemes). Details of these methods may be found in Ref. [39]. The step-by-step integration may be carried out after transforming the system equation in normal coordinates which considerably reduces computational effort.

For the implementation of integration schemes the time histories of loading function are required as opposed to the description of the spectral density function. This introduces an additional effort to efficiently simulate sample functions of multi-variate and/or multi-dimensional random processes. These sample functions may be generated by utilizing FFT-based techniques, or alternatively, by means of parametric time series approach, e.g., ARMA (auto-regressive moving average) [40-45]. The ARMA representation entails weighted recursive relations that connect the random quantity being simulated at successive increments. This procedure utilizes a recursive relationship in which the coefficients are ascertained from the given covariance of the random field. The desired sample functions are generated using the recursive relationship. Unlike FFT-based techniques this approach does not require storage of large amounts of data, rather only limited information, e.g., coefficient matrices are stored and long time histories may be simulated through recursive relationships.

The time histories of response components and their covariance may be effectively obtained in a recursive form for a multi-degree-of-freedom system subjected to random wind loading utilizing a z-transform. The input to the system may be described at discrete time intervals or be represented by an ARMA model. The associated response is expressed in terms of either an ARMA model or time histories which may be utilized to derive the response covariance [46]. The recursive techniques reduce the comptational effort required for the analysis, thus offers an efficient procedure for response analysis. It is noteworthy that this approach, like other direct methods, does not require the damping matrix to be classical.

The response of a multi-degree of freedom system to multi-correlated random wind loads may be carried out by employing a theory of stochastic differential equations. By applying Ito's stochastic differential equations, the multi-correlated stochastic process is represented by a white noise filtered by a set of first order linear differential equations. The calculation of response variances may be simplified and computational time reduced compared to methods using power spectral densities. The description of wind velocity vector by stochastic differential equation makes it possible to also simulate these random processes numerically [47-49].

APPLICATIONS

Typically a building is modelled by means of a computer code, e.g., ETABS which formulates mass and stiffness matrices. Based on these matrices mode shapes and natural frequencies are computed. At this stage depending on the building characteristics the analysis of wind effects could take several different courses. For relatively stiffer buildings falling in the category of mid-rise buildings, designer may elect to use code specified procedures for computing dynamic response. The ANSI 82 [11] is limited to the procedure for computing the alongwind response, whereas, Canadian [7] code provides empirical relationships for estimating across-wind and torsional responses. These response estimates are utilized to check for strength and serviceability. Necessary changes warranted by these strength and performance checking criteria are implemented to meet the code recommendations. However, designer has the option to conduct a wind tunnel study involving pressure and force balance measurements and use the results obtained from these tests for design.

For taller buildings with unconventional shapes and a combination of structural systems that may lead to coupled lateral and torsional motions it is customary in design practice to undertake a wind tunnel testing program. Such a study ensures appropriate modelling of the surrounding terrain, influence of any adjacent buildings, and special features of building shape and its envelope. In this case, in addition to pressure model for cladding design purposes, either a base-balance study or a complete aeroelastic test is conducted. Once the building has coupled mode shapes, the use of spectra derived from a base-balance study becomes more approximate. However, utilization of a multi-level second generation balance alleviates this shortcoming. For more slender buildings with highly coupled modes, an aeroelastic model offer both the inclusion of the effects of coupled mode shapes and the motion induced load effects, but not without extra cost and time needed for design and fabrication of the model.

In the following section an analysis of a special class of torsionally coupled tall building modelled as a lumped mass system is presented

[22,50]. The center of mass of the floors lie on one vertical axis so that center of resistance of the stories lie on another vertical axis so that ecdentricities are the same for all the stories, and all the floors have the same radius of gyration about the vertical axis through the center (Fig. 1).

Consider the equations of motion of an undamped n-degree-of-freedom model of this building system

$$\begin{bmatrix} m & 0 & 0 \\ 0 & m & 0 \\ 0 & 0 & m \end{bmatrix} \begin{Bmatrix} \ddot{\underset{\sim}{x}} \\ r\ddot{\underset{\sim}{\theta}} \\ \ddot{\underset{\sim}{y}} \end{Bmatrix} + \begin{bmatrix} K_{xx} & K_{xt} & \\ K_{xt}^T & K_{tt} & K_{ty}^T \\ 0 & K_{yt} & K_{yy} \end{bmatrix} \begin{Bmatrix} \underset{\sim}{x} \\ r\underset{\sim}{\theta} \\ \underset{\sim}{y} \end{Bmatrix} = \begin{Bmatrix} f_{\underset{\sim}{x}}(t) \\ \frac{1}{r} f_{\theta}(t) \\ f_{\underset{\sim}{y}}(t) \end{Bmatrix} \tag{14}$$

in which $\underset{\sim}{x}, \theta, \underset{\sim}{y}$ = response subvectors; $f_{\underset{\sim}{x}}(t), f_{\theta}(t), f_{\underset{\sim}{y}}(t)$ = loading subvectors in the x, y, and θ directions, respectively; r_k = radius of gyration of kth floor deck; and m, K_{xx}, K_{xt}, K_{yt}, K_{yy}, K_{tt} = n x n submatrices of mass and stiffness, respectively.

Damping can be introduced into this system conveniently at a later stage as viscous damping in each natural mode of vibration. Following the approach in Ref. 50 this $3n$-degrees-of-freedom system is reduced to an uncoupled system with n-degrees-of-freedom in each respective direction. Now the analysis is reduced to solving the nth order eigenvalue problem for first J modes in each direction. This is followed by defining a mode-generalized system in jth mode associated with the jth group of three uncoupled vibration modes as having the properties given by the following relationship for $j = 1, J$:

generalized mass $\quad m_j$

generalized stiffness $\quad K_{xj} = w_{xj}^2 m_j$

$$K_{yj} = w_{yj}^2 m_j$$

$$K_{tj} = w_{\theta j}^2 m_j$$

$$\frac{e_x}{r} = \frac{\{\psi_{yj}^T\}\,[K_{yt}]\,\{\psi_{\theta j}\}}{w_{yj}^2 m_j} \quad \text{and}$$

$$\frac{e_y}{r} = \frac{\{\psi_{xj}^T\}\,[K_{xt}]\,\{\psi_{\theta j}\}}{w_{xj}^2 m_j} \tag{15}$$

in which ψ_{xj}, ψ_{yj} and $\psi_{\theta j}$ are jth mode shapes in each respective direction. Now the jth mode-generalized system has three degree-of-freedom and the corresponding equations of motion are given by

$$\begin{bmatrix} m_j & & \\ & m_j & \\ & & m_j \end{bmatrix} \begin{Bmatrix} \ddot{x}^{(j)} \\ r\ddot{\theta}^{(j)} \\ \ddot{y}^{(j)} \end{Bmatrix}$$

$$+ \begin{bmatrix} K_{xj} - \frac{e_y}{r} K_{xj} & & 0 \\ \frac{-e_y K_{xj}}{r} & K_t & \frac{e_x K_{yj}}{r} \\ 0 & \frac{e_x K_{yj}}{r} & K_{yj} \end{bmatrix} \begin{Bmatrix} x^{(j)} \\ r\theta^{(j)} \\ y^{(j)} \end{Bmatrix}$$

$$= \begin{Bmatrix} f_x^{(j)}(t) \\ \frac{1}{r} f_{\theta}^{(j)}(t) \\ f_y^{(j)}(t) \end{Bmatrix} \tag{16}$$

in which $x^{(j)}, \theta^{(j)}$, and $y^{(j)}$ = response of jth mode-generalized system in respective directions, and $f_x^{(j)}(t), f_\theta^{(j)}(t), f_y^{(j)}(t)$ = jth mode generalized forcing function in x, θ, and y directions. According to Ref. 51, the contribution to the response of a wind-excited high-rise building in modes higher than the fundamental is insignificant in practice except for higher derivatives of response. Since the wind loads are stochastic in nature, the dynamic response of the system can be obtained using frequency domain methods or step-by-step integration using the time domain approach. The frequency domain approach is followed in the analysis presented here. Following the classical random vibration approach, the relationship between the forcing function and response cross-spectral density functions is given by

$$[S_R(f)] = [H(i2\pi f)]\,[S_F(f)]\,[H^*(i2\pi f)]^T;$$

$$[H(i2\pi f)] =$$

$$(2\pi f)^p\,[-(2\pi f)^2[M] + i2\pi f[C] + [K]\,]^{-1};$$

$$[S_F(f)] = \begin{bmatrix} S_{F_x}(f) & S_{F_{x\theta}}(f) & S_{F_{xy}}(f) \\ S_{F_{\theta x}}(f) & S_{F_\theta}(f) & S_{F_{\theta y}}(f) \\ S_{F_{yx}}(f) & S_{F_{y\theta}}(f) & S_{F_y}(f) \end{bmatrix} \quad (17)$$

in which $[H(i2\pi f)]$ = matrix of frequency response function with * denoting the complex conjugate; $[S_R(f)]$ = cross spectral density matrix of response, and $[S_F(f)]$ = cross spectral density matrix of forcing function; [M], [K] = mass and stiffness matrices (Eq. 16); [C] = damping matrix, which can be obtained from critical damping ratios associated with each mode; and p = higher derivatives; i.e., p equal to 1, 2, 3 represents velocity, acceleration, and jerk response components, respectively. The cross spectral density matrix can be obtained from a base-balance study in the wind tunnel which also provides the level of correlation. The forcing functions in the alongwind and acrosswind directions are generally independent for typical prismatic buildings. However, there exists correlation between the algonwind and torsion as well as acrosswind and torsion (17, 23, 52).

The matrix of mean square response is given by

$$[\sigma_R^2] = \left[\int_0^\infty [S_R(n)]\,dn\right];$$

$$[S_R(n)] = \begin{bmatrix} S_x(n) & S_{x\theta}(n) & S_{xy}(n) \\ S_{\theta x}(n) & S_\theta(n) & S_{\theta y}(n) \\ S_{yx}(n) & S_{y\theta}(n) & S_y(n) \end{bmatrix} \quad (18)$$

A computer program was developed to evaluate the lateral-torsional response of typical tall buildings conforming the the dynamic characteristics discussed earlier in the formulation of the equations of motion. As indicated earlier, the corners of a building experience considerable effects of torsional response, i.e., higher stresses and human discomfort. Therefore it is important to consider response at the top of the building, which for the corner $x = D/2$, $y = B/2$ is obtained from the response at the center of the building as

$$x_c^{(p)}(t) = x^{(p)}(t) - \frac{B}{2}\theta^{(p)}(t);\quad y_c^{(p)}(t)$$

$$+\frac{D}{2}\theta^{(p)} \tag{19}$$

in which $x_c(t)$ = x-component of corner response; $y_c(t)$ = y-component of corner response; $x(t)$ = x-component of the center; $y(t)$ = y-component of the center of the building response; and superscript p = higher derivatives of response, i.e., 0, 1, 2, 3 for displacement velocity, acceleration, and jerk, respectively. The rms value of the building corner response is given by

$$\sigma_{x_c}^{(p)} = \sqrt{\sigma_x^2 + \frac{B^2}{4}\sigma_\theta^2 - B\sigma_{x\theta}^2};$$

$$\sigma_{y_c}^{(p)} = \sqrt{\sigma_y^2 + \frac{D^2}{4}\sigma_\theta^2 + D\sigma_{y\theta}^2} \tag{20}$$

in which $\sigma_x^2, \sigma_y^2, \sigma_\theta^2$ = mean square values of the alongwind, acrosswind, and torsional response at the center of the building and $\sigma_{x\theta}^2 = \int_0^\infty S_{x\theta}(n)\,dn$, and $\sigma_{y\theta}^2 = \int_0^\infty S_{y\theta}(n)\,dn$ The response estimates obtained in this manner correspond to the response of the building top corners in the fundamental mode. The response at any other level can be obtained using the mode shapes of the torsionally coupled building. The coupled mode shapes are obtained from the product of the uncoupled mode shapes of the building and the mode shapes associated with the jth mode-generalized system in respective directions ($\alpha_{xj}\ \psi_{xj};\ \alpha_{yj}\ \psi_{yj};\ \alpha_{\theta j}\ \psi_{\theta j}$).

The procedure outlined above was utilized to study the dynamic response characteristics of a torsionally coupled building. Detailed results can be found in Ref. 22]. Aerodynamic loading spectra for the alongwind, acrosswind and torsional directions obtained from wind tunnel measurements were utilized for this example problem [Fig. 2]. The ratios of the torsion-induced to the acrosswind and alongwind acceleration are given in Fig. 3. The comparison of response levels emphasizes the importance of torsional response in the design of high-rise buildings. Ignoring this important component of overall response could result in underestimation of the building response to windloads. A parameter study was conducted in which the eccentricity between the mass and resistance centers was systematically varied. It is noted that certain combinations of eccentricities in both lateral directions tend to amplify the overall response [Ref. 22].

For design purposes the response statistics of a building are synthesized with meteorological characteristics of the local estimate. This requires definition of both the dependence of particular response parameters on wind speed and direction and statistical characteristics of the local climate, i.e., wind speed and direction. Typically a rate of up-crossing of a threshold response values, which are a function of both wind speed and direction is utilized. This requires information on the joint probability density function of wind speed and direction. From the up-crossing rate the mean recurrence interval for a specified response level is obtained. Subsequently the risk of exceeding the response level associated with a given mean recurrence interval for a specified time duration is determined. Utilizing the preceding information plots between the peak base moments and shears, and peak or rms top floor acceleration versus average number of events/year are plotted. These plots serve as limit state checking against preestablished thresholds for strength and serviceability. For example, a typical criterion for limiting tall building motion is given below [53].

"The mean recurrence interval for storms causing an rms acceleration of 8-10 mg at the building top shall not be less than 10 years".

Applications of probability based approaches for design of buildings are addressed in the literature [e.g., 6, 54-57].

DAMPING IN TALL BUILDINGS

Estimation of damping in structural systems poses the most difficult problem in structural dynamics. Unlike mass and stiffness characteristics of a structural system, damping does not refer to a unique physical phenomenon; increasingly damping is being recognized as a key factor in the design of structures that are sensitive to wind. Damping is of particular interest to the designers of high-rise buildings where it plays an important role to help meet serviceability limit-states from human comfort considerations. Damping is often required rather than engineered in a structural system, therefore, the estimates of damping in a structural system have intrinsic variability. The availability of damping values accurately, at the design stage, would certainly alleviate a major source of uncertainty experience in the design of dynamically sensitive structures. As presently the selection of an appropriate damping value is a subject of controversy. The present state of this in the design of tall buildings is such that it is difficult to predict structural damping closer than plus or minus 30% until the building construction is completed. Currently, the damping is essentially ascertained on the basis of the knowledge gleaned from existing buildings of similar material and structural systems on which tests have been conducted. Although it is a general consensus that damping values change with amplitude of motion, their functional descriptions are rather limited [58-61]. Besides the complex nature of damping, the methods employed to ascertain damping of full-scale structures, and the analysis and interpretation of data introduce additional uncertainty. In a recent study by Davenport and Hill-Carroll [59], the variability of damping was analyzed by carefully selecting data obtained from available full-scale studies. Although the COV of damping estimates based on the available data ranged from as low as 33% to as high as 78%, they suggested a value of 40%. Based on a selected group of measurements, meeting sufficient quality criterion, Jerry [58] has proposed a theoretical model that provides damping values which are in good agreement with measured values. A similar expression is also suggested in ESDU [61]. Kareem and Sun [62] have presented a second-order perturbation technique to study and quantify the effects of damping variability on the transient and steady-state dynamic response of structural systems.

Damping is being currently augmented in structural systems directly by means of viscoelastic dampers, or in some cases indirectly through active or passive dynamic vibration absorbers or tuned mass dampers (TMDs) [47, 63-68]. Such damping systems offer promising alternatives to improve the damping capacity of structures in view of the limited control over traditional parameters that help to reduce structural motion, e.g., mode shapes, stiffness and mass. A viscoelastic damper is another passive discrete damping device that is capable of dissipating large amounts of energy in shear. However, the viscoelastic material properties are a function of frequency and temperature, which requires these features to be included in the performance evaluation of such a system. Currently, viscoelastic dampers have been successfully used in the World Trade Center, New York, and Columbia Center Building, Seattle [66]. One of the most promising passive motion mitigate motion device. In such a device, the response of structures with tuned sloshing liquid is mitigated when one of the sloshing modes of the secondary fluid appendage is tuned to the fundamental mode of the primary system [69-71].

CONCLUDING REMARKS

This paper reviews the techniques to estimate the wind-induced loads on buildings utilizing boundary layer wind tunnels. These methods include pressure field measurements, pneumatic averaging of pressure field, base-balance, multi-level second-generation force transducer, aeroelastic models. The procedures for dynamic analysis of tall buildings are discussed. An example of a torsionally coupled building is presented which demonstrates the need for including torsional response of buildings for evaluating building performance from human comfort considerations. The potential of the foregoing technique is fully realized by synthesizing the results with meterological statistics of the local climate. This provides a probabilistic framework to examine the building performance based on pre-established limit states concerning serviceability and servivability.

The importance of damping in reducing building motion is emphasized. Issues concerning damping variability and its efforts on the response analysis are discussed. Sources to augment damping in buildings are highlighted in view of the limited control over traditional parameters that help to reduce structural motion. These techniques include passive devices such as viscoelastic dampers, tuned mass dampers, and tuned sloshing dampers. The active control of wind excited structures offers great promise, but applications are at present rather limited to research and development phases.

ACKNOWLEDGEMENT

The support for this work was provided by the NSF-PYI-84 award to the author by the National Science Foundation under Grant No. BCS90-96274 and matching funds provided by the American Institute of Steel Construction. Their support is gratefully acknowledged. Any opinion, findings, conclusions, or recommendations expressed in this paper are those of the author and do not necessarily reflect the views of the sponsors.

REFERENCES

1. Davenport, A.G., Gust Loading Factors, *Journal of the Structural Division, ASCE,* June 1967, 93(ST3).

2. Vickery, B.J., On the Assessment of Wind Effects on Elastic Structures, Civil Engineering Transactions, Inst., Aust., 1966.

3. Vellozi, J. and Cohen, E., Gust Response Factors, *Journal of the Structural Division, ASCE,* 1968, 94(ST6).

4. Simiu, E., Revised Procedure for Estimating Alongwind Response, *Journal of the Structural Division, ASCE,* 1980, 106(ST1).

5. Solari, G., Along-Wind Response Estimation: Closed Form Solutions, *Journal of the Structural Division, ASCE,* 1982, 108(ST1).

6. Simiu, E. and Scanlan, R.H., Wind Effects on Structures: An Introduction to Wind Engineering, Wiley and Sons, NY, 1986.

7. Canadian National Building Code, *Supplement No. 4*, Commentaries on Part 4, 1980.

8. Kareem, A., Synthesis of Fluctuating Along Wind Loads on Buildings, *Journal of the Engineering Mechanics Div., ASCE,* 1986, 112(2).

9. Kareem, A., "Nonlinear Wind Velocity Term and Response of Offshore Compliant Structures," *Journal of Engineering Mechanics, ASCE,* Vol.. 110, No. 10, 1984.

10. Kareem, A., Fluctuating Wind Loads on Buildings, *J. Engr., Mech. Div., ASCE,* December 1982, 108(EM6).

11. American National Standard, Building Code Requirements for Minimum Design Loads in Buildings and Other Structures, ANSI A58.1-1982, American National Standards Institute, Inc., 1430 Broadway, New York, NY 10018.

12. Murakami, S., "Computational Wind

Engineering," *Proceedings of the Sixth U.S. National Conference on Wind Engineering,* (Editor: Kareem), University of Houston, Houston, TX, March 1989.

13. Kareem, A., Mapping and Synthesis of Random Pressure Fields, *Journal of Engineering Mechanics, ASCE,* Vol. 115, No. 11, 1989.

14. Reinhold, T.A., Distribution and Correlation of Dynamic Wind Loads, *Journal of the Engineering Mechanics Division, ASCE,* 1983, 109(6).

15. Surry, D. and Stathopoulos, T., An Experimental Approach to the Economical Measurement of Spatially Averaged Wind Loads, *J. Eng. Aerodynamics,* 1977, 2.

16. Fuji, K., Hibi, K., Ueda, H., Shimada, K., "Visualization of Fluctuating Surface Pressure Distribution on Bluff Body Using Electronically Scanning Pressure Sensors," 5th International Symposium on Flow Visualization, 8, 1989.

17. Kareem, A., Acrosswind Response of Buildings, *Journal of the Structural Division, ASCE,* 1982, 108(ST4).

18. Kareem, A., Model for Predicting the Acrosswind Response of Buildings, *Engineering Structures,* April 1984, 6.

19. Vanmarcke, E.H., Random Fields, MIT Press, Cambridge, Mass., 1983.

20. Vanmarcke, E.H. and Grigorio, M., Stochastic Finite Element Analysis of Simple Beams, *Journal of Engineering Mechanics, ASCE,* October 1983, 109(5).

21. Kareem, A. and Sun, Wei-Joe, Probabilistic Finite Element Analysis of Structures with Parametric Uncertainties, University of Houston, Department of Civil Engineering, Research Report, in preparation, 1987.

22. Kareem, A., Lateral Torsional Motion of Tall Buildings to Wind Loads, *Journal of the Structural Division, ASCE,* 1985, 111(11).

23. Kareem, A., Aerodynamic Load Effects on Prismatic Structures, Department of Civil Engineering, University of Notre Dame, in progress.

24. Kareem, A. and Cermak, J.E., Wind-Tunnel Simulation of Wind-Structure Interactions, 1979, 18(4).

25. Tschanz, T. and Davenport, A.G., The Base Balance Technique for the Determination of Dynamic Wind Loads, *Journal of Wind Engineering and Industrial Aerodynamics,* 1983, 13(1-3).

26. Vickery, P.J. *et al.*, The Effect of Model Shape on the Wind-Induced Response of Tall Buildings, *Proceedings of the Fifth U.S. National Conference on Win Engineering,* Lubbock, Texas, November 1985.

27. Boggs, D.W. and Peterka, J.A., "Aerodynamic Model Tests of Tall Buildings," *Journal of Engineering Mechanics, ASCE,* Vol. 115, No. 3, 1989.

28. Reinhold, T.A. and Kareem, a., Wind Loads and Building Response Predictions Using Force Balance Techniques, *Proceedings of the Third ASCE Engineering Mechanics Conference - Dynamics of Structures,* UCLA, 1986.

29. Saunders, J.W. and Melbourne, W.H., Tall Rectangular Building Response to Cross-Wind Excitation, *Proceedings of the Fourth International Conference on Wind Effects on Buildings and Structures,* Cambridge University Press, Cambridge, Mass., 1977.

30. Cermak, J.E., Aerodynamics of Buildings, Annual Review of Fluid Mechanics, 1976, 8.

31. Steckly, A., Vickery, B.J. and Isyumov, N., "On the Measurement of Motion-Induced Forces in Turbulent Shear Flow," *Proceedings of the Sixth U.S. National Conference on Wind Engineering* (Editor: A. Kareem), Houston, TX, March 1989.

32. Isyumov, N., Aeroelastic Modeling of Tall Buildings, Wind Tunnel Modeling for Civil Engineering Applications, (ed. T.A. Reinhold), Cambridge University Press, 1982.

33. Davenport, A.G., Perspectives on Full-Scale Measurement of Wind Effects, *Journal of Industrial Aerodynamics,* 1973, 1(1).

34. Williams, D., "Dynamic Loads in Aeroplanes Under Given Impulsive Loads with Particular Reference to Landing and Duct Loads on a Large Flying Boat, Great Britain Royal Aircraft Estiablishment Reports SME3309 and 3316, 1945.

35. Hansen, R.J. *et al.*, Human Response to Wind-Induced Motion of Buildings, *Journal of the Struct. Division, ASCE,* July 1973, 99(ST7).

36. Reinher, H. and Meister, F., The Sensitivity of Human Beings to Vibrations, *Die Empfinlichkeit des Menschengegen Ershuttterugen Forschung aufdem Gebiete des Ingenieu and Wesens,* November 1931, 2(11), (English Translation).

37. Chen, P.W. and Robertson, L.E., Human Perception Threshold of Horizontal Motion, *J. of Struct. Divs., ASCE,* 1972, 98(ST8).

38. Kareem, A., Mitigation of Wind Induced Motion of Tall Buildings, *Journal of Wind Engineering and Industrial Aerodynamics,* 1983, 11.

39. Bathe, K.J., Finite Element Procedures in Engineering Analysis, Prentice-Hall, Inc., New Jersey, 1982.

40. Shinozuka, M., Monte Carlo Solution of Structural Dynamics, *Computers and Structures,* 1972, 2, 855-874.

41. Wittig, L.E. and Sinha, A.K., Simulation of Multi-correlated Random Processes Using the FFT Algorithm, *Journal of the Acoustal Society of America,* 1975, 68.

42. Li, Yonsun, and Kareem, A., "ARMA Systems in Wind Engineering," *Probabilistic Engineering Mechanics*, Vol. 5, No. 2, June 1990.

43. Samaras, E.F., Shinozuka, M. and Tsurui, A., ARMA Representation of Random Vector Processes, *Journal of Engineering Mechanics, ASCE,* March 1985, 111(3).

44. Reed, D.A. and Scanlan, R.H. "Autoregressive Representation of Longitudinal, Lateral, and Vertical Turbulence Spectra, *Journal of Wind Engineering and Industrial Aerodynamics,* 1984, 17.

45. Spanos, P.-T.P and Schultz, K.P., "Two-Stage Order- of Magnitude Matching for the von Karman Turbulence Spectrum," *Proceedings of the Fourth International Conference on Structural Safety and Reliability,* ICOSSAR, 1985, 1.

46. Li, Yonsun, and Kareem, A., "Recursive Modelling of Dynamic Systems," *Journal of Engineering Mechanics, ASCE,* Vol. 116, No. 3, 1990.

47. Suhardjo, J. and Kareem, A., "Active Control of Wind Excited Structures," Research Report in Preparation, Dept. of Civil Engineering, University of Notre Dame, 1990.

48. Gobmann, E. and Waller, H., "Analysis of Multi-Correlated Wind-Excited Vibrations of Structures Using the Covariance Methods," *Engineering Structures,* 5, 1983.

49. Muscolino, G., "Stochastis Analysis of Linear Structures Subject to Multicorrelated Filtered Noises, *Engineering Structures,* 8, 1986.

50. Kan, L., Christopher, Chopra, A.K., "Elastic Earthquake analysis of Torsionally Coupled Multi-Story Buildings," *Earthquake Engineering and Structural Dynamics,* Vol. 5, 1977.

51. Kareem, A., Wind Excited Response of Buildings in Higher Modes, *Journal of Structural Engineering, ASCE,* 1981, 107(ST4).

52. Isyumov, N. and Poole, M., Wind Induced Torque on Square and Rectangular Building Shapes, *Journal of Wind Engineering and Industrial Aerodynamics,* 1983, 13.

53. Kareem, A. and Allen, R.H., "WISER: A Knowledge-Based Expert System For the Design Modification of High-Rise Buildings for Serviceability," Preprints - *Proceedings of the Seventh International Conference on Wind Engineering,* Vol. 2, Aachen, W. Germany, 1987.

54. Davenport, A.G., "Note on the Distribution of the Largest Value of a Random Function with Application to Gust Loading," *J. Inst. Civ. Eng.*, 24, 1964.

55. Hart, G.C. *et al.*, Structural Design Using a Wind Tunnel Test Program and Risk Analysis, *Journal of Wind Engineering and Industrial Aerodynamics*, 14(1-3).

56. Kareem, A., "Wind Effects of Structures: A Probabilistic Viewpoint," *Probabilistic Engineering Mechanics*, Vol. 2, No. 4, 1987.

57. Davenport, A.G., "The Relationship of Reliability to Wind Loading," *Journal of Wind Engineering and Industrial Aerodynamics*, Vol. 13, No. 1-3, 1983.

58. Jeary, A.P., Damping in Tall Buildings - A Mechanism and a Predictor, *Earthquake Engineering and Structural Dynamics*, 1986, 14.

59. Davenport, A.G. and Hill-Carroll, P., Damping in Tall Buildings; Its Variability and Treatment in Design, Building Motion in Wind, Proceedings of a Seszsion, ASCE Convention, Seattle, Washington, 1986.

60. Hart, G.-C., "Estimation of Damping for Building Design," Tall Building, Vanderbilt University.

61. Damping of Structures, Part 1: Tall Buildings, Engineering Science Data Units, Item No. 83009, Sept. 1983, London.

62. Kareem, A., and Sun, W.J., "Dynamic Response of Structures With Uncertain Damping," *Engineering Structures*, Vol. 11, 1989.

63. McNamara, R.J., "Tune Mass Dampers for Buildings," *Journal of the Structural Division, ASCE,* Vol. 103, No. ST9, 1977.

64. Samali, B., Yang, J.N.; and Yeh, C.T., "Control of Lateral-Torsional Motion of Wind-Excited Buildings, *Journal of Structural Engineering, ASCE,* Vol. 111, No. 6, 1985.

65. Abdel-Rohman, M. and Leipholz, H.H., "Active Control of Tall Buildings," *Journal of Structural Division, ASCE,* Vol. 109, No. 3, 1983.

66. Keel, C.J. and Mahmoodi, P., "Design of Viscoelastic Dampers for Columbia Center Building," Building Motion in Wind, ASCE Convention, Seattle, Washington, April 8, 1986.

67. Wiesner, K.B., "Taming Lively Buildings," Civil Engineering, ASCE, June 1986.

68. Bergman, L.A., McFarland, D.M., Hall, J.K., Johnson, E.A. and Kareem, A., "Optimal Distribution of Tuned-Mass Dampers in Wind Sensitive Structures, *Proceedings of the 5th International Conference on Structural Safety and Reliability,* ICOSSAR '89, San Francisco, Ca, 1989.

69. Kareem, A. and Sun, W.J., "Stochastic Response of Structures with Fluid-Containing Appendages," *Journal of Sound and Vibration*, Vol. 119, No. 3, 1987.

70. Fujino, Y., et al., "Fundamental Study of Tuned Liquid Damper (TLD) - A New Damper for Building Vibrations," *Proceedings of the Symposium/Workshop on Serviceability of Buildings*, 16-18, May 1988.

71. Kareem, A., "Reduction of Wind Induced Motion Utilizing a Tuned Sloshing Damper," *Proceedings of the Sixth U.S. National Conference on Wind Engineering*, (Editor: Kareem), March 8-10, 1989.

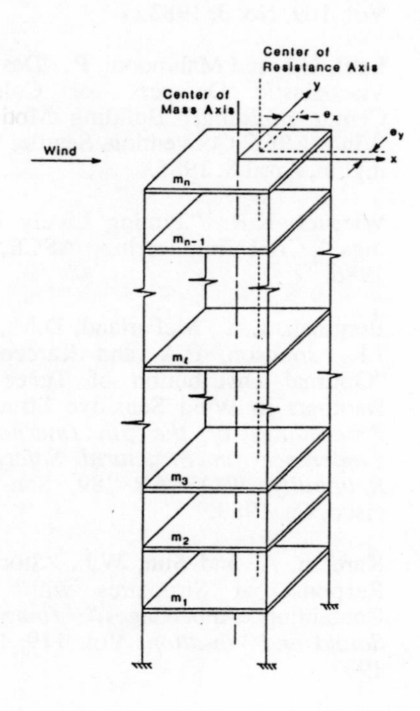

Fig. 1 - Idealized Building

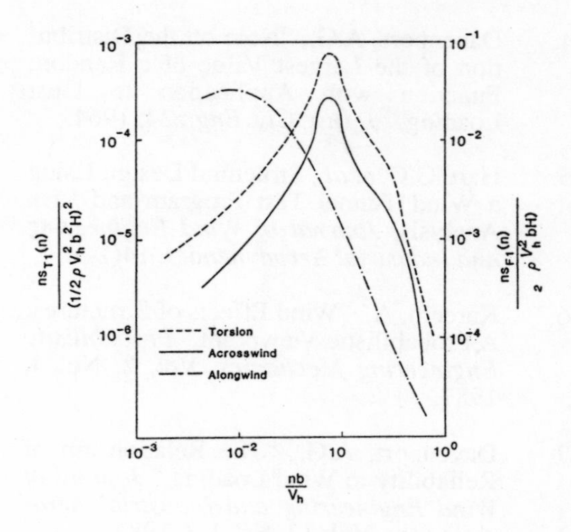

Fig. 2 - Normalized Reduced Modal Spectra of Torsional, Acrosswind and Alongwind Forces in Urban Environment

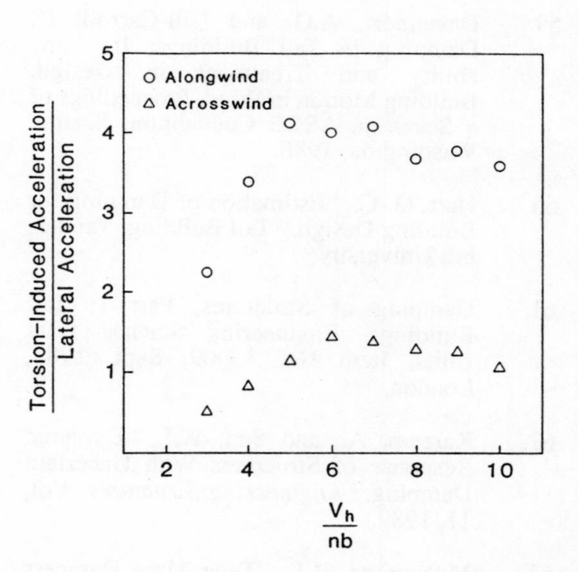

Fig. 3 - Comparison of Torsion-Induced and Lateral Accelerations

WIND LOADS ON TOWERS AND CHIMNEYS

B.J. Vickery

Boundary Layer Wind Tunnel Laboratory
Faculty of Engg. Science
University of Western Ontario
London,
Ontario, N6A 5B9, CANADA

SYNOPSIS

The two-dimensional nature of the flow past slender towers and chimneys permits theoretical approaches to response prediction provided that the necessary aerodynamic data are available. The paper defines the data that must be obtained by experiments and describes experimental techniques suited to this end. Special attention is paid to the circular cross-section for which data must be obtained from full-scale observations. Attention is also given to motion-induced forces which play a much stronger role in determining the response than is the case for most buildings.

1.0 INTRODUCTION

From a general view point the aerodynamic behaviour of chimneys and towers differs little from that of tall buildings. There are however differences which permit a somewhat different approach to be taken in the evaluation of wind loads on the more slender towers and chimneys. Buildings are generally of comparatively low aspect ratio (a height to breadth ratio less than 7 and commonly much smaller). Buildings are commonly located in an environment which contains numerous structures of comparable height. The flow past buildings is strongly three-dimensional and strongly dependent on nearby structures. In such circumstances the only reliable means of prediction of response is wind tunnel testing with complete modelling of the environs. In contrast, chimneys and towers are commonly slender and the flow is only weakly three-dimensional. They often rise well above the surrounding structures and are not subject to strong interference effects. A further and significant difference between the tower or chimney and the tall building lies in the design criteria. For tall buildings the controlling criteria are most often associated with deflection and acceleration limitations rather than strength. For chimneys and unoccupied towers the controlling criterion is normally strength.

With only weak three-dimensionality and in the absence of interference effects, the wind loading on chimneys and towers can be estimated by theoretical methods provided the necessary aerodynamic data is available. This possibility opens the way to the use of sectional model tests rather full models. Because acceleration levels are not normally a design consideration the so-called "critical speed" may well fall within the design range. In such circumstances motion-induced forces are likely to play an important role.

In Section 2 of this paper the principal dynamic forces acting on chimneys and towers are described. In Section 3 the theoretical approach to response prediction is outlined with the aim of defining the aerodynamic data essential to the prediction calculations. Section 4 is concerned with experimental techniques suited to chimney and towers. Data for the prediction of wind-induced response is presented in Section 5.

2.0 DYNAMIC WIND FORCES

In broad terms, the wind forces resulting in dynamic excitation can be divided into three groups as follows:

(1) Forces induced by the turbulent fluctuations in the incident flow.

(2) Forces induces by the unsteady nature of the wake and particularly to vortex shedding.

(3) Forces induced by motion of the structure.

2.1 Forces Induced By Turbulence

The forces induced by atmospheric turbulence cover a wide frequency band with significant energy existing from frequencies as low as 0.01 Hz to about 2 Hz. Such forces are most commonly the chief source of excitation in the along-wind or drag direction but the lateral components of turbulence may induce significant across-wind forces. The turbulent velocity fluctuations may arise from the general nature of the earth's turbulent boundary layer or from the unsteady wakes of one or more nearby structures. In the latter case the excitation is referred to as buffetting and the frequency band is often narrow with most of the energy being within ±30% of a central frequency determined by the size and shape of the upstream structure and the wind speed.

2.2 Wake Effects

The wakes of bluff bodies are unsteady and, for slender structures particularly, across-wind forces are induced by the shedding of vortices. For very slender structures in comparatively smooth flow the spectrum of the vortex induced is narrow and centred on a shedding frequency, f_s, given by;

$$f_s = S.\frac{U}{B}$$

where U = mean wind speed
B = breadth of structure
S = Strouhal Number

The Strouhal Number is a function of the cross-sectional shape the aspect ratio and (for curved shapes) the Reynolds Number but is typically between 0.1 and 0.2. For lower aspect ratios and in high intensity turbulence the Strouhal Number may change slightly but there is a distinct broadening of the spectrum. The general features of the spectra of turbulence-induced and wake-induced forces are shown in Figures 1 and 2 (Davenport [1]). Spectra associated with buffeting from a nearby structure would have characteristics similar to those shown in Figure 2. The influence of aspect ratio is shown in Figure 2a, where the hatched vertical line represents a typical design speed. For slender towers and chimneys the wake excitation is more peaked or narrow-banded and reaches a maximum at some critical speed which is below the design speed. For buildings, the critical speed will almost invariably fall above the design speed since achieving acceptable acceleration levels is likely to prove difficult or impossible at or near critical speed.

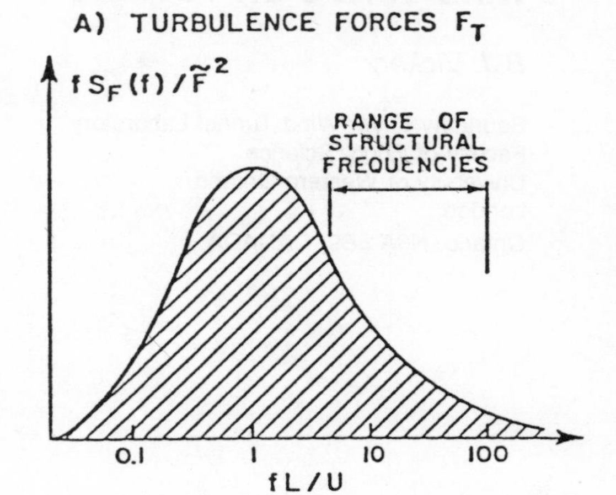

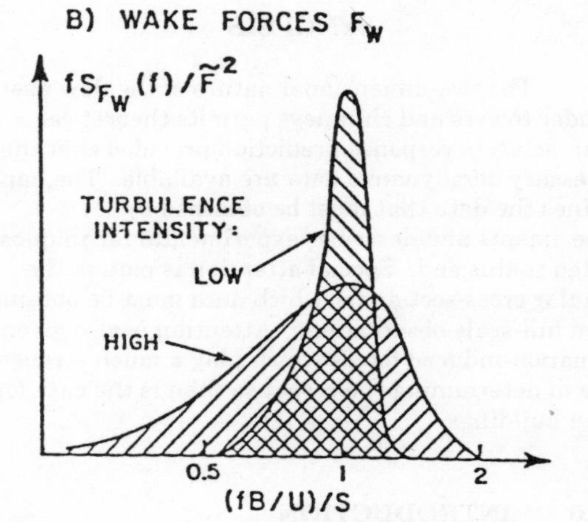

Fig. 1: Typical spectra for forces due to (A) atmospheric turbulence and (B) wake vortex shedding (Davenport [1])

2.3 Motion-Induced Forces

Forces induced by motion include those in phase with displacement or acceleration (aerodynamic stiffness or mass) and those in phase with velocity (negative aerodynamic damping) or 180° out of phase (positive aerodynamic damping). In the along-wind direction, the aerodynamic damping is almost always positive but for the across-wind motions of more slender structures it may well be negative and of the same order as the positive structural damping. The positive drag damping tends to increase monotonically with wind speed whereas for across-wind motions there is commonly negative damping for a range of $U^*(= U/f_oB)$ (f_o = frequency of vibration) commencing just above a value of $1/S$. A schematic representation of the role of aerodynamic damping is shown in Figure 3.

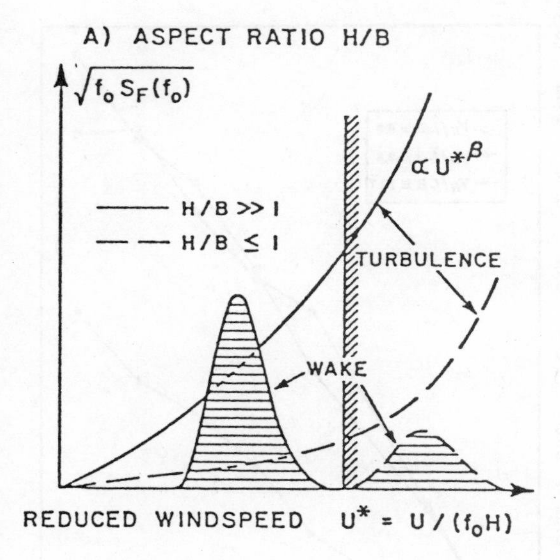

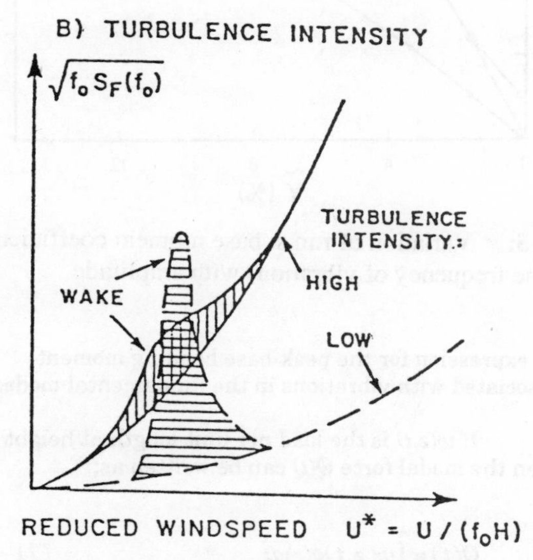

Fig. 2: The influence of (A) aspect ratio and (B) turbulence intensity on excitation by turbulence and by shedding (Davenport [1])

In addition to inducing phase related forces the motion of a structure may influence the nature of those forces which already exist on the stationary body. Steckley [2] noted an increase in correlation of the forces associated with shedding (even at frequencies removed from the frequency of vibration of the structure) and consequent increases in the base moment induced by shedding. Figure 4 shows the variation of the base moment coefficient (excluding forces at the vibration frequency) with tip amplitude while Figure 5 shows the variation of the component at the frequency of vibration. The results in Figures 4 and 5 are for a square prism with an aspect ratio of 13.3 tested in turbulent shear flow. The R.M.S. Moment Coefficient (σ) is defined as;

$$\sigma = M / \tfrac{1}{2} \rho_a V^2 B H^2$$

M	=	RMS Base Moment
B	=	width
H	=	height
V	=	mean speed at height H
ρ_a	=	air density
and Y	=	Δ / B
where Δ	=	R.M.S. tip motion

The almost linear growth of the phase related aerodynamic stiffness and damping is clear in Figure 5 while the influence of motion on the existing forces is evident but not of great significance at amplitudes of interest for most structures.

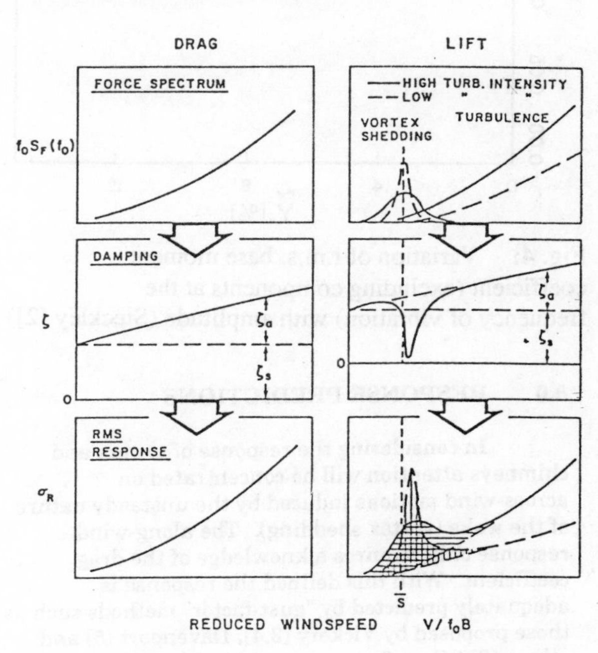

Fig. 3: Schematic representation of the resonant response of a tower

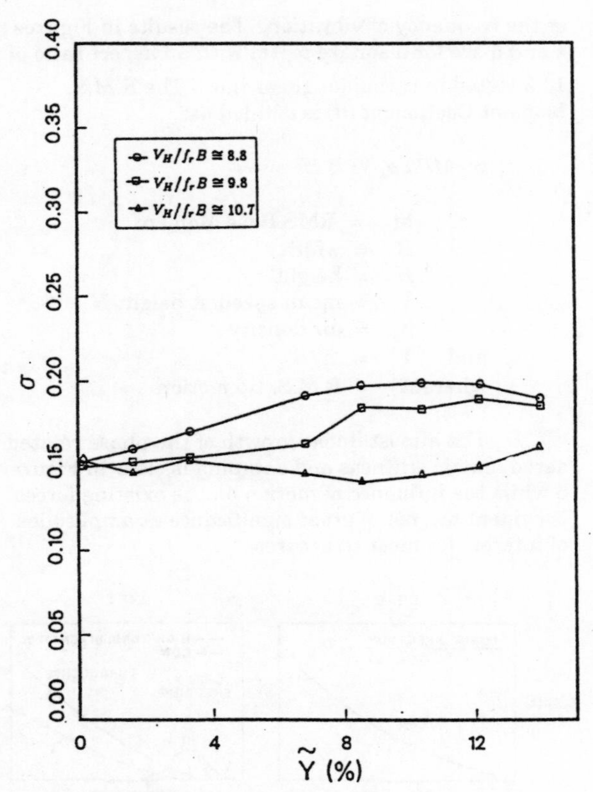

Fig. 4: Variation of r.m.s. base moment coefficient (excluding components at the frequency of vibration) with amplitude (Steckley [2])

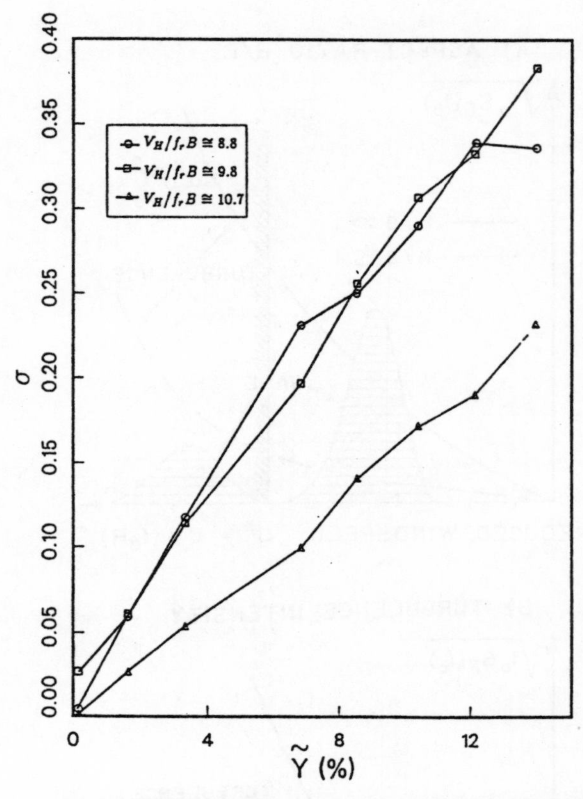

Fig. 5: Variation of r.m.s. base moment coefficient (at the frequency of vibration) with amplitude

3.0 RESPONSE PREDICTIONS

In considering the response of towers and chimneys attention will be concentrated on across-wind motions induced by the unsteady nature of the wake (vortex shedding). The along-wind response only requires a knowledge of the drag coefficient. With this defined the response is adequately predicted by "gust-factor" methods such as those proposed by Vickery [3,4], Davenport [5] and others [6]. Gust-factor approaches have now been incorporated into many national building codes and into specialist codes such as the CICIND model codes for chimneys. Across-wind response is somewhat more difficult to handle and it is not at all uncommon that these motions control the design.

Across-wind response is dominated by the resonant amplification near the natural frequency of the fundamental mode. Only for very slender self-supporting towers or guyed towers is it necessary to consider higher modes. In the derivation that follows attention is restricted to the development of an expression for the peak base bending moment associated with vibrations in the fundamental mode.

If $w(z,t)$ is the load per unit length at height z then the modal force $Q(t)$ can be written as;

$$Q(t) = \int_o^h w(z,t)\phi(z)dz \tag{1}$$

where $\phi(z)$ is the mode shape. The r.m.s. modal amplitude

$\sigma_a = \sqrt{\overline{a}^2}$, is given by;

$$\sigma_a = \frac{\sqrt{\frac{\pi}{4\beta} f_o S_{QQ}(f_o)}}{K} \tag{2}$$

where $S_{QQ}(f)$ is the spectrum of $Q(t)$, f_o is the natural frequency and K the modal stiffness where;

$$K = (2\pi f_o)^2 \int_o^h m(z)\phi^2(z)dz$$

$m(z)$ = *mass per unit height*

β = damping, including both structural and aerodynamic

$$S_{QQ}(f) = \int_o^h \int_o^h S_w(z_1 z_2 f)\phi(z_1)\phi(z_2)dz_1 dz_2 \qquad (3)$$

where $S_w(z_1\, z_2\, f)$ is the co-spectrum of the loads at positions z_1 and z_2. The r.m.s. moment is;

$$\sigma_M = \sigma_a \int_o^h (2\pi f_o)^2 m(z) z\phi(z)dz \qquad (4)$$

It is clear from equations (1) to (4) that the key aerodynamic data relate to the co-spectrum, $S_w(z_1\, z_2\, f)$, and the aerodynamic contribution to the damping, β.

The co-spectrum may be expressed as;

$$S_w(z_1, z_2, f) = R(z_1, z_2, f).\sqrt{S_w(z_1, f) S_w(z_2, f)}$$

For slender structures, the exact form of the narrow-band correlation function is not as important as its integral, the correlation length (L),

$$L = \int_o^\infty R(z_1, z_2, f) d(z_1 - z_2)$$
$$= L.D$$

The spectrum, $S(z_1, f)$, for vortex shedding forces is adequately represented by a Gaussian distribution of the form;

$$S_w(z_1 f) = \left\{\tilde{C}_L . \frac{1}{2}\rho\bar{\mu}^2(z_1).D(z_1)\right\}^2$$
$$\frac{1}{B\sqrt{\pi}} \exp\left\{-\left(\frac{1 - f/f_s}{B}\right)^2\right\}$$

where;

C_{LN} = r.m.s. lift force coefficient

$\mu(z_1)$ = mean speed at z_1

$D(z_1)$ = breadth of structure at z_1

B = a bandwidth parameter

f_s = shedding frequency

= $S\ \mu(z_1)/D(z_1)$

S = Strouhal Number

In order to predict the response a knowledge of the following aerodynamic parameters is essential;

Strouhal Number;	S
Lift Coefficient;	C_L
Correlation Length;	L
Bandwidth;	B

for a uniform structure of width D in uniform flow the rms base moment is then;

$$\sigma_M \neq \sigma C_L . \frac{1}{2}\rho\bar{\mu}^2 D H^2 \left\{\frac{\sqrt{\pi} L\lambda}{2B\beta}\right\}^{1/2}$$

$$x \exp\left\{-\frac{1}{2}\left(\frac{1 - f_o D / S\bar{\mu}}{B}\right)\right\}^2 x\left(\int_o^H \phi^2(z)dz\right)^{1/2}$$

$$x\left(\int_o^H m(z)(z/H)\phi(z)dz\right) / \left(\int_o^H m(z)\phi^2(z)dz\right)$$

where

$$\lambda = H/D$$

$$\beta = \beta_s + \beta_{aero}$$

$$\beta_{aero} = \phi\{\bar{\mu}/f_o D, \sigma_a/D\}$$

The values of σC_L, S, L, and β_{aero} can be evaluated from sectional tests. B can be estimated from the intensity of turbulence as described by Vickery & Basu [7].

4.0 EXPERIMENTAL TECHNIQUES

Experimental techniques in common use may be broken into two broad classes although a variety of approaches exist within each class. The two classes are;

(i) **Aeroelastic;** the models are dynamically scaled and deflections, moments etc. may be directly measured and transposed to prototype scale. The models may be of the full structure or of a section of the structure but in the latter case the overall performance cannot, in general, be obtained directly from the test results.

(ii) **"Rigid" models;** rigid or very stiff models satisfy only geometric scaling and the measurements are restricted to forces and moments measured by a balance (usually at the base of the structure) or by integration of surface pressures. The motions of the prototype are then predicted by theoretical methods with the motion-induced forces either being ignored or estimated from existing data. Since only geometric scaling is required the influence of mass, stiffness and damping on the response form part of the predictive phase and a range of values can be examined or predictions made interactively with the designer as the design progresses.

4.1 Aeroelastic Tests

There are a variety of aeroelastic models in common use. The simplest to appreciate are "replica" models that reproduce the prototype such in virtually all respects. An example of such a model, that of the 550m CN Tower in Toronto [8], is shown in Figure 6. The concrete tower is modelled geometrically with a metal filled plastic representing the concrete. The plastic and concrete have essentially the same density but the lower modulus of the plastic results in a reduced velocity scale ie;

$$V_M / V_P = \sqrt{(E/\rho)_M} \; / \sqrt{(E/\rho)_P}$$

Replica models such as that shown in Figure 6 are only possible for comparatively simple structural systems and only then if model construction materials are available to produce a velocity scale that can be attained in the tunnel. A more common approach is to model the structural system by a spine which has the correct stiffness properties and then to attach to this spine, segments which reproduce the external geometry and the overall mass without adding to the stiffness.

Other forms of aeroelastic models include rigid base-pivotted models, sectional models and taut-strip models. The use of base pivotted or "stick" models is discussed by Isyumov [9]. Sectional models are suitable for long slender structures with constant cross-section eg. bridges and towers.

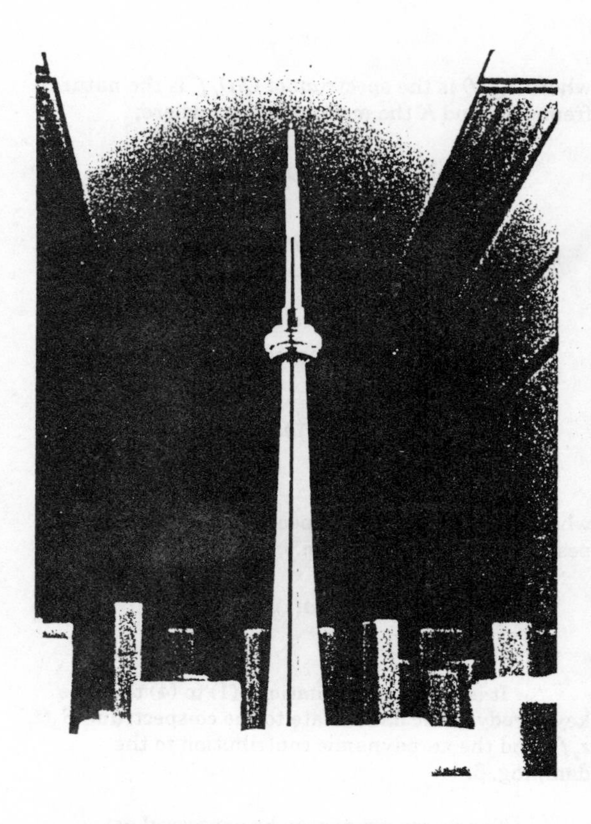

Fig. 6: Replica model of the CN Tower, Toronto (Isyumov et al [8])

Two-dimensional sections can be elastically mounted to produce the correct torsional, vertical and horizontal frequencies and basic data such as aerodynamic derivatives and turbulence admittance functions derived for inclusion in a theory designed to predict behaviour of the complete structure.

4.2 Rigid Models (Force and Pressure Measurements)

As stated, the attractions of force measuring systems are the reduced cost, the more rapid response to the designer and the ability to investigate rapidly the influence of changes in stiffness, damping, mass and mode shape. The prime disadvantage is that motion-induced forces are not measured and this deficiency must be accounted for in the response predictions or, if these forces are ignored, there must be a clear understanding of the magnitude and the sense of the errors so induced. The most critical situation is that of negative aerodynamic damping which, if ignored, could produce severe underestimates of the response. Rigid body force measurements can be supplemented by forced vibration tests in which the motion-induced forces are measured directly and then incorporated into the response prediction. This technique has been employed since the 1960's for bridge sectional models and is described in reference [10] but has not as yet been employed extensively in the study of towers. Steckley [2] has explored the possibility of this technique for buildings and has successfully measured aerodynamic stiffness and damping on a rigid base-pivotted model. The equipment employed by Steckley is shown in Figure 7. The load cells 1 and 2 measure the base moment while the cells 3 and 4 produce a signal used to remove the driving inertial force component from the signal from cells 1 and 2. Steckley demonstrated that reliable measurements of aerodynamic damping and mass (or stiffness) could be obtained by driving the model at small amplitudes with a frequency swept sine wave. The technique would increase the wind tunnel test time by a factor of three but would be an invaluable aid in circumstances where negative aerodynamic damping was a factor.

Some results obtained by Steckley are presented in Figure 8. The results presented show the dimensionless stiffness and damping (α and β) for a 1:1:13 prism as a function of the dimensionless wind speed (V/f_oB) and the r.m.s. amplitude (as a percentage of B). The parameters α and β are defined such that;

$M(t)$ = motion-induced base moment

$$= \frac{2}{3}\rho_a B^2 H^3 \omega_o \{\alpha\theta + \beta\dot{\theta} / \omega_o\}$$

where θ = base rotation (radians)

The aerodynamic damping as a fraction of critical for a uniform prismatic structure is;

$$\eta_{aero} = \beta\rho_a / \rho_b$$

and the aerodynamic mass as a fraction of the actual mass is;

$$M_{aero} / M_{bldg} = 2\alpha\rho_a / \rho_b$$

The same technique was employed by Steckley to examine the influence of turbulence and aspect ratio on α and β and these results are shown in Figures 9 and 10.

The strong negative aerodynamic damping at values of V/f_oB just beyond $1/S$ are evident and predictions which did not include this motion-induced force would be seriously in error. At lower values of V_H/f_oB the aerodynamic damping is positive and to ignore it would produce conservative response estimates. At a value of V_H/f_oB of 5 the value of

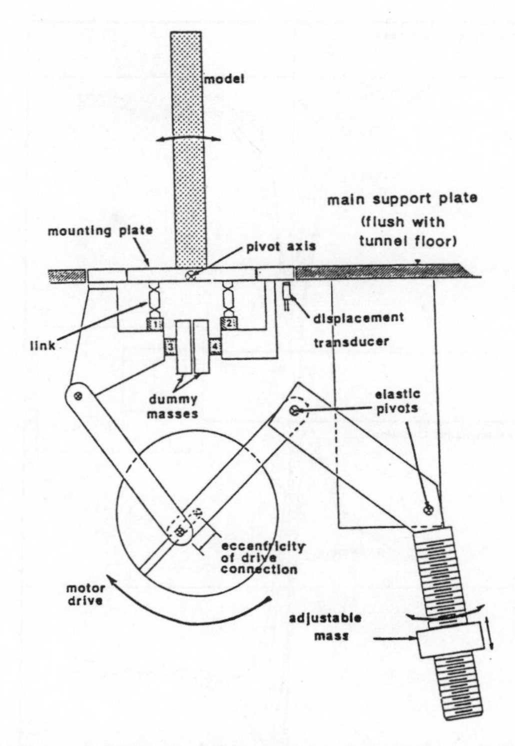

Fig.7: Schematic diagram of a forced base-pivotted model for the determination of motion-induced forces (Steckley [2])

β is about 0.6 and the aerodamping for a "typical" ρ_a/ρ_b 1:150 would be about 0.4% and would reduce the resonant response by perhaps 20%.

4.2.1 Base balance methods

The two rigid-body techniques in use are the base-balance approach and an approach based on surface pressure integration. The former is by far the simpler and is a widely used and accepted technique for many buildings and towers. In concept, the approach is simple; a rigid model is mounted on a stiff balance and the spectra of the base reactions are measured over the frequency range of interest. The measured spectra (and cross-spectra) may then be used to predict the response. In practice, the design of a suitable balance and model involves a difficult compromise between sensitivity and stiffness and the use of base reactions requires that assumptions be made as to the form of the load distribution (unless the true mode shape is linear for sway and uniform for twist).

The design requirements for a suitable balance have been discussed in detail by Tschanz [ref. 9, pp. 296-312] while methods for dealing with

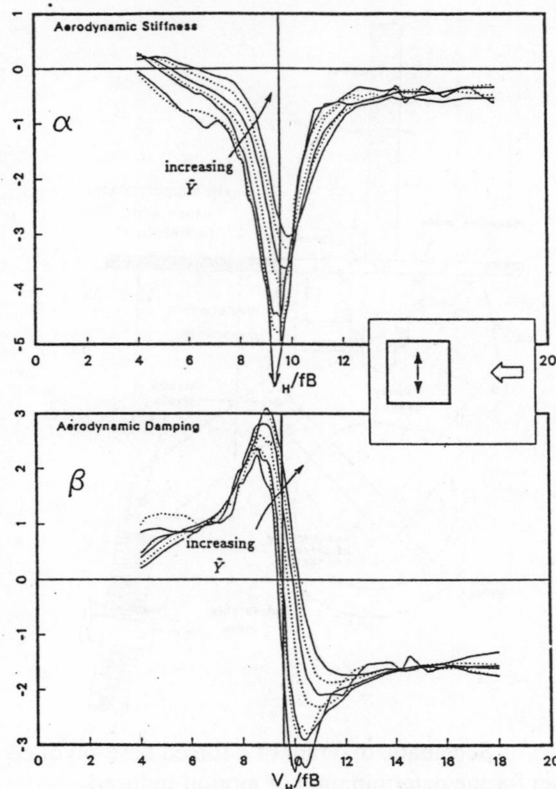

Fig. 8: Aerodynamic stiffness and damping coefficients for a 1:1:13 prism in turbulent flow (Steckley [2])

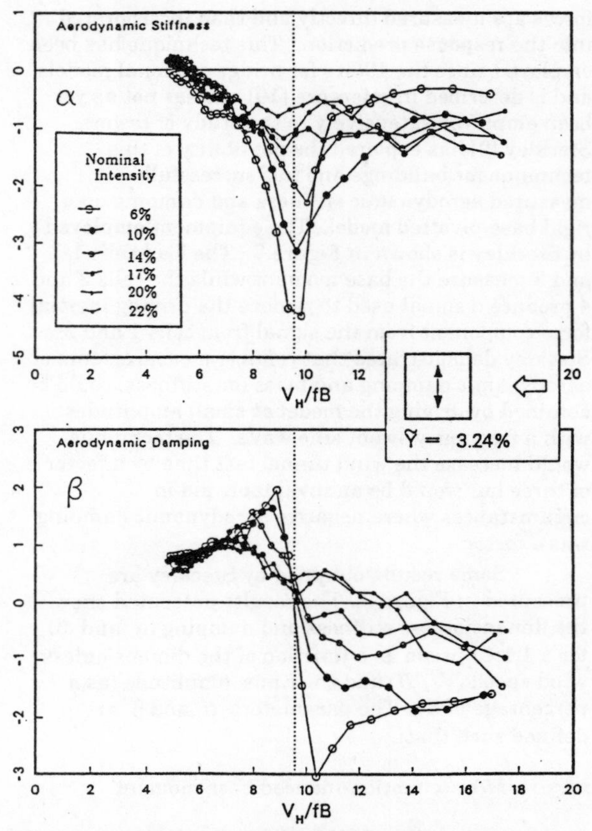

Fig. 9: Influence of turbulence on aerodynamic mass and damping (Steckley [2])

non-linear mode shapes have been discussed by Vickery et al [11] and by Holmes [12].

At the present state of development the base balance technique provides the most efficient method of obtaining response estimates for many buildings and towers. The limitations in respect to motion-induced forces have been discussed earlier and it is clear that the technique should be restricted to values of V/f_oB significantly less than $1/S$. Since the predictions rely only on a knowledge of base reactions, assumptions must be made as to load distribution. For nearly linear sway modes the sensitivity to such assumptions is weak but for complex mode shapes, particularly those involving sway in two directions and torsion, it has yet to be shown that reliable methods exist to move from base reaction to response. In such instances the base balance technique can still provide valuable estimates in the early stages of the design but recourse to other methods (such as full aeroelastic testing) would be a prudent final check.

4.2.2 Surface pressure integration

The prediction of response from integrated surface pressures is, again, simple in principle but complicated by practical requirements and equipment cost. The method is more flexible than the base-balance technique but models are more complex with a consequent increase in cost and construction time.

If modal analysis is used to predict the response then the modal force, $Q_i(t)$, for the ith mode is given by,

$$Q_i(t) = \int_A p(t)\overline{\eta}_o \overline{\delta}_i dA$$

where $p(t)$ = surface pressure at (x,y,z)
$\bar{n}$ = unit inward vector normal to the surface at (x,y,z)
δ_i = displacement vector in the ith mode at (x,y,z)

and the integral is taken over the surface area of the structure.

If the pressures are measured at discrete points then an approximation to $Q_i(t)$ is;

$$Q_i(t) \neq \sum p_j(t)\overline{\eta}_j . \overline{\delta}_{ij}\delta A_j$$

where $p_j(t)$ is the pressure at the jth tap, δA_j, is the finite elemental area attributed to the measuring point and $\bar{n}$, and δ_{ij} are the unit normal vector and the mode displacement vector at the jth tap.

If a system were available in which;

(i) a separate pressure transducer was fitted to each pressure point

(ii) a scanning device could sample each tap at essentially the same time

(iii) a computer could compute a number of summations of equation 4.2 in a time less than the required time between sample sets ie, $1/2f_m$ where f_m is the maximum frequency of interest

(iv) the frequency response characteristics of the tubing system were consistent with the required f_m

then, $Q_i(t)$ could be computed in real time and the data for response predictions could be readily derived provided that the spacing of the points was sufficiently fine (a spacing equal to or less than the lesser of about $B/6$ or $V/5f_m$ horizontally and about $H/6$ or $V/5f_m$ vertically).

For tall buildings the value of f_m is unlikely to exceed 0.5 Hz and for a simple prismatic building the minimum number of points is likely to be of the order of 100 per side.

Such systems are presently available but with a cost of the order of U.S. $300,000 or more they are not presently an accepted tool for commercial testing but have been employed in research projects [13]. The potential of such systems for complex modal analysis and also for non-linear dynamic analysis is clear but before being widely used as a tool for commercial testing there will probably be a development period of a year or two.

The use of surface pressures to predict response has been an accepted technique for many years but the lack of high speed systems has forced the use of less powerful but very useful approaches to the evaluation of the summation of Equation 4.2. The prime restriction of most pressure systems in common use is the limited number of readings that can be

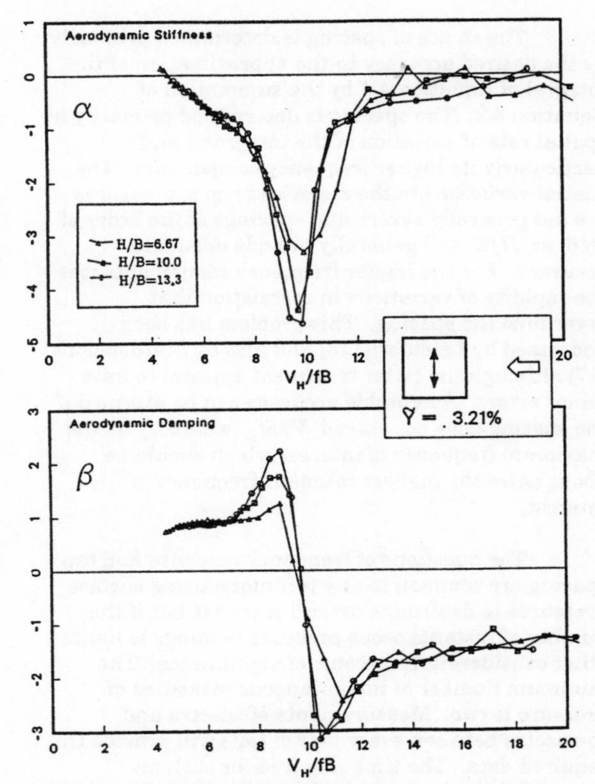

Fig. 10: Influence of aspect-ratio on aerodynamic mass and damping (Steckley [2])

obtained at one instant (typically 8 or 16 but often less).

The key factors to be considered when determining overall loads from surface pressure measurements are the frequency response of the pressure measuring system and the spacing of the pressure measuring points. Frequency response is not normally a problem. At typical linear scales of 1:400 and velocity scales of about 1:2 the frequency scaling is about 200:1. For large structures the frequency range of interest rarely exceeds 0.5 Hz and a model frequency response which is essentially flat to 100 Hz will be adequate. Frequencies of this order are readily attained by modern transducers and the limitations on response are determined almost entirely by the tubing system from the model to the transducers. The design of suitable pressure tubes has been described thoroughly by Gumley [14, 15]. The work of Gumley will not be discussed here but it can be concluded that with careful design it is feasible to achieve satisfactory performance up to frequencies of the order of 0.8 a_o/L where a_o is the speed (adiabatic) of sound and L the tube length. For typical tube lengths of 1m to 2m, this suggests that an acceptable response should be attainable for frequencies less than about 200 Hz.

The choice of spacing is determined primarily by the desired accuracy in the approximation of the integral of Equation 4.1 by the summation of Equation 4.2. The spacing is determined primarily by spatial rate of variation of the integrand and particularly its higher frequency components. The spatial variations in the mean and r.m.s. pressures are not generally severe and spacings of the order of $B/6$ or $H/6$ will generally provide adequate accuracy. For the higher frequency components it is the rapidity of variations in correlation that determine the spacing. This problem has been addressed by Letchford [16] and also by Stathopolous [17] although the latter treatment appears to have minor errors. Acceptable accuracy can be attained if the spacing does not exceed $V/5f_m$ where f_m is the maximum frequency of interest which should be about twice the highest resonant frequency of interest.

The questions of frequency response and tap spacing are common to any technique using surface pressures to determine overall response but if the number of instantaneous pressure readings is limited other considerations become of significance. The minimum number of instantaneous measures of pressure is two. Measurements of spectra and co-spectra between every pair of taps will provide the required data. The time required for such an approach is however prohibitive. The technique is discussed by Kareem [ref. 9, pp. 275-295].

In order to reduce the need for a multitude of transducers many experimenters have made use of "pneumatic averaging". Since the flow in pressure tubing is commonly laminar the pressure in a manifold linked to a number of taps by a series of identical tubes will be the average of the pressures at the connected taps. By varying the spacing, and hence the attributable area to each tap, the pneumatic average becomes a weighted average that can be made proportional to a modal force, an induced structural reaction or a Fourier component of the load. Such approaches have been used by many, Draisey [18] used pneumatic averaging of over 100 taps to produce signals proportional to the modal forces of a circular flat plate while Osborne [19] used the technique to produce signals proportional to the Fourier components of the pressure distribution on a circular cylinder.

In order to avoid problems associated with pressures at discrete points a technique using spatially continuous pneumatic averaging has been employed. With this technique a shallow chamber is formed behind a skin of porous polyethylene which forms the exterior of the model. Flow through the skin is laminar and the pressure in the chamber is the average over the covering polyethylene. The frequency response of such a system is acceptable to frequencies in excess of 100 Hz [20].

5.0 Aerodynamic Data

The principal data requirements for response prediction are the values of C_L, S, B, β and L. The four basic parameters are functions of the Reynolds Number, a turbulence parameter, the cross-sectional shape and the height to breadth ratio, λ. For sharp edged cross-sections the Reynolds Number is not important above a value of perhaps 1000 but for circular cross-section its importance is paramount and data must be determined from full-scale observations or from tests conducted at Reynolds Numbers above about 2×10^6. In this section, attention is given to the data presently available and attention is concentrated on the circular and square cross-sections.

5.1 Aspect Ratio or Tip Effects

The finite aspect ratio of towers and chimneys has two effects on the lateral forces. Flow over the tip reduces the base pressure and weakens the shedding over the structure as a whole. In addition to this general reduction there will be a tip effect due to the fact that the lift producing circulation cannot persist to the tip. The influence of aspect ratio has not been studied extensively but the data available suggests that both effects are significant but the tip effect is particularly important since reductions in lateral load near the tip have a very strong influence on the response of cantilever type structures.

Measurements of the lateral fluctuating force on a square cross-section have been reported by Steckley [2] and his observations are shown in Figures 11A and 11B for the force removed from the frequency of motion and the motion-induced forces respectively. Both show a strong tip effect. For the forces on a stationary body the tip effect is reasonably represented by the function;

$$C_L(x) = C_{L_o}(1 - e^{-x/B})$$

where x is the distance from the tip and C_{LoN} is the r.m.s. lift coefficient well away from the tip. The reduction in the motion-induced force and the damping parameter, β is also reasonably represented by;

$$\beta(x) = \beta_o(1 - e^{-x/B})$$

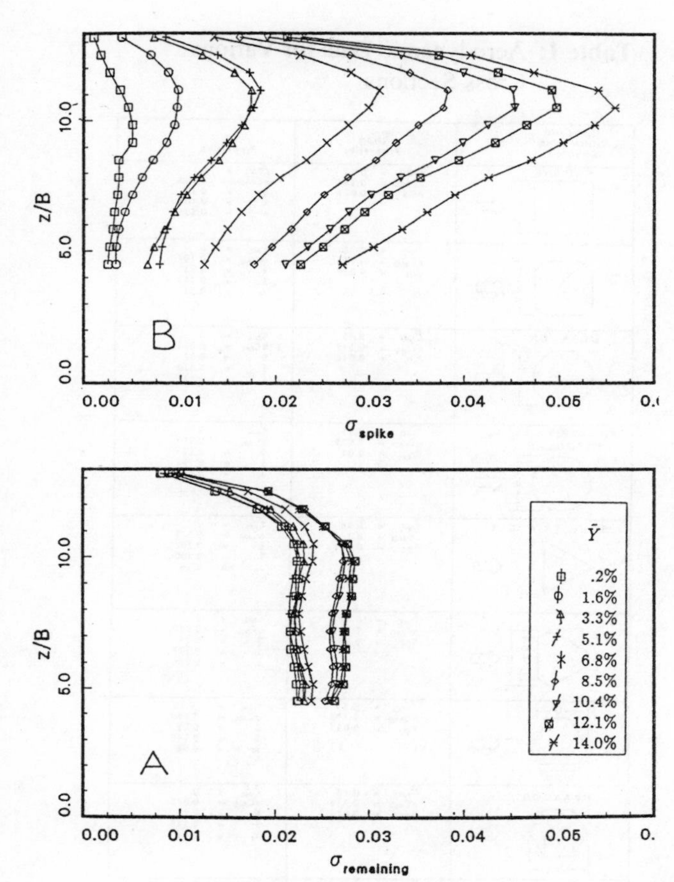

Fig. 11: Distribution of vortex shedding and motion-induced forces

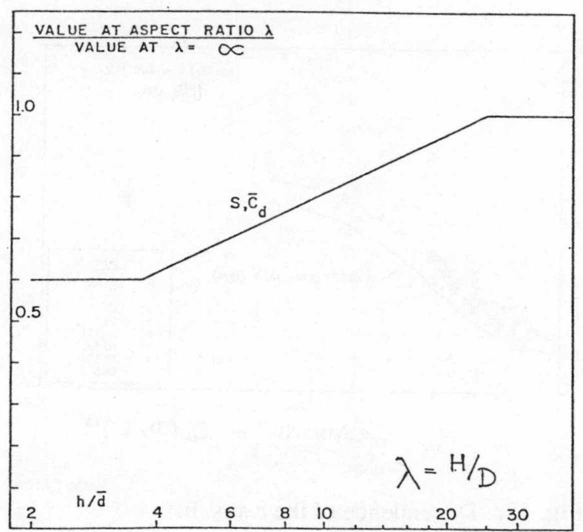

Fig. 12: Influence of aspect ratio on the Strouhal Number and Drag Coefficient

The reduction in C_{Lo} and α_o due to weaker base pressure is not well defined but models of the vortex street, as discussed by Vickery (21), suggest that the reduction in C_{Lo} is very similar to the reduction in the mean drag. The reduction in the mean drag has received considerable attention and advice is contained in many codes of practice. The Strouhal Number is also dependent on aspect ratio and it reduces with decreasing H/B ratio. Variations of the Strouhal Number and Drag Coefficient suggested by Vickery & Basu (22) are shown in Figure 12.

5.2 Circular Cross-Sections

The strong dependence of the aerodynamic properties of the circular cross-section precludes the use of model tests except in very large high speed tunnels capable of attaining values in excess of 2 x 10^6. For the most part data must be gleaned from full-scale observations. Vickery & Daly [23] noted the strong dependence of C_L on turbulence and this has been confirmed by recent full-scale measurements by Waldeck [24]. The relationship suggested by Vickery & Daly is shown in Figure 13 together with the observations by Waldeck. In low levels of turbulence the r.m.s. lift coefficient is about 0.05 but attains values in excess of 0.2 for values of $I_u(D/L)^{1/3}$ greater than 0.1

I_u = turbulence intensity

D = diameter

L = integral scale of turbulence.

The bandwidth, B, is also strongly dependent on turbulence and exhibits considerable scatter. The full-scale data suggests that B is reasonably well predicted by the relationship;

$$B = 0.15 + 1.5 I_w$$

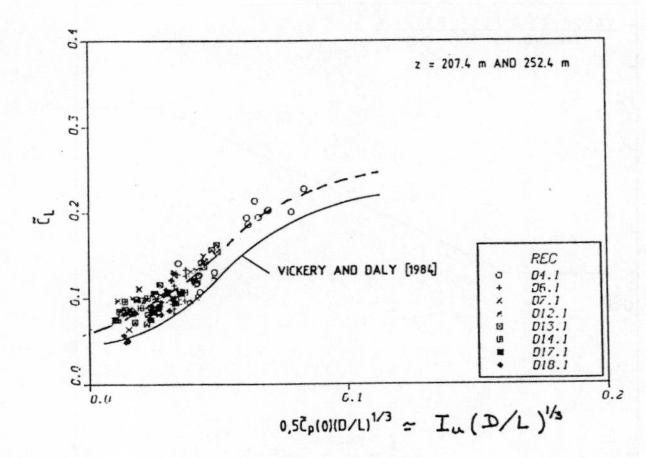

Fig. 13: Dependence of the r.m.s. lift force coefficient on the turbulence parameter

In turbulent flow the lift coefficient is due in part to varying wind direction at frequencies much lower than shedding frequency. The values of C_L shown in Figure 13 include these low frequency components. The value of C_L associated only with shedding is typically 80% of the full value.

The correlation length is not well defined but full-scale measurements indicate it to be from one to two diameters. Waldeck's data suggest a value of 1.5D. The Strouhal number is not markedly influenced by turbulence and the data of Waldeck and others suggests a value close to 0.20 for height to diameter ratios of about 15.

The damping parameter, β, is not well defined. No direct full-scale measurements have been made and it must be deduced from observations. From a study of the instability of steel chimneys Daly [25] concluded that near the critical speed and in turbulent flow the value of β in the transcritical range was 0.43 with a coefficient of variation of 40%.

5.3 Square Cross-Sections

The square cross-section has been studied in some depth and the basic aerodynamic properties are reasonably well defined. It has been shown [16] that turbulence has a strong influence on both the mean drag and the fluctuating lift force. In turbulent flow the value of C_L is about 0.65 for two-dimensional conditions, the correlation length is about 3.5 times the width and the Strouhal Number about 0.14. For

Table 1: Aerodynamic Data for Various Cross-Sections

Shape and Critical Wind Direction	Drag Properties	Lift Properties
SQUARE (side b)	$\bar{C}_{MD}$ = 0.52 σC_{MD} = 0.154 R_{OS} = 0.021	$\bar{C}_{ML}$ = 0 σC_{ML} = 0.140 B = 0.22 S = 0.103 a_r = 0.51
(b)	$\bar{C}_{MO}$ = 0.38 σC_{MO} = 0.114 R_{OS} = 0.029	$\bar{C}_{ML}$ = 0 σC_{ML} = 0.107 B = 0.33 S = 0.120 a_r = 0.54
OCTAGON (b)	$\bar{C}_{MO}$ = 0.45 σC_{MO} = 0.117 R_{OS} = 0.018	$\bar{C}_{ML}$ = 0 σC_{ML} = 0.061 B = 0.22 S = 0.130 a_r = 0.32
SUB-CRITICAL R_e CIRCULAR (b)	$\bar{C}_{MO}$ = 0.35 σC_{MO} = 0.089 R_{OS} = 0.01	$\bar{C}_{ML}$ = 0 σC_{ML} = 0.053 B = 0.25 S = 0.157 a_r = 0.49
EQUI. TRIANGLE (b)	$\bar{C}_{MO}$ = 0.58 σC_{MO} = 0.143 R_{OS} = 0.022	$\bar{C}_{ML}$ = 0 σC_{ML} = 0.072 B = 0.26 S = 0.100 a_r = 0.56
(3/4 b; b)	$\bar{C}_{MO}$ = 0.52 σC_{MO} = 0.137 R_{OS} = 0.032	$\bar{C}_{ML}$ = 0 σC_{ML} = 0.086 B = 0.24 S = 0.127 a_r = 0.90
(b; 2/3 b)	$\bar{C}_{MO}$ = 0.54 σC_{MO} = 0.133 R_{OS} = 0.025	C_{ML} = -0.16 σC_{ML} = 0.098 B = 0.25 S = 0.140 a_r = 0.94
HEXAGON (b/2; b)	$\bar{C}_{MO}$ = 0.46 σC_{MO} = 0.119 R_{OS} = 0.013	$\bar{C}_{ML}$ = 0 σC_{ML} = 0.073 B = 0.21 S = 0.135 a_r = 0.61
(b; 0.50 b)	$\bar{C}_{MO}$ = 0.56 σC_{MO} = 0.137 R_{OS} = 0.056	$\bar{C}_{ML}$ = 0 σC_{ML} = 0.064 B = 0.31 S = 0.098 a_r = 0.72
(0.50 b; b)	$\bar{C}_{MO}$ = 0.21 σC_{MO} = 0.063 R_{OS} = 0.080	$\bar{C}_{ML}$ = 0 σC_{ML} = 0.112 B = 1.57 S = 0.068 a_r = 1.0
(b; 0.64 b; 0.08 b)	$\bar{C}_{MO}$ = 0.52 σC_{MO} = 0.130 R_{OS} = 0.025	$\bar{C}_{ML}$ = 0 σC_{ML} = 0.088 B = 0.21 S = 0.124 a_r = 0.85
(0.64 b; b)	$\bar{C}_{MO}$ = 0.23 σC_{MO} = 0.064 R_{OS} = 0.06	$\bar{C}_{ML}$ = 0 σC_{ML} = 0.102 B = 1.6 S = 0.054 a_r = 1.0

finite aspect ratios the Strouhal Number is reduced and has a value of about 0.11 for H/B ratios near 15. Values of the damping parameter β have been presented in Section 4 of this paper.

An alternate method of defining the aerodynamic data for towers is to define the nature of the dynamic component of the base momentum rather than the actual distributed loads which produce this moment. Table 1 shows aerodynamic data for various shapes. These data are useful for obtaining preliminary response estimates of chimneys and towers with H/B ratios from about 10 to 20 but are not a substitute for wind tunnel testing if reliable loads are required. The data in Table 1 include;

(i) the mean and r.m.s. base moment coefficient for drag loads;

(ii) the mean and r.m.s. base moment coefficient for across-wind loads;

(iii) the spectral parameters S, B and a_r which define the spectral shape. S (Strouhal Number) and B (bandwidth) have been discussed. The parameter a_r is the fraction of total spectrum caused by shedding as opposed to lateral turbulence.

6.0 CONCLUSION

The aims of this paper were to;

(i) describe the nature of the wind forces acting on chimneys and towers;

(ii) outline theoretical methods of predicting dynamic response;

(iii) describe experimental techniques particularly suited to chimneys and towers; and

(iv) to present aerodynamic data suitable for preliminary estimates of the wind-induced dynamic response.

Full coverage of all these aspects in a single paper is not possible but it is believed that the more important considerations have been discussed.

REFERENCES

1. Davenport, A.G. "The representation of the dynamic effects of turbulent wind by equivalent static loads", Proc. AISC/CISC Symposium on Structural Steel, Chicago U.S.A., May 1985.

2. Steckley, A. "Motion-induced Wind Forces on Chimneys and Tall Buildings", Ph.D. Thesis, The University of Western Ontario, London, Canada, June 1989.

3. Vickery, B.J. "On the assessment of wind effects on elastic structures", Civil Eng. Trans., Inst. Eng. Aust., pp. 183-192, 1966.

4. Vickery, B.J."On the reliability of gust-loading factors", Civil Eng. Trans., Inst. Eng. Aust., Vol. CE13, No. 1, 1971.

5. Davenport, A.G. "Gust Loading Factors", J. Struct. Div. Proc. A.S.C.E., Vol. 93, No. ST3, 1967.

6. Velozzi, J. and Cohen, E. "Gust response factors", J. Struct. Div., Proc. A.S.C.E., Vol. 94, No. ST6, 1968.

7. Vickery, B.J. and Basu, R.I. "Across-wind vibrations of structures of circular cross-section - Parts I and II", J. Wind Eng'g & Indust. Aero., Vol. 12, No. 1, pp. 49-97, 1983.

8. Isyumov, N., Davenport, A.G. and Monbaliu, J. "CN Tower, Toronto - Model and full scale response to wind", Proc. 12th Congress, IABSE, Vancouver, Canada, Sept. 1984.

9. Reinhold, T.A. (editor) "Wind Tunnel Modelling for Civil Engineering Applications", Proc. Int. Workshop on Wind Tunnel Modelling Criteria and Techniques in Civil Engineering Applications, Gaithersburg, U.S.A., Cambridge University Press 1982.

10. Ukeguchi, N., Sakata, H. and Nishitani, H. "An investigation of aeroelastic instability of suspension bridges", Proc. Int. Symp. on Suspension Bridges, Lab. Nac. de Engen., Lisbon, pp. 273-284 (1966).

11. Vickery, P.J., Steckley, A., Isyumov, N. and Vickery, B.J. "The effect of mode shape on the wind-induced response of tall buildings", Proc. 5th U.S. National Conference on Wind Engineering, Lubbock, Texas, pp. 1B/41-48, November 1985.

12. Holmes, J.D. "Mode shape corrections for dynamic response to wind", Engineering Structures, Vol. 9, pp. 210-212, 1987.

13. Fujii, K., Hibi, K., Uead H. and Shimada, K. "Visualization of fluctuating pressure distribution on bluff body using electronically scanned pressure sensors", Inst. of Technology, Shimizu Construction, Etchujima 3-4-17, Kot-ku, Tokyo 135, Japan (VIDEO TAPE).

14. Gumley, S.J. "A detailed design method for pneumatic tubing systems", J. Wind Eng'g & Industrial Aero., Vol. 13, No. 1, pp. 441-452, 1983.

15. Gumley, S.J. "Tubing systems for the pneumatic averaging of fluctuating pressures", J. Wind Eng'g and Industrial Aero., Vol. 12, No. 2, pp. 189-228, 1983.

16. Letchford, C.W. "On the discrete approximation in pneumatic averaging", Recent Advances in Wind Eng'g (ed. T.F. Sun), Int. Academic Publishers, Beijing, Vol. 2, pp. 1159-67, June 1989.

17. Stathopoulos, T. "Wind pressure loads on flat roofs", BLWT-3-75, The University of Western Ontario, London, Canada, 1975.

18. Draisey, S. "The influence of wall openings on the dynamic behaviour of large span roofs", M.Eng.Sc. Thesis, The University of Western Ontario, London, Canada, May 1987.

19. Osborne, C. "Wind loading on chimneys", B.Eng.Sc. Thesis, The University of Western Ontario, London, Canada, March 1981.

20. Rosales, M.B. "A novel technique for measuring spatially averaged pressures", M.Eng.Sc. Thesis, The University of Western Ontario, London, Canada, December 1983.

21. Vickery, B.J. "Fluctuating Lift and Drag on a Long Cylinder of Square Cross-Section in a Smooth and In a Turbulent Stream", J. Fluid Mech., Vol. 25, Pt. 3, pp. 481-494, 1966.

22. Vickery, B.J. "The Response of Reinforced Concrete Chimney to Vortex Shedding", J. Eng'g Struct., Vol. 6, p. 324, Oct. 1984.

23. Vickery, B.J. and Daly, A. "Wind Tunnel Modelling as a Means of Predicting the Response of Chimneys to Vortex Shedding", J. Eng. Struct., Vol. 6, pp. 363-368, Oct. 1984.

24. Waldeck, L. "Measurement and Analysis of the Dynamic Effects of Atmospheric Wind on a 300m Concrete Chimney", Ph.D. Thesis, University of Witwatersrand, Johannesburg, South Africa, 1990.

25. Daly, A. "The Response of Chimneys to Wind-Induced Loads and the Evaluation of the Resulting Fatigue Damage", Ph.D. Thesis, The University of Western Ontario, London, Canada, 1986.

THE ASSESSMENT OF WIND LOADS ON COOLING TOWERS

H.-J. Niemann *M. Kasperski*

Building Aerodynamics Laboratory
Ruhr-Universitat Bochum, GERMANY

Summary

For cooling towers, the dynamic action of the natural wind normally is taken into account by an equivalent static wind load, based on the gust response factor of the meridional tension force. Other stress distribution resultants such as hoop or bending forces may be subject to considerably larger gust response factors. In this paper, a new approach with individual equivalent static loads for each design situation - i.e. the design of the reinforcement in the meridional and circumferential direction and the design against buckling - is presented. For the reinforcements, these equivalent static loads are based on the tensile stresses in the reinforcement layers, i.e. this approach accounts for the net effect of membrane force, shear and bending.

General

The action of the natural wind induces static and dynamic pressures on the outer and inner surface of the shell. The complete load process can be understood as a highly complicated multi-dimensional stochastic process which conveniently is described by statistic means. To describe the complete response process, the most powerful method is a combination of the covariance analysis for the low-frequent part of the responses and a spectral analysis for the high-frequent part. Since these methods themselves and the sets of data needed to describe the load process lead to an enormous amount of experiment and computation, generally this procedure is used for research purposes. From the obtained results, equivalent static loads are developed for the design. Since cooling towers mostly are acting quasi-statically - i.e. the main part of the responses is due to the low-frequent wind action, the resonant part is small - it is convenient to split up the equivalent static loads into two parts: a load distribution for calculating the quasi-static responses, which is based on the mean pressure distribution times the gust response factor of the reaction under consideration, and a resonance factor which takes into account the resonant amplification. Considering economics, a general demand for the development of equivalent static wind loads should be: as simple as possible, as safe as required, as economically as possible.

Background of equivalent static loads

A further step of simplification generally is used if the quasi-static behaviour of a structure is described. Instead of individual gust response factors the equivalent steady gust model is used, which forms the basis of many design codes of practice. The steady gust can be understood as a gust which envelops the whole building. All pressures induced by this gust will vary in time simultaneously proportional to their mean values. The equivalent static load now is defined by the mean pressure distribution times the gust velocity pressure of the enveloping gust. The dimension of the gust or the averaging time respectively can be obtained using the TVL-formula:

$$t \cdot v = 4.5 \cdot L \qquad (1)$$

with t - averaging time of the enveloping gust

v - mean velocity at the top of the building

L - characteristic dimension of the building, for cooling towers the height H

For cooling towers with heights above 100 m equ. (1) leads to averaging times of 10s to 20s. For the sake of simplicity, mostly an uniform averaging time of e.g. 5s is used in design codes, neglecting the influence of the building's dimensions. Generally, this simplification is also used for cooling towers.

So, the remaining problem is to describe the mean pressure distribution for the outer and inner surface. Most results we have today have been obtained in wind tunnel experiments. But there are also some full-scale tests, used for calibration and control of the wind tunnel tests [e.g. 1,2,3] (see fig. 1).

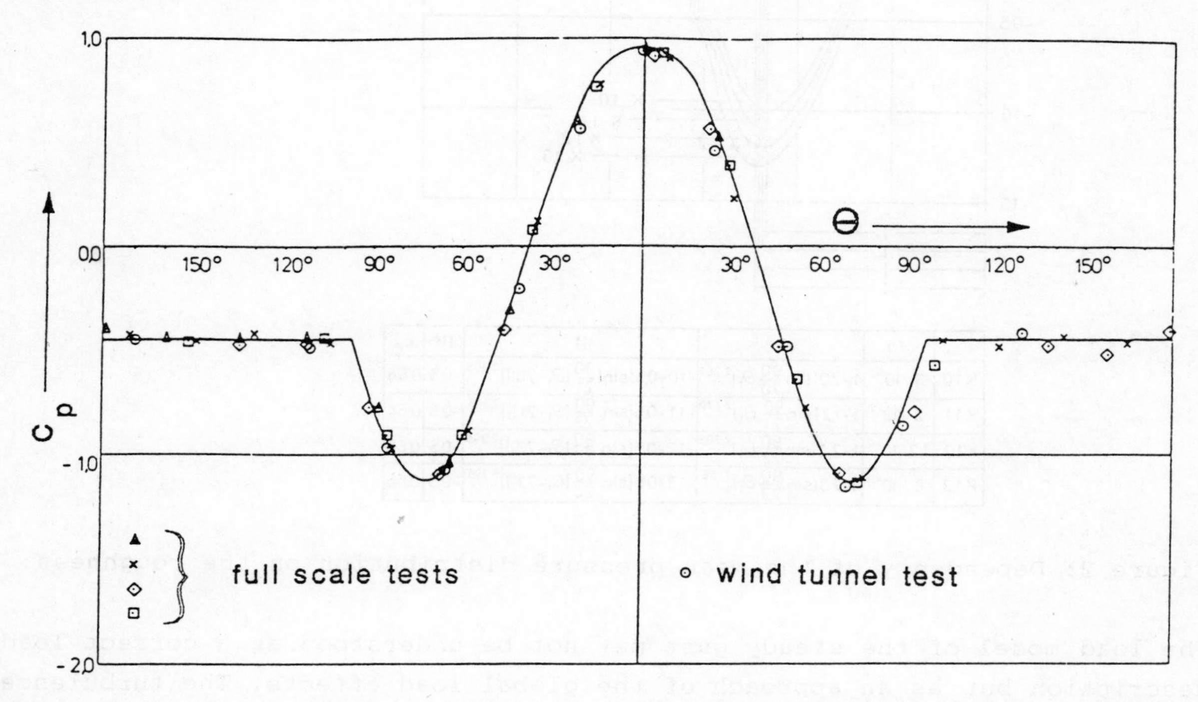

Figure 1: Mean pressure distribution around a cooling tower. Comparison of full-scale experiment and wind tunnel test

In wind tunnel tests, the main problem is due to the simulation of the Reynolds number Re. For scaling factors less than 1 to 10, the similarity of Re can not be realized using the same fluid (i.e. air) in the model test as in nature. Fortunately, in nature the pressure distribution of cooling towers is in the transcritical region, so the distribution will not directly depend on Re. In wind tunnel tests, generally a technical roughness is used to ensure a transcritical state, using either sand-grain roughness or ribs. Since the pressure distribution strongly varies with varying roughness (fig. 2), it has become design practice to use ribs in nature to control for example the maximum of the suction. So in full-scale, there are more or less smooth cooling towers as well as cooling towers, roughened artifically by vertical ribs in different arrangements.

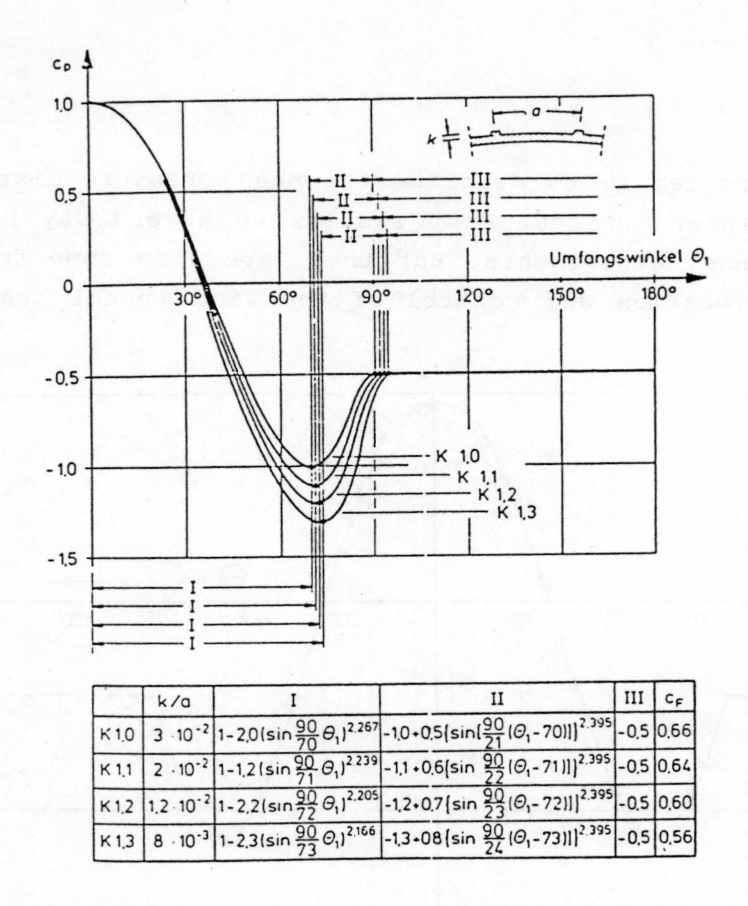

	k/a	I	II	III	c_F
K 1.0	$3 \cdot 10^{-2}$	$1-2.0(\sin\frac{90}{70}\Theta_1)^{2.267}$	$-1.0+0.5\{\sin[\frac{90}{21}(\Theta_1-70)]\}^{2.395}$	-0.5	0.66
K 1.1	$2 \cdot 10^{-2}$	$1-1.2(\sin\frac{90}{71}\Theta_1)^{2.239}$	$-1.1+0.6\{\sin[\frac{90}{22}(\Theta_1-71)]\}^{2.395}$	-0.5	0.64
K 1.2	$1.2 \cdot 10^{-2}$	$1-2.2(\sin\frac{90}{72}\Theta_1)^{2.205}$	$-1.2+0.7\{\sin[\frac{90}{23}(\Theta_1-72)]\}^{2.395}$	-0.5	0.60
K 1.3	$8 \cdot 10^{-3}$	$1-2.3(\sin\frac{90}{73}\Theta_1)^{2.166}$	$-1.3+08\{\sin[\frac{90}{24}(\Theta_1-73)]\}^{2.395}$	-0.5	0.56

Figure 2: Dependency of the mean pressure distribution on the roughness

The load model of the steady gust may not be understood as a correct load description but as an approach of the global load effects. The turbulence of the oncoming flow is more or less well decribed by this approach, fluctuations induced by the tower and fluctuations near locations where the mean changes in sign from pressure to suction are not well modeled. This may lead to problems, if a reaction is more sensitive to a smaller non-uniform load distribution than to a greater uniform load.

In Figure 3, typical instantaneous gust pressure distributions (i.e. pressure distribution minus mean pressure) and the upper and lower limit of the local probable extreme gust pressures are compared to the load distribution obtained using the concept of the steady gust. While shear and meridional tension are less sensitive to non-uniform asymmetric loadings, the extreme reactions in hoop direction are due to load effects with a relatively smaller amplitude but an unfavourable asymmetric distribution, leading to a higher gust response factor for the reactions in hoop direction.

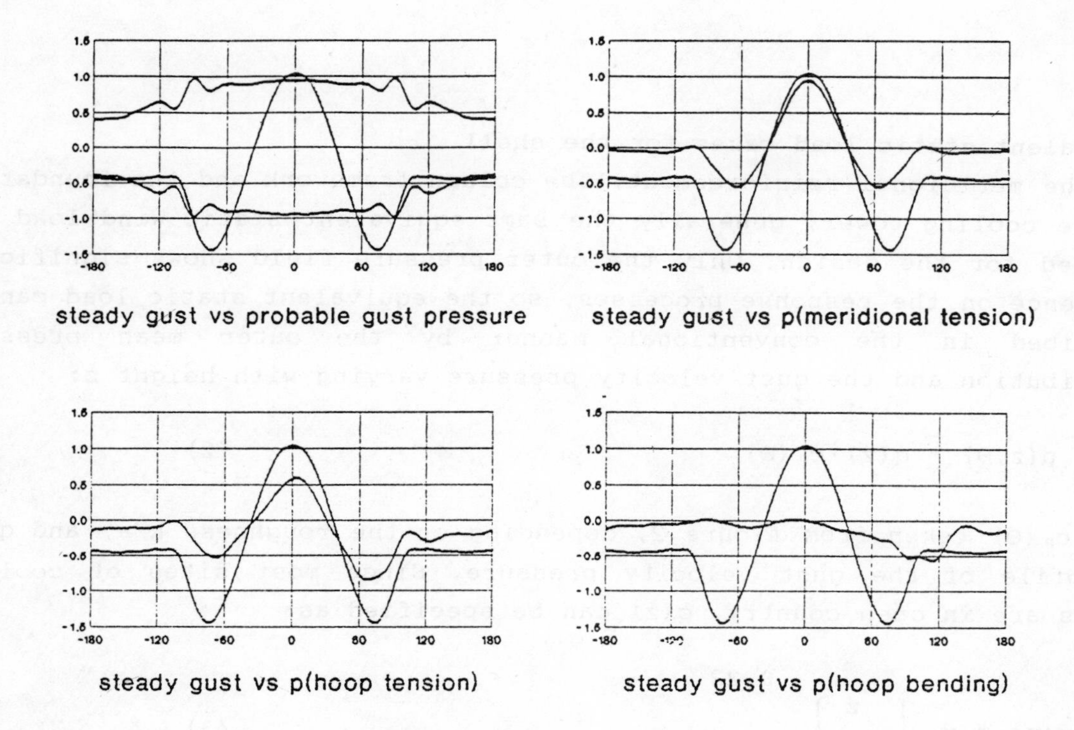

Figure 3: Instantaneous gust pressure distribution which induce an extreme reaction and local probable extremes compared to the load distribution of the steady gust model

In table 1 the typical results of gust response factors for different reactions as well as their resonance factors are listed. A great inhomogeneity concerning both the gust response factor and the resonance factor has to be stated, so generally more than one equivalent static wind load is needed for the design.

reaction	gust response factor	resonance factor
tensile force meridional direction	1.80	1.04
tensile force hoop direction	7.50	1.10
bending moment hoop direction	3.10	1.50
pressure force meridional direction	2.00	1.02

Table 1: Typical inhomogeneity of gust response factors and resonance factors for different reactions of a cooling tower

Equivalent static load cases for the shell

For the meridional reinforcement, the column framework and the foundation of the cooling tower, generally the same equivalent static wind load can be used for the design. Only the outer pressure field shows significant influence on the response processes, so the equivalent static load can be described in the conventional manner by the outer mean pressure distribution and the gust velocity pressure varying with height z:

$$p(z,\Theta) = q(z) \cdot c_p(\Theta) \qquad (2)$$

with $c_p(\Theta)$ taken from figure 2, depending on the roughness k/a, and q(z) = profile of the gust velocity pressure. Since most sites of cooling towers are in open country, q(z) can be specified as:

$$q(z) = q_H \left(\frac{z}{H} \right)^{0.22} \qquad (3)$$

with q_H - gust velocity pressure at cooling tower's height

Generally, the same load is used for the design against buckling. Additionally, an inner pressure of

$$p_i = -0.4 \cdot q_H \qquad (4)$$

has to be taken into account. The buckling safety can be estimated using for example the Mungan criterion for the buckling interaction [4]:

$$\nu_B = 0.8 \left[\frac{\sigma_{11}}{\sigma_{110}} + \frac{\sigma_{22}}{\sigma_{220}} \right] + 0.2 \left[\frac{\sigma_{11}}{\sigma_{110}}^{2} + \frac{\sigma_{22}}{\sigma_{220}}^{2} \right]^{2} \qquad (5)$$

with σ_{11}, σ_{22} - membrane stress in hoop direction and meridional direction respectively

σ_{110}, σ_{220} - critical buckling stress in hoop and meridional direction

In this criterion, the effect of the bending moments in the sense of a pre-deformation is not included, furthermore the equivalent static load described as above will not lead to the real interactive pair of buckling forces. Therefore, the buckling safety generally is fixed higher than the safety for a failure due to tensile stresses, e.g. $\nu_B = 5$.

As mentioned above, for the design of the hoop reinforcement a second equivalent static load is needed. A final concept is not yet available, although in [5] a proposal for a more rougher shell has been made. A conservative estimation can be obtained using the same pressure distribution as for the meridional reinforcement times an additional factor of about 1.5 and taking into account a mean internal pressure of:

$$p_i = -0.4 \cdot \overline{q}_H \qquad (6)$$

with $\overline{q}_H$ - mean velocity pressure at cooling tower's height.

Resonance factors should be individual factors for the three design situation. For the meridional reinforcement a new and very realistic concept uses the resonance curve of figure 4, where the resonance factor φ_{22} depends on the mean velocity at the cooling tower's height H, the lowest eigenfrequency of the shell, the height of the cooling tower H and some geometrical parameters combined in a factor A:

$$A = ({}^{7}/_{3} - {}^{10}/_{3}\ F) \cdot A_{TL} \qquad (7)$$

with F - form parameter

$= \overline{D} / 2 \cdot H$

$\overline{D}$ - mean diameter of the shell

A_{TL} - throat's location parameter

$$A_{TL} = \begin{cases} 1.0 & \text{for} \quad T_L < 0.8 \\ 0.8 & \text{for} \quad 0.8 \le T_L < 0.85 \\ f(T_L) & \text{for} \quad T_L \ge 0.85 \end{cases}$$

$$f(T_L) = {}^{10}/_{3}\ (T_L^2 - T_L + \tfrac{1}{2})$$

T_L - normalized location of the throat

$= (H - H_o) / H$

H - total height of the cooling tower

H_o - height above throat

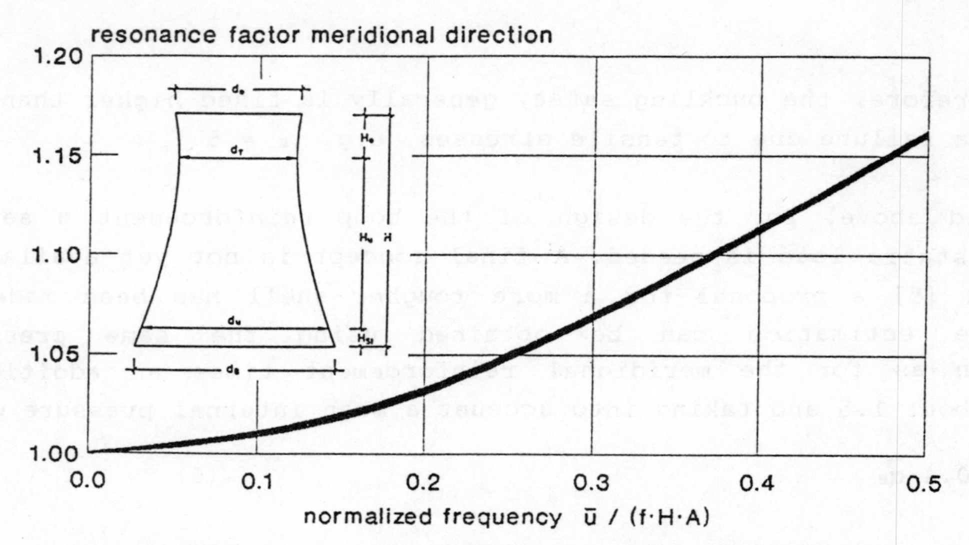

Figure 4: Resonance curve for the meridional reinforcement

For the buckling analysis, a resonant influence can be neglected, so the resonance factor $\varphi_B = 1.0$.

An approach for the resonance factor φ_{11} of the hoop reinforcement is given in [5], leading to a relative complicated set of diagrams. For simplicity, a conservative estimation may be used as follows:

$$\varphi_{11} = [1 + 2\ (^{H}/_{100} - 1.5)]\cdot\varphi_{22} \quad \text{for } H \geq 150 \text{ m} \quad (8)$$
$$\varphi_{11} = \varphi_{22} \quad \text{else}$$

Safety aspects in design and construction

For the design, the wind action has to be taken into account for two different states - the working state and the limit state - each for tensile and buckling failure.

For the working state, an interaction between dead load, wind load and temperature loads due to solar radiation and load differences between inside and outside during operation is considered:

$$w + g + t \leq p$$

with w - stress due to wind action

g - stress due to wind load

t - stress due to temperature effects

p - permissible stress

For the limit state design, a combination of wind load and dead load each with partial safety factors should be considered:

$$\gamma_w \cdot w + \gamma_g \cdot g \leq u$$

with γ_w - safety factor for the wind action should be between 1.5 - 1.75

γ_g - safety factor for the dead load should be between 0.9 - 1.0

u - ultimate stress

Additionally, a minimum reinforcement is required to ensure a sufficient stiffness of the shell. Mostly, a minimum uniform reinforcement of e.g. 0.3% is chosen for both the meridional and circumferential direction. If however the shape of the cooling tower and or its height are unusual, the minimum reinforcement for the hoop direction should be increased for the upper third of the shell to e.g. 0.5%.

Interference aspects

Generally, cooling towers are not erected as free standing buildings, but they are integrated in a more or less complex form of a power plant probably with other cooling towers or high boiler houses etc. (fig.5).

5. Arrangement of a power plant with three cooling towers

The neighbourhood of other buildings may influence both the static and the fluctuating wind loads, leading to favourable shielding effects or to unfavourable load amplifications. Without any special investigation of the problem, at least some rather unfavourable position of neighbouring buildings must be avoided. An example of such a distance rule is given in figure 6: neighbouring buildings with a similar height as the cooling tower should be outside a critical region which is given as 2-times the mean diameter of the cooling tower [6] or as the maximum of 2-times the upper diameter of the shell and 1.5-times the base diameter [7].

If however this critical distance can not be avoided, a considerable interaction effect must be taken into account for a safe design. Although generally interaction factors are specified to be in the range of 1.1 to 1.15, there are a few arrangements which will induce a much higher increase of the responses, especially for the hoop direction. Examples of these arrangements and observed interaction factors are given in fig. 7.

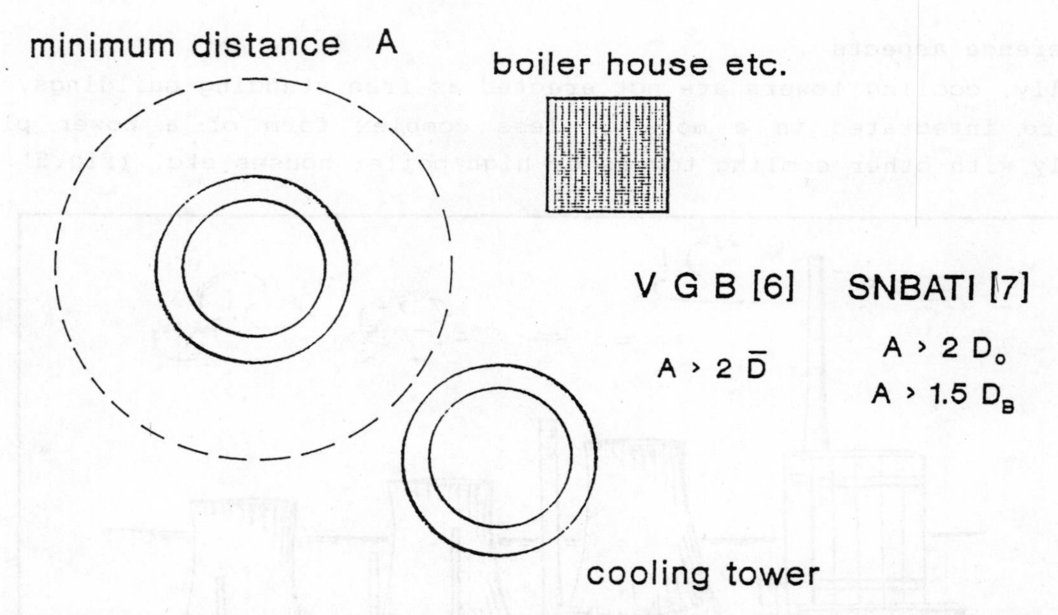

Figure 6: Most critical distances for unfavourable interaction effects

Obviously, a checking of the critical distance as described above will not be sufficient for all possible situations. If interaction is believed to be critical for a special situation, the best instrument for estimating the effects is a wind tunnel experiment using an aeroelastic model. The technique itself is rather complicated: For a geometrical scale of 1 : 400 a wall thickness of less than 1 mm in the model has to be fabricated, a stiffening of the shell due to the applicated strain gauges has to be taken into account leading to a series of calibration experiments, temperature effects should be well observed etc. The main advantages of this technique are its completeness in both the structural and the fluidmechanical behaviour and in its direct insight into critical stress amplifications at differing flow directions.

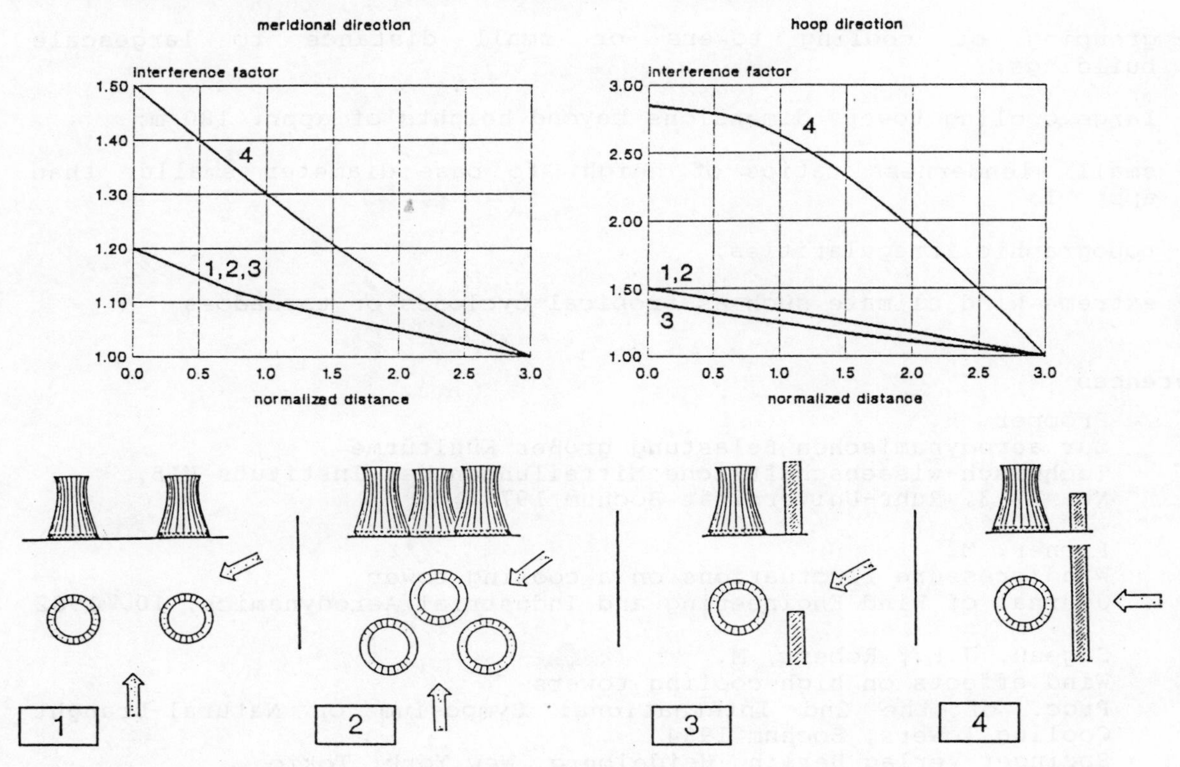

Figure 7: Examples of critical arrangements and interference factors

If interference effects are considerably, it may be useful to take into account a refined model of the local wind climate characterized by a wind speed rosette. Now the aim of the investigation is to avoid unfavourable flow directions, i.e. the alignment of the configuration has to be optimized: The most critical flow direction with respect to stress amplification should be the flow direction with the lowest probability of high wind velocities.

Conclusions

In recent decades, a great amount of theoretical knowledge and practical experience about the structural behaviour of natural draught cooling towers have been obtained world wide, leading to a well established and highly developed standard. However, there are some situations in which the established framework of considering the dynamic wind action by equivalent static loads may become inappropriate. Such situations are:

- grouping of cooling towers or small distance to largescale buildings;
- large cooling towers dimensions beyond heights of appr. 180 m;
- small slenderness ratios of height to base diameter smaller than appr. 1;
- topographic irregularities;
- extreme wind climate such as tropical cyclones or tornadoes

References

[1] Pröpper, H.
Zur aerodynamischen Belastung großer Kühltürme
Technisch-wissenschaftliche Mitteilungen des Instituts KIB, Nr. 77-3, Ruhr-Universität Bochum 1977

[2] Pirner, M.
Wind pressure fluctuations on a cooling tower
Journal of Wind Engineering and Industrial Aerodynamics, 10, 1982

[3] Sageau, J.F.; Robert, M.
Wind effects on high cooling towers
Proc. of the 2nd International Symposium on Natural-Draught Cooling Towers, Bochum 1984
Springer Verlag Berlin, Heidelberg, New York, Tokio

[4] Mungan, I.
Buckling of reinforced cooling towers shells
BBS Approach American Concrete Institute Journal, Sept.-Oct. 1982

[5] Kasperski, M.; Niemann, H.-J.
Refined equivalent static wind loads for cooling towers
Proc. of the 3rd International Symposium on Natural-Draught Cooling Towers, Paris 1989

[6] VGB - Technische Vereinigung der Großkraftwerksbetreiber
Bautechnik bei Kühltürmen, Essen 1980

[7] SNBATI - Syndicat National du Béton Armé et des Techniques Industrialisées
Regles professionelles applicables à la construction des refrigerants atmospheriques en béton armé
Juni 1986

Wind Loads on Structures, 1990

WIND ENGINEERING OF BRIDGES: AN EXAMINATION OF THE STATE OF THE ART

Robert .H. Scanlan

Dept. of Civil Engineering
The Johns Hopkins University
Baltimore, MD 21218-2699, U.S.A.

SYNOPSIS

The paper addresses the state of the art in bridge design against wind. Review of the extensive existing literature on the subject is not attempted. Rather, leading issues in bringing the problems of bridge dynamic response to wind under close analytical definition are raised. Emphasis is placed on the required minimum of experimental results upon which reliable analysis can be based. The paper closes with critical comment on methods of investigation presently employed.

INTRODUCTION

This paper does not aim at a review of the extensive existing literature on bridge aerodynamics and aeroelasticity. Rather, it discusses certain aspects of the present state of the art, which is advancing out of empiricism toward more incisive analytical formulations based on essential measured aerodynamic parameters.

The wind engineering of all types of bridges is in principle based on actual or inferred knowledge of the expected highest velocity of wind at the site that will strike the bridge within a defined time interval. Assessment of this wind velocity is a well-defined problem in extreme value statistics based on meteorological data for the bridge site that are known or inferred. Pressures derived from a steady horizontal wind velocity U are routinely described by a formula like

$$p = \frac{1}{2}\rho U^2 C_p \qquad (1)$$

ρ being being air density and C_p a pressure coefficient not exceeding unit value. Forces derived through the integration of such pressures over the bridge structure may be expressed as

$$F = \frac{1}{2}\rho U^2 A C_D \qquad (2)$$

where A is some reference area and C_D is a drag coefficient that may range from a value as low as 0.3 to a number larger than unity, since leeward pressures are usually suctions (negative). Detailed wind tunnel model investigations are involved for specific bridge deck forms if refined values of C_D are sought. Alternate approaches used quite successfully for many years have, more simply, chosen conservative values of static pressures to be applied to windward bridge surfaces. A more modern practice is to increase the design wind velocity by a factor like 1.25 to allow for gusting.

When additional refinements beyond this are sought they must usually be based upon wind tunnel model investigation; it is common then to employ a geometrically faithful model of the desired structural element such as deck, pier, tower,etc. Under such conditions observations not only of drag, but of uplift and overturning moment are made, each with its appropriate coefficient. Approaches of this general kind are "standard" and to be found in most construction code and wind textbook sources [1–3]. In general, the steady-state pressures employed in modern bridge design have long since been proven to be adequate and are no longer to be considered topics for research or special investigation except in rare cases of particular geometry or site peculiarities. Focus in the present paper will therefore be upon the more recondite aspects of bridge wind design, namely *dynamics*.

SOURCES OF DYNAMIC ACTION

Wind velocity, when measured, invariably exhibits time histories with stochastic aspects. Thus while the wind has potential for dynamics in its steady flow, its gusting, caused by mechanical stirring due to upstream objects, is an additional source of dynamic effects. When a very smooth and steady airflow passes over a bluff object like an element of bridge structure, the result is not steady but in fact is a fluctuating net effort on the structure. Thus so-called *steady* force coefficients are themselves only mean values of fluctuating components, even when the oncoming flow is smooth. Thus eqs. (1) and (2), if their coefficients C_p and C_D are taken as constants, must properly be interpreted as means or averages only. Alternately, eqs. (1) and (2) must be replaced by more adequate representations when time-dependency is to be taken into account. In general, bridge structures can be *buffeted* by two turbulence effects: the oncoming turbulence present in the wind and the signature (or self-induced) turbulence initiated by the structure.

Further, the structure itself, being elastic, will respond more or less to the fluctuating pressures acting on it. When and if this response becomes appreciable, the possibility of *flow-structure interaction* or *aeroelastic* effects, arises. In this circumstance, structural motions directly influence the formation of flow-induced pressures over the structure. The possibilities of "force feedback" are thus opened up. In most cases such feedback tends merely to damp the structural motion, but cases do arise (usually at higher wind speeds) where the feedback causes net effective damping to become negative, i.e. such that energy flows into the structure during the motion, instead of away from it. Cases of this sort take on different aspects, depending on certain particular features of either the structure or the oncoming flow. The most commonly identified dynamic bridge responses to wind may be characterized as "buffeting", "vortex-induced", and "flutter".

PHYSICAL MODELING FOR DYNAMIC ACTIONS

Given a new bridge design, particularly one with a long or flexible span, its susceptibility to wind-induced dynamic action should be assessed as part of the design procedure. The most vulnerable portion being the deck, it is usual to begin with a geometrically faithful wind tunnel model of a typical deck *section*. Such sections are invariably studied for the longest and most flexible bridges— the classical catenary suspension and the cable-stayed types. However, notable examples of other flexible cases, such as those with any type of extremely long span, are also wind-susceptible. Also employed are appropriately scaled full-bridge (and occasionally partial-bridge) models.

As the stages of broad empiricism in testing have progressed toward a more scientific view, questions have begun to be posed as to the *aerodynamic mechanisms* involved in bridge response to wind and the *essential minima* of results needed to assure a stable design. Many of these can in fact be inferred from the section model. Today such models are best looked upon as basic *analog sources* for the data necessary to identify (and possibly correct) the mechanisms affecting a given bridge design under wind. The present paper is particularly concerned with the analytical formats whereby the effects of these mechanisms are captured.

PRINCIPAL MECHANISMS OF BRIDGE RESPONSE

Broadly speaking, the key parameters influencing bridge dynamic performance under wind are deck cross-sectional shape, natural frequencies in the degrees of freedom permitted, structural mass and structural damping.

Vortex-Induced Oscillation. The geometric form of a deck cross-section will influence the wind cross flow about it in such a way that the forces induced either agitate the structure or tend to quiet it. When the structure is particularly bluff— i.e. unstreamlined— it is very often observed that at a fairly low wind speed the cross flow agitates the structure at one of its natural frequencies. This generally low-speed and limited-amplitude phenomenon is somewhat loosely termed "vortex-induced" oscillation.

Decks approaching the "H-section" form have long been known to be susceptible to vortex-induced oscillation. Such shapes are generally avoided in modern design. The primary palliative against this type of oscillation is appropriate readjustment of the cross-sectional shape during the design stage. This process generally requires the attention of a skilled aerodynamicist. The present paper will not deal with details of this very interesting design-related activity.

When a design having a particular deck cross-sectional shape is projected it must be subjected to verification as to its performance. It is thus not desirable to incorporate the projected form immediately into a full-bridge model. Rather, the inexpensive section model is resorted to. Below is briefly outlined a method for extrapolating the wind-tunnel action of such a model under vortex-induced oscillation from section to full-scale, full-bridge performance.

Typically, the physical model will exhibit vertical (h) degree-of-freedom oscillations, which representative action will be discussed here. A representative mathematical model of this is given by

$$m[\ddot{h} + 2\zeta\omega_h\dot{h} + \omega_h^2 h] = F(h,\ \dot{h},\ \ddot{h},\ U,\ t) \tag{3}$$

where

$$F(h, \dot{h}, \ddot{h}, U, t) = \frac{1}{2}\rho U^2(2B) \tag{4}$$

$$\times \quad \left[Y_1(1 - \epsilon\frac{h^2}{B^2})\frac{\dot{h}}{U} + Y_2\frac{h}{B} + \frac{1}{2}C_L\sin(\omega t + \theta)\right]$$

In the above equations

- m = mass per unit span
- h = vertical displacement
- $\dot{h}$ = dh/dt
- ζ = mechanical damping ratio
- ω_h = natural mechanical radian frequency
- ρ = air density
- U = cross-wind velocity
- B = a reference structural dimension (typically, deck width)
- ϵ, C_L, Y_i $(i = 1,\ 2)$ = aerodynamic coefficients, functions of K
- K = $B\omega/U$ = reduced frequency
- ω = radian frequency of oscillation

In several applications it was found that the value of C_L could be taken as negligible; C_L therefore will be omitted in the subsequent discussion.

Further defining

$$\eta = h/B$$
$$s = Ut/B$$
$$\eta' = d\eta/ds$$
$$K_h = B\omega_h/U$$
$$m_r = \rho^2/m = \text{mass ratio}$$

Equations (3) and (4) are put into the form

$$\eta''(s) \quad + \quad 2\zeta K_h\eta'(s) + K_h^2\eta(s)$$
$$= \quad m_r\left\{Y_1[1 - \epsilon\eta^2(s)]\eta'(s) + Y_2\eta(s)\right\} \tag{5}$$

To gain the necessary analog input data, the smooth wind flow over the model is first adjusted until maximum steady-state oscillatory amplitude occurs at the condition called "lock-in".

When this "resonant" condition has been established the model is displaced to an initial amplitude A_o that is 2 or 3 times greater than the steady-state amplitude, and the oscillation is allowed to "decay to resonance". An approximate solution to eq.(5) under these conditions (obtained by the method of slowly varying parameters) is

$$\eta(s) = \frac{\beta}{\sqrt{1 - \frac{A_o^2 - \beta^2}{A_o^2}\exp\frac{-\alpha\beta^2 s}{4}}} \times$$
$$\cos\left[Ks - \frac{1}{2K}(m_r Y_2 + K^2 - K_h^2)s - \psi\right] \tag{6}$$

where

$$\beta = \text{steady-state amplitude of } \eta\ (s \to \infty)$$
$$= \frac{2}{\sqrt{\epsilon}}\left[1 - \frac{2\zeta K_h}{m_r Y_1}\right]^{\frac{1}{2}}$$

$A_o =$ initial displacement of η

$K = \frac{B\omega}{U}$

$=$ reduced frequency at experimentally observed radian frequency ω

$\alpha = m_r Y_1 \epsilon$

$\psi =$ a phase angle

Identifying the full experimental response with eq.(6) permits Y_1, Y_2 and ϵ to be determined.

This sectional-model method is then extended to the three-dimensional prototype bridge of span length L by replacing η by the expression

$$\eta = \xi_o \phi(x) \cos \omega t \quad (7)$$

where $\phi(x)$ is the bridge mode shape normalized at unity at its largest amplitude, and ξ_o is calculated from the result

$$\xi_o = 2\left[\frac{\Phi_2}{\epsilon\Phi_4}\left(1 - \frac{2\Phi_{20}\zeta(K_h)}{m_r Y_1 \Phi_2}\right)\right]^{\frac{1}{2}} \quad (8)$$

with, for $p = 2$ or 4

$$\Phi_p = \int_o^L \phi(x)\phi^p(x)\frac{dx}{L} \quad (9)$$

and

$$\Phi_{20} = \int_o^L \phi^2(x)\frac{dx}{L} \quad (10)$$

For sinusoidal modes, numerical values of Φ_2, Φ_4 and Φ_{20} are:

$$\Phi_2 = 0.4244$$
$$\Phi_4 = 0.3395$$
$$\Phi_{20} = 0.5$$

Estimates by this method yielded a computed maximum amplitude of vortex-induced response for the Deer Isle Bridge of 0.119 m, whereas field observations yielded 0.096 m. Details on the method are provided in [4,5].

Flutter. At higher wind speeds generally the possibility arises of bridge deck oscillatory response that may grow without bound in amplitude. This possibility was singularly demonstrated at Tacoma Narrows in 1940. Two commonly observed characteristics of the high-speed fluid-structure interaction may be noted:

- It most often occurs in or near a single natural mechanical mode of the deck structure. Typically this mode has a prominent torsional component.

- Even when the presence of other modes is observable in the response, a single principal mode often appears to be the "driving" agent.

- Classical coupled vertical/torsional motion can occur in some instances.

In seeking, via theoretical formulation, to identify the fundamental mechanisms involved in bridge-deck instability [6] the classical model of airfoil flutter [7] was at first evoked. In this latter model, the essential criteria for flutter susceptibility reside in the pre-calculated *flutter derivatives* of the thin airfoil. It was clear, however, from the earliest stages that appropriate flutter derivatives for the bridge deck problem could not similarly be theoretically derived from first principles, since bridge decks, aerodynamically viewed, are quite bluff bodies that give rise to separated flow. Any hope for assimilating bridges to airfoils [8] or for calculating the oscillating aerodynamic forces acting on them starting from first theoretical principles, was not rewarded [9]. Even today, with high-capacity computer help available, the calculation of flutter forces on bluff bodies by exclusively computational means remains a formidable enterprise, to date not highly recompensed with results. Thus the more immediately appealing idea of identifying the flutter derivatives experimentally has been pursued. The section model has been the natural analog source for this purpose.

The flutter derivatives obtained under laminar approach flow remain the central descriptors of how the bridge will respond to wind. Their values under turbulent approach flow have not to date been extracted under wholly appropriate conditions of similitude. In fact, this remains an area of active research interest. An important consideration therein is that active, rather than passive, means may have to be resorted to to establish turbulence of the appropriate scale lengths [10].

Experience with bridge section models reveals that their strongest tendencies toward instability generally reside in the torsional motions. The damping (velocity-related) flutter derivatives clearly reveal these tendencies. The cross-coupling flutter derivatives appear in general to be much less influential. Their effects can largely be negated, for example, by distinct separation of the vertical and torsional mechanical frequencies of a given deck structure— in which circumstance aerodynamic effects usually cannot force a clearly coupled response.

Following the format (but not the detail) of the airfoil flutter problem a linear model of the obviously self-excited system was early formulated as reviewed below. Assume a bridge deck section free to move in vertical (h) and torsional (α) motions. As with the airfoil, lift L and torsional moment M were first posited in a general form, which later was reduced to one containing dimensionless coefficients. From this, the commonly recognized forms have evolved:

$$L = \frac{1}{2}\rho U^2 B \times \left[K H_1^* \frac{\dot{h}}{U} + K H_2^* \frac{B\dot{\alpha}}{U} + K^2 H_3^* \alpha + K^2 H_4^* \frac{h}{B} \right] \tag{11}$$

$$M = \frac{1}{2}\rho U^2 B^2 \times \left[K A_1^* \frac{\dot{h}}{U} + K A_2^* \frac{B\dot{\alpha}}{U} + K^2 A_3^* \alpha + K^2 A_4^* \frac{h}{B} \right] \tag{12}$$

where H_i^*, A_i^* are dimensionless forms of aerodynamic coefficients, commonly called "flutter derivatives".

The equations of motion of a centrally balanced bridge deck section are then

$$m[\ddot{h} + 2\zeta_h \omega_h \dot{h} + \omega_h^2 h] = L \tag{13}$$

$$I[\ddot{\alpha} + 2\zeta_\alpha \omega_\alpha \dot{\alpha} + \omega_\alpha^2 \alpha] = M \tag{14}$$

where

m = mass per unit span
I = mass moment of inertia per unit span
ζ_h = damping ratio, h–motion
ζ_α = damping ratio, α–motion
ω_h = natural radian frequency, h–motion
ω_α = natural radian friquency, α–motion

Eqs. (13) (14) have been generalized to three dimensions for use on full-span prototype bridges [11]; the stability of the system defined by them is a direct function of both the mechanical parameters and the flutter derivatives.

One of the most common characteristics of flutter-prone bridge designs is the tendency of their torsional damping flutter derivative (A_2^*) to reverse sign at some point in the range of the parameter K. Under the common conditions of single-mode flutter driven by mode i the flutter criterion can be shown [12] to be

$$H_1^* G_{h_i} + A_2^* G_{\alpha_i} \geq \frac{4\zeta_i I_i}{\rho B^4 L} \left(1 + \frac{\rho B^4 L}{2 I_i} A_3^* G_{\alpha_i} \right)^{\frac{1}{2}} \tag{15}$$

where

$G_{q_i} = \int_o^L q_i^2(x)dx/L$
q_i = dimensionless modal component $h(x)$ or $\alpha(x)$ of deck mode i
ζ_i = mechanical damping ratio in mode i
I_i = generalized inertia of mode i

Eq.(15) is basically a comparison of net available aerodynamic damping to available mechanical damping. The problem of *obtaining* the flutter derivatives remains essentially an identification problem based on experimental results, and it is still an objective attacked variously from several quarters. Some of the means employed are illustrated in [6, 13–16].

Buffeting. Aerodynamic efforts (per unit span) of this type have been written in the quasi-steady form [11]

$$\text{Lift}: \quad L_b(t) = -\frac{1}{2}\rho U^2 B\left[C_L\left(\frac{2u(t)}{U}\right) + \left(\frac{dC_L}{d\alpha} + C_D\right)\frac{w(t)}{U}\right] \quad (16)$$

$$\text{Drag}: \quad D_b(t) = \frac{1}{2}\rho U^2 B\left[C_D\left(\frac{2u(t)}{U}\right)\right] \quad (17)$$

$$\text{Moment}: \quad M_b(t) = \frac{1}{2}\rho U^2 B^2\left[C_M\left(\frac{2u(t)}{U}\right) + \frac{dC_M}{d\alpha}\frac{w(t)}{U}\right] \quad (18)$$

where $u(t)$, $w(t)$ are horizontal and vertical gust velocities. In these expressions C_L, C_D, C_M are steady-state force coefficients. In view of the fact that $u(t)$ and $w(t)$ are varying in time, the question arises as to the validity of steady-state derivatives in this context. This observation has been responded to by including an experimental multiplier, the *aerodynamic admittance*, to correct the observed *spectra* of L_b, D_B and M_b [17]. These effects have been included in three-dimensional analyses [18].

CRITICAL REVIEW

An aim of the writer is to place the wind engineering of bridges— especially their dynamics— on a sound analytical footing, based on the minimum essential set of aerodynamic and aeroelastic facts. The art of bridge dynamics under wind and its aeroelastic aspects has been undergoing evolution for half a century. During much of this period an ad hoc or empirical approach has been common. This has manifested itself primarily in the building and direct testing of reduced-scale wind tunnel models considered in one or another sense to "represent" the prototype full-scale bridge, both dynamically and aerodynamically. Progress in the art has occasioned step-by-step departures from this early approach.

The *section model* was early-on the major investigative device and may be considered so now. However, uncritically accepting its stability or instability as evidence of analogous prototype action should now be considered unwarranted. Other more sophisticated reactions to it are in order. Its response requires *interpretation.*

It has the major advantage of being the largest convenient model representation of a typical part of the prototype bridge structure; yet, its scale will still be such that it will necessarily involve some possible Reynolds number discrepancies. This will be particularly true for the finer structural details such as grids, slots, and fences, and for the vortex and turbulence effects associated with them. Insufficient attention has been paid by the profession to date to the full implications of these aerodynamic niceties.

The section model cannot conveniently (or correctly) be endowed with the identical freedoms that a comparable section of the prototype will have. In a modern full-scale bridge such as a cable-stayed structure the natural vibration modes are quite complex, and each deck section will undergo its own local part or manifestation of the total modal action of the full structure. Thus no section model can actually act "typically", except insofar as its geometry is typical.

It is in taking these facts fully into account that the impetus was initially gained to employ the section model in a manner that emphasizes its sole real attribute: its geometric form. This line of reasoning leads inevitably to viewing its primary role to be an analog source of aerodynamic and aeroelastic data, not of stability or instability information per se. The aerodynamic data obtainable from it are, first, the mean or static values of C_L, C_D, and C_M and, second, their associated spectra (related to the aerodynamic admittance) under both laminar and (correctly scaled) turbulent approach flow. When model motion in specified degrees of freedom is properly controlled and interpreted the third set of aeroelastic information that can be extracted is the flutter derivatives.

Experimental evidence to date on three-dimensional models has shown that the presence of turbulence in the approach flow tends to delay to higher values the velocity at which dynamic instability occurs. This fact, coupled with the observation that many long-span bridges are in fact approached by unusually smooth wind flow and/or wind flows characterized by very long turbulence length scales ($> 100m$) at the level of their decks, suggests that first obtaining the flutter derivatives under smooth flow conditions is the fundamental and prudent design investigation. If a given design is rendered stable up to a given wind speed according to its smooth flow flutter derivatives, the likelihood of its stability under appropriate turbulent wind is considerably enhanced. It is, also, an incorrect design inference to count upon the presence of turbulence to assure any kind of bridge stability, including that against vortex-induced oscillation.

The question of the effects of turbulence upon bridge stability has recently been approached with considerable analytic vigor. At issue is the question of just how turbulence in the approach flow affects the flutter derivatives and how to reflect these effects analytically. In this context recourse is had alternately to the use of indicial response functions [19], resulting in certain formulations [20–22] of the equations of motion that possess stochastic coefficients. Under these conditions structural stability may have to be differently interpreted. However, the physical bases for certain of the analytic models postulated to date have not been fully substantiated.

The problem of experimental assessment of the effects of turbulence upon the mean force and flutter derivatives remains open as an area of research that has not to date been addressed to the best possible simulation capability. In the cases where it has been addressed so far, the spectra and scales of turbulence developed have not been wholly appropriate. It appears as if active (driven) turbulence generation [10] may be a necessary recourse if turbulence scale lengths appropriate to the relatively large scales of section models are to be realized. This avenue of research still offers promise if only to confirm that it may eventually prove to be an unnecessary or redundant exercise, given the primary usefulness of flutter derivative data taken under laminar approach flow.

Because of the open questions concerning the effects of turbulence, full-bridge models (up to as large as 1:100 scale) have been built and tested in well-simulated turbulent boundary layer flow. Also, extended-span models of various sorts short of complete full-bridge simulation have been built. The shortcomings to guard against in full-bridge models will be discussed first. Unless exquisitely (and therefore expensively) executed, their characteristics may not adequately duplicate the necessary attributes of the prototype. Full aeroelastic models require extreme attention to mass placement and appropriate (usually alternate) structure for elastic scaling— and, in fact, to scaling laws in general. Their vibration modes should faithfully reproduce those of the prototype to a number as high as 10 or 15, or more, since in the prototype it is not unusual for a very high mode (the author has seen the 13th and 17th respectively in two designs) to become the eventual key mode for flutter instability. When, on the other hand, the scale of the model is quite small (below 1:300, for example) the model inevitably loses both aerodynamic and elastic faithfulness, and its damping values are likely to be unrelated to expected prototype values. An impressively "stable" performance of such a model in the wind tunnel cannot be considered a guarantee of prototype action, since the particular mechanisms responsible for the observed stability are not made evident in this context and may in fact differ from those of the prototype.

Small-scale full-bridge models are often considered to be unavoidable because of the size of the available wind tunnel and the short turbulence length scales achievable by passive means therein. In such cases both the presence of turbulence and the model's geometric, structural and damping inadequacies may tend to mask important facts to be concluded relative to the prototype. The same criticisms apply with more cogency to abbreviated models of various sorts, for example those endowed with over-simplified structure favoring certain pre-selected modal responses.

Alternatives to inadequate models are a) recourse to the best available aerodynamic data from geometrically accurate section models: static and flutter derivatives and aerodynamic admittances; b) acquisition of a well-calculated set of full-bridge modes (for example from advanced finite element codes); and c) the calculation of stability and buffeting response based on sound theory. Under such calculational control, many realistic simulations can be carried out, including desired effects of special terrain, live loads, etc. Where funds permit, useful corroboration of calculational techniques may be accomplished through the parallel study of high quality full-bridge models.

It is hoped through these comments to suggest that the era of empiricism in the investigation of long-span bridge response to wind can be superseded— as in the analogous case of aeronautical design— by one where sound calculations, based on a minimal set of accurate observations and a substantive analytic rationale, will replace ad hoc investigations.

REFERENCES

1. Draft (1985) Indian Standard Criteria for Design of Structures for Wind Effects (Revision of IS:875 pertaining to wind loads) Doc: BDC 37(3571)

2. American National Standard: "Minimum Design Loads for Buildings and Other Structures" ANSI A58.1 (1982). Revised as ASCE Document 7-88 (1990) ASCE, New York

3. Simiu, E. and Scanlan, R.H.: *Wind Effects on Structures* 2nd Ed. (1986) Wiley, N.Y.

4. Ehsan, F. and Scanlan, R.H.: "Vortex-Induced Vibrations of Flexible Bridges" *Jnl. Engineering Mechanics, ASCE* Vol.116, No.6 (Jun.1990) pp.1392–1411

5. Ehsan, F.: "Vortex-Induced Response of Long, Suspended-Span Bridges", Doctoral dissertation, Dept. of Civil Engineering, Johns Hopkins University, Baltimore, MD (Jun.1988)

6. Scanlan, R.H. and Tomko, J.J.: "Airfoil and Bridge Deck Flutter Derivatives" *Jnl. Engineering Mechanics Div, ASCE*, Vol.97, No.6 (Dec.1971) pp.1717–1737

7. Theodorsen, T. "General Theory of Aerodynamic Instability and the Mechanism of Flutter", National Advisory Committee for Aeronautics Report No.496 (1934)

8. Bleich, F.: "Dynamic Instability of Truss-Stiffened Suspension Bridges under Wind Action" *Proc. ASCE*, Vol.74, No.8 (1948) pp.1269–1314; Vol.75,No.3 (Mar.1949) pp.413–416; Vol.75,No.6 (Jun.1949) pp.855-865

9. Selberg, A. and Hjorth-Hansen, E.: "The Fate of Flat Plate Aerodynamics in the World of Bridge Decks", *Proc. Theodorsen Colloquium, Det Kongelige Norske Videnskabers Selskab* (1976) pp.101–113

10. Huston, D.R.: "The Effects of Upstream Gusting on the Aeroelastic Behavior of Long Suspended-Span Bridges", Doctoral dissertation, Dept. of Civil Engineering, Princeton University (May 1986)

11. Scanlan, R.H.: "On Flutter and Buffeting Mechanisms in Long-Span Bridges" *Probabilistic Engineering Mechanics*, Vol.3,No.1 (Mar.1988) pp.22–27

12. Scanlan, R. H., and Jones, N.P.: "Aeroelastic Analysis of Cable-Stayed Bridges." *Jnl. Structural Engineering, ASCE*, Vol.116, No.2 (Feb.1990) pp.279–297

13. Shinozuka, M., Imai, H., Enami,Y. and Takemura, K.: "Identification of Aerodynamic Characteristics of a Suspension Bridge Based on Field Data", *Stochastic Problems in Dynamics*, B.L. Clarkson, Ed., Pitman, London (1977) pp.214–236

14. Shinozuka, M., Chung-Bang Yun, and Imai, H.: "Identification of Linear Structural Dynamic Systems", *Jnl. Engineering Mechanics Div., ASCE*, Vol.108, No.EM6 (Dec.1982) pp.1371–1390

15. Xie, Jiming: "CVR Method for Identification of Nonsteady Aerodynamic Model", *Proc. 7th Int'l Conf. on Wind Engrg.*, Aachen, W. Germany (July 1987), reprinted in *Jnl. of Wind Engineering and Industrial Aerodynamics*, Vol.29 (1988) pp.389–398

16. Imai, H., Yun, C.-B., Maruyama, O. and Shinozuka, M.: "Fundamentals of System Identification in Structural Dynamics", *Probabilistic Engrg. Mechanics*, Vol.4,No.4 (1989) pp.162–173

17. Jancauskas, E.W.: "The Cross-Wind Excitation of Bluff Structures and the Incident Turbulence Mechanism", Doctoral dissertation, Dept. of Mechanical Engineering, Monash University, Australia (1983)

18. Kumarasena, T.: "Wind Response Prediction of Long-Span Bridges", Doctoral dissertation, Dept. of Civil Engineering, Johns Hopkins University, Baltimore, MD (Oct.1989)

19. Scanlan, R. H.: "Role of Indicial Functions in Buffeting Analysis of Bridges", *Jnl. Structural Engineering, ASCE*, Vol.110,No.7 (Jul.1984) pp.1433-1446

20. Lin, Y.K.: "Motion of Suspension Bridge in Turbulent Winds", *Jnl. Engineering Mechanics Div., ASCE*, Vol.105,No.EM6 (Dec.1979) pp.921–932

21. Lin, Y.K. and Yang, J.N.: "Multimode Bridge Response to Wind Excitations", *Jnl. Engineering. Mechanics Div., ASCE*, Vol.109,No.2 (Apr.1983) pp.586–603

22. Bucher, C.G. and Lin, Y.K.: "Effect of Spanwise Correlation of Turbulence Field on the Motion Stability of Long-Span Bridges", *Jnl. Fluids and Structures*, Vol.2 (1988) pp.437–451

Wind Loads on Structures, 1990

WIND LOADS ON CABLE-STAYED BRIDGES

Naruhito Shiraishi
Professor

Masaru Matsumoto
Associate Professor

School of Civil Engineering
Kyoto University
Kyoto 606, JAPAN

ABSTRACT

This paper presents some basic aero-static and aerodynamic forces acting on cable-stayed bridges and their responses. Stress is particularly placed on aerodynamic characteristics of cable-stayed bridges, including aerodynamic behaviors of such individual structural elements as freely standing pylons and cables under erection. And a method of wind safety diagnosis of bridges is also considered based on the data of wind characteristics at the site.

INTRODUCTION

It is well known that after the failure of the Tacoma Narrows Bridge in Washington aerodynamic examination of so flexible structure as suspension bridge is primarily important to secure its safety. Tremendous efforts have been made to analyze aerodynamic behaviors of suspension bridges in various countries. In early sixties of this century, it was said that truss type stiffening girder was much safer for aerodynamic instability, but appearance of streamline like closed box section of Severn Bridge in Britain was so attractive for bridge designers that box type girders have been employed for a large number of suspension bridges as well as cable-stayed bridges.

In wind resistant design, there are number of factors common to both suspension bridges and cable-stayed bridges, but at the same time it should be noted that cable-stayed bridge was so different from suspension bridge. The sag ratio for suspension bridge usually remains from 1/8 to 1/12, while the ratio of height of pylon to center span of cable-stayed bridge remains from 1/3 to 1/5. In other words, the height of pylon of suspension bridge in structural mechanical sense is comparatively low to compare with the height of cable-stayed bridge. This indicates that the pylon of cable-stayed bridge is structurally much flexible so as to be examined aerodynamically. In suspension bridge, main cable is connected with stiffening girder by hangers, so that the displacement of cable is assumed to be same as the displacement at the connecting point of girder. This means that it is not so important to examine the vibrations of cable of suspension bridge aerodynamically. However cables used for cable-stayed bridges are associated with completely different structural and mechanical feature. They tend to vibrate independently from pylons and girders with remarkably small damping characteristic. Aerodynamically, cable aerodynamics is considered as one of the newly realized engineering problems. The problem will be described in detail in subsequent paragraph. As long as the aerodynamic problem of girder is concerned, there is such similarity between suspension bridge and cable-stayed bridge that past knowledge on aerodynamic behaviors of girders of suspension bridges can be utilized quite effectively.

In connection with aerodynamic safety evaluation, an attention should be placed on the different vibrational characteristics of suspension bridges and cable-stayed bridges. For very few cases one has to pay attention on coupling of pylon and stiffening girder of suspension bridge, but for cable-stayed bridge major vibrational modes are coupled modes of girder and pylon.

In this paper firstly some fundamental discussion will be made on coupling of vibrational modes of stiffening girder and pylon. The aerodynamic behavior of pylon, particularly at construction stage, will be discussed. And then so-called rain vibrations of cables will be introduced.

VIBRATIONAL CHARACTERISTICS OF CABLE-STAYED BRIDGES

In order to examine the aerodynamic behaviors of cable-stayed bridges, it is necessary to clarify the vibrational characteristics, namely eigenfrequencies and corresponding modes. Since the vibrational characteristics depend on mass and inertia distribution and structural rigidities including bending and torsional resistances of each members, an analysis is required to be performed by taking all structural members into an account simultaneously. Detailed discussions can be found in the reference [1] and note that equivalent mass and equivalent inertia take an important role for clarification of aerodynamic responses of cable-stayed bridges. Generally speaking, the larger the mass, the more resistant the structure in dynamic sense, one is required to examine which vibrational mode is associated with the least value of equivalent mass and the least value of equivalent inertia. In other words the least equivalent mass is considered to correspond to the most possible flexural vibrational mode of girder and the least inertia is also considered to correspond to the most possible torsional vibrational mode of girder, respectively. In free vibrations of cable-stayed bridges, complexity is noticed mainly by coupling of modes of stiffening girder and pylon, so that an attention should be placed on whether the mode of stiffening girder or that of pylon is dominant. For example, for prestressed concrete cable-stayed bridge consisting of two plane cable system, torsional modes of girder are frequently obtained at comparatively lower degree of vibrational modes of low natural frequencies, but if equivalent inertia are so large it stands for that the out of phase vibrational modes of two pylons are dominant. In other words, the torsional mode of girder has less meaning in connection with vortex-induced vibrations of girders in torsional mode but an attention should be placed on the out of phase mode of pylon induced by winds.

WIND-INDUCED OSCILLATION OF PYLON OF CABLE-STAYED BRIDGE

As already mentioned above, wind-induced oscillations of pylon of long span bridges including suspension bridge and cable-stayed bridge were frequently observed when they stand without any connection with cables under construction stage. Since most of pylons of this kind of structures are so-called rectangular cross section, vortex-induced oscillation in either bending mode or torsional mode can be generated as well as galloping oscillation when wind velocity exceeds the critical wind speed corresponding to this phenomenon. However, it should be mentioned that wind-induced oscillation for pylon of suspension bridge would occur when wind blows perpendicular to the bridge axis, but for cable-stayed bridge of single cable plane system, oscillation of pylon at construction completed stage is induced by wind parallel to bridge axis.

Effective counter measures for this kind of vibrations are of course introduction of mechanical dampers to increase damping force, but sometime it becomes more effective to install a space or a slit at section of pylon parallel to bridge axis or to install such aerodynamic devices as cowling and corner cut at edges of cross section of pylon. [2]

Since the longer span the cable-stayed bridge, the higher the pylon, aerodynamic characteristics of pylons at construction stage are known to take an important role in safety evaluation. As the cross section of a majority of pylons of cable-stayed bridges are more bluff to compare with those of girders, it is required to pay a particular attention for its aerodynamic behaviors for both construction stage and completed stage.

INCLINED CABLE AERODYNAMICS

In 1979, J. Wianecke [3] reported cable vibrations of the Brotonne Bridge in France. Based on observation of severe vibrations of Brotonne Bridge, wind tunnel test was carried out to disclose possibility of cable vibrations and to characterize the mechanism of this sort of vibrations. The test was reported that cables showed considerably large amplitude of vibrations and wind tunnel test results appeared to be similar vibrations to those observed at the site.

In Japan, Y. Hikami [4,5] made the first attempt to analyze observed cable vibrations of the Meikonishi Bridge located at the inlet of Nagoya Harbor in central part of the main land of Japan. The time history of cable vibrations of the Meikonishi Bridge, as indicated in [6], are considered clearly that the vibrations are closely related with rainfall incidental to wind blowing. The records of amplitudes of all 24 cables for two typical wind directions were obtained. In the case of NNW wind which had wind component blowing from the center span to side span, the vibrations appeared only in the cables in side span. On the other hand, for SSE wind direction in which wind blew from the side to center span, only the cables in the center span vibrated. Succeeding wind tunnel test indicated that main cause of cable vibrations was due to formation of rivulet tracked down along the lower surface of cables by rain incident to wind.

Recently further investigations on cable aerodynamics have been performed by number of researchers. And similar kinds of cable vibrations were observed at a number of cable-stayed bridges throughout nations. The mechanism to cause cable vibrations is explained as follows; namely circular cylinder inclined to wind direction can vibrate due to formation of secondary axial flow generated along cable at both windward and leeward stagnation point. And the formation of rivulet at upper and lower surface of cables tends to enhance the vibration. More detailed discussions on cable vibrations are described in the succeeding paragraph.

In 1988, the committee for cable vibrations of cable-stayed bridges was organized by the Japan Institute of Construction Engineering in order to survey recent state of wind-induced vibrations of cables for cable-stayed bridges in Japan. The results of survey by this committee [7] confirmed the already reported characteristics of cable vibrations as follows;

(1) This kind vibration was only observed for cable-stayed bridges which locate at so smooth terrain as inlet of river.

(2) The vibrations were observed in heavy, medium and light rain, even drizzle or misty rain and sleet, together with wind.

(3) Rain-wind induced vibrations were only observed for poly-ethylene coated cables of diameter of 150 to 200 mm.

(4) This kind of vibrations were only observed at the certain limited region of wind velocity, namely 6 m/sec to 18 m/sec, which corresponds to the reduced velocity (V/fD where V stands for wind velocity, f for natural frequency and D for diameter of cable) of approximately 20 to 90 and subcritical Reynolds number of 60,000 to 2x100,000.

(5) The frequency of cable vibrations vary from 0.6 Hz to 3 Hz approximately.

(6) The amplitudes of vibrations reached to an amount of twice as much as the cable diameter for most of cases. Exceptionally much more amplitude of vibrations were observed as much as 5.4 times diameter (approximately 2m).

(7) Cables in downstream region from pylon were observed to be apt to vibrate sensitively to compare with those in upstream region from pylon. However, there were some cases that both cables in downstream and upstream regions from pylon vibrated simultaneously or individually.

(8) For most of cases, cable vibrations were in-plane modes but exceptionally there were out of plane modes of vibrations.

One of the important characteristics of cable vibrations is to define cable or cylinder orientations relative to the wind direction, namely definition of yaw angle (the angle between wind direction and projection of cable to the horizontal plane) and definition of pitch angle (the angle between longitudinal axis of cable and its projection to horizontal plane). However, the configuration of inclined cable can be alternatively expressed by introducing the relative angle of yaw (the angle between longitudinal axis of cable and wind direction). This means that wind-induced cable vibrations are sensitively dependent on this relative angle of yaw.

As mentioned already, cable vibrations observed at number of cable-stayed bridges were at the first place thought due to simultaneous effects of both wind and rain. But for some of bridges it was known that cable could vibrate without rain, so an attention was placed on analysis of factors to cause for cable to vibrate without rain. The investigations for inclined cylinder were made by Shiragashi et al [8], Nakagawa [9] and Matsumoto et al [10] to indicate the existence of secondary flow behind yawed and inclined cylinders and this secondary flow was experimentally observed. Note that the flow crosses the cylinder normal to its axis, thereby inducing the secondary axial flow in the near wake. It was postulated that this secondary flow acts in a manner reminiscent of base bleed or a splitter plate placed in the wake of a cylinder so that galloping type of aerodynamic instability is generated by formation of inner circulatory vortices in region surrounded by leeward side of cable and secondary axial flow. The vibration mechanism of yawed cable or cylinder without rivulet can be characterized experimentally as follows;

(a) The angle of yaw between inclined cable or cylinder and perpendicular axis to wind direction (see **Fig.5**) characterize the stability of wind-induced vibrations of inclined cable. This is graphically shown in **Fig.6,** indicating that for small angle of yaw only buffeting type (mainly rolling) vibration is generated but for large angle of yaw, more than the critical angle of yaw (namely, 22.5 to 25 degree), divergent type of galloping is generated.

(b) As indicated in **Fig. 7,** axial velocity of secondary flow generated behind yawed cable tends to increase as angle of yaw increases until the angle of yaw reaches to 45 degree. The secondary axial flow can be easily visualized by placing number of small flags near to the inclined cable in the wake. (see **Fig. 8**). The hypothesized significance of secondary axial flow along inclined cylinder was examined by a series of wind tunnel tests whether circular cylinder mounted perpendicularly to wind direction could vibrate by externally applied secondary flow or not. The experiment system was schematically shown in **Fig. 9,** in which the nozzle-scoop system was installed along the rear stagnation line of spring-supported cylinder to supply artificial secondary flow. As indicated in the response curve, an introduction of artificial secondary flow was considered sufficient enough to unstabilize the vibration of cylinder in wind.

(c) Note that even weak intensity of turbulence tends to stabilize the wind-induced vibrations of inclined cylinder. (see **Fig. 10**) Though more detailed discussions are necessary, it should be mentioned that the inclination of lift coefficient of cylinder with separate splitter plate with very small spacing tends to change drastically from negative value in smooth flow to positive value in turbulent flow.

(d) Note that vibrations of cable caused by secondary axial flow can not be stabilized so significantly by increase of the Scruton number. (see **Fig. 12 & 13**)

As mentioned above, firstly severe vibrations of cables of cable-stayed bridges were observed in both rain-falling and wind-blowing. But later there were several reports that cables of cable-stayed bridges vibrated in windy days without rain. The above-mentioned experimental works could thus lead to the conclusions that the angle of yaw was responsible for vibrations of cables on an account of generation of secondary axial flow, along cable. Since this vibration of cable caused by secondary axial flow can be stabilized by incidental turbulence, it is thought to be rarely observed for existing cable-stayed bridges.

A series of wind tunnel tests for inclined cylinders or yawed cables with rivulets on upper and lower surfaces of cross section were performed. The results can be summarized as follows;

(1) At the angle of yaw of 45 degree, cables or cylinders tend to vibrate either with or without rivulets. (see **Fig. 14** & **Fig. 15**)

(2) There is no secondary axial flow along cables to be generated when cable subjects to no angle of yaw. However, introduction of rivulet causes for cable to vibrate as shown in **Fig. 16.** The result indicates typical example of response curves consisting of vortex-induced vibration in rather narrowly region of low wind velocity and divergent galloping type of vibration at high wind velocity region. And note that the location of rivulet on the surface of cable plays an important role to characterize unstabilized vibrations of yawed cables as shown in **Fig. 17.** This unstabilized dynamic response for yawed cable corresponds fairly well to the result postulated based on gradient of lift coefficient for cable with rivulet. (see **Fig.**

18).

(3) The vibrations of yawed cable with rivulet are greatly influenced by turbulence of air flow. As the intensity of turbulence increases, the restricted vibrations for this case tend to diminish but the divergent type of oscillations tend to be more enhanced as shown in **Fig. 19.** It should be noted that for non-yawed cable with rivulet the response was so stabilized to diminish as the intensity of turbulence increased, as shown in **Fig. 20.**

(4) The effects of the Scruton number are considered to stabilize the vibrations of yawed cable as shown in **Fig's 21** to **25.**

DIAGNOSIS OF STRUCTURAL SAFETY FOR STRONG WINDS

In order to design a wind resistant structure or to evaluate safety of existing structures against strong winds, probabilistic characteristics of both natural winds at the site and static and dynamic behaviors of structure should be analyzed. In this paper, probabilistic characteristics of winds are classified into three types, namely

(1) parent distribution $p_P(U)$ of wind velocity U, which is usually expressed by Weibull distribution function,

(2) extreme value distribution function $p_{ex}(U)$ of velocity U in design life time T of structure, which is usually expressed by Gumbel distribution function,

(3) probability function of $p_{1,R}(U)$ of velocity U, which can occur more than once at least during the design life time.

The parent distribution of wind can be accurately estimated if wind velocity measurement is made at the site. However, this is scarcely possible and in many cases the parent distribution of wind velocity can be evaluated by use of measured data at neighboring meteorological observatories to identify the Weibull parameters, c(scale parameter) and k (shape parameter), in the regressive formula as shown as

$$c_j=(\sum_i a_{ci} X_i+b_c)_j$$

$$k_j=(\sum_i a_{ki} X_i+b_k)_j$$

where a and b indicate regressive coefficients, X indicates characteristic parameter and j indicates wind direction (j = 1 to 16). The extreme value distribution of wind velocity for each of 16 divided wind directions can be evaluated as follows.

STEP 1 The extreme value distribution function is obtained by Gomes and Vickery method using above mentioned Weibull distribution in the following fashion,

$$p'_{ex}(U)=\exp\{\exp(-a_E<U-U'_R>)\}$$

$$U'_R=U_1+\left(\frac{1}{a_E}\right) lnR$$

$$U_1=c\,(lnN)^{\frac{1}{k}}\left\{1+\frac{k-1}{k^2}\frac{ln(lnN)}{lnN}\right\}$$

$$\frac{1}{a_E}=\frac{c}{k}\,(lnN)^{\frac{1}{k}-1}\left\{1+\frac{(k-1)}{(klnN)}-(1-\frac{1}{k})^2\frac{ln(lnN)}{lnN}\right\}$$

$$N=2\pi\,\nu_U\,\beta_U\,\sigma_u\,k/c$$

STEP 2 The R year extreme value distribution function $p_{ex}(U)$ is obtained from $p'_{ex,R}$ in STEP 1 by modifying by use of the "typhoon factor" which is assumed independent from wind direction in typhoon prone area. The typhoon factor is given as

$$\phi(R)=\frac{U_{ex}(R)_{\text{measured}}}{U_{ex}(R)_{\text{G\&Vmethod}}}$$

where the numerator indicates the expected value for R year in extreme value distribution obtained from annual maximum wind velocity, while the denominator indicates the expected value of R year extreme value distribution obtained by use of Gomes & Vickery method. Using this typhoon factor, the R year extreme value distribution function $P_{ex,R}$ is expressed as

$$P_{ex,R}(U)=\exp\{\exp(-a_E<U-U_R>)\}$$

where $U_R=\phi(R)\times U'_R$

$$\phi(R=1)=1.0$$

The probability function $p_{1,R}$ of wind velocity U which can occur more than once at least during the return period R is obtained from the parent distribution p_U as follows,

$$p_{1,R}(U)=1-\{1-f(U)\,dU\}^n$$

where

$$f(U)\,dU=\left(\frac{k}{c}\right)\left(\frac{U}{c}\right)^{k-1}\exp\left\{-\left(\frac{U}{c}\right)^{k}\right\}\tilde{p}_U\,dU$$

$$n=365\times24\times6\times R$$

and $\tilde{p}$ stands for the occurrence probability of reference wind direction.

CONCLUDING REMARKS

This paper presents recent investigations on some fundamental problems for aerodynamic behaviors of cable-stayed bridges. Cable-stayed bridge is generally speaking one of the most attractive bridge structures for bridge engineers to design because of variety of structural combinations of cables, pylon and girders. However an attention should be placed on dynamic behaviors on an account of its flexibility. Based on recent aerodynamic investigations on cable-stayed bridges the following concluding remarks can be mentioned;

1. As one of basic information for aerodynamic safety of cable-stayed bridge, an attention should be placed on vibrational characteristics of cable-stayed bridges. It is considered that main vibrational features of cable-stayed bridges are coupling modes of pylon and girder so that one should note the least equivalent mass for fundamental flexural mode and the least equivalent inertia for fundamental torsional mode for both vibrations of pylon and girder.

2. To compare with structural configuration of suspension bridges, the ratio of height of pylon to center span length of cable-stayed bridge is comparatively so large that an aerodynamic investigation for pylon of cable-stayed bridge is necessary. There are number of effective counter-measures in order to decrease aerodynamic responses of pylon and girder by means of mechanical devices as well as aerodynamic improvement.

3. A particular attention should be placed on dynamic behaviors of cables because of low capacity of damping capacity and length of cables of small diameters. Very large amplitudes of cable vibrations were reported for number of cable-stayed bridges throughout nations [11]. As mentioned above, important characteristics for this type of vibrations of cables are due to geometrical arrangement for cable axis and wind direction and formation of rivulet appeared at upper and lower surface of cables. In other words, the so-called rain-wind induced vibrations of inclined cables are associated with two different kinds of aerodynamic instability factors, namely, (a) secondary axial flow generated in near wake of inclined cable and (b) formation of rivulet on upper and lower surfaces of cables by rain.

4. It is concluded that cable or circular cylinder becomes aerodynamically unstable to vibrate in either restricted type or divergent galloping type if certain amount of angle of yaw is introduced. However this type of vibrations tend to diminish in turbulent flow. But formation of rivulet on surface of cables tends to enhance aerodynamic instability even in turbulent flow if rivulet is formed at particular position on surface of cables.

5. It is interesting to note that the most effective wind direction to girder is perpendicular to the bridge axis and the one to independently standing pylon for single plane type of cable system is parallel to the bridge axis, while the most effective wind directions to cables are skew to the bridge axis in the horizontal plane.

6. The commonly used wind resistant design method is based on drag coefficient for structural members and design wind velocity to define so-called wind loads on the structure. Note that the drag coefficient is not directly associated with geometrical shape of cross section but it varies by flow pattern around the section if some neighboring structure is located.

7. Aerodynamic safety for any flexible structures can be assured by employing the feasible probabilistic evaluation of failure of probability based on statistical data of wind velocity at site. One of the safety evaluation methods is presented in this paper. But there still remain number of unsolved problems to have to be clarified by future investigations.

ACKNOWLEDGEMENT

The authors would like to express their sincere thanks to Dr. H. Shirato and our other colleagues for their cooperation and assistance in collection of data, numerical calculation and experimental works. This investigation could not have been made without their support.

REFERENCES

1 N.Shiraishi & M.Matsumoto; Aerodynamic Characteristics on Cable-stayed Bridges, Proc. 1st South-East Asia Conf. on Structural and Construction Engineering, Bangkok, 1986

2 N. Shiraishi, M. Matsumoto, H. Shirato & H. Ishizaki; On Aerodynamic Stability Effects for Bluff Rectangular Cylinders by their Corner-cut, Proc. 7th International Conference on Wind Engineering, Vol.2, pp 263-272, Aachen, Germany, 1987

3 J. Wianecki; Cables Wind Excited Vibrations of Cable-stayed Bridges, Proc. 5th International Conference of Wind Engineering, Colorado, USA, pp 1381-1393, Oxford-New York, Pergamon Press, 1979

4 Y. Hikami; Rain Vibrations of Cables of a Cable Stayed Bridge, Jour. of Wind Engineering, No. 27, pp 17-28, 1986 (in Japanese)

5. Y. Hikami & N. Shiraishi; Rain-wind induced Vibrations of Cables in Cable-stayed Bridges, Jour. Wind Engineering & Industrial Aerodynamics, 29, pp 409-418, Elsevier Science Publishers, 1988

6. N. Shiraishi & M. Matsumoto; Aerodynamic Response of Structures Subjected to Strong Wind and its Safety Evaluation, Proc. Japan-China (Taipei) Joint Seminar on Natural Hazard Mitigation, Kyoto, Japan, 1989

7. M. Matsumoto et al; Cable Aerodynamics of Cable-stayed Bridges, Proc. 3rd Kyoto University-KAIST Joint Seminar/Workshop on Civil Engineering, Seoul, Korea, 1990

8. M. Shiragashi, A. Hasegawa & S. Wakiya; Effect of Secondary Flow on Karman Vortex Shedding, Bulletin of JSME, Vol. 29, No.250, pp 1124-1128, 1986

9. K. Nakagawa, K. Kishida & K.Igarashi; Aerodynamic Vibration and Wake Characteristics of a Yawed Circular Cylinder, Wind Tunnel Report of Osaka University, pp 31-37, 1983 (in Japanese)

10. M.Matsumoto, C.W.Knisely, N.Shiraishi, M. Kitazawa & T.Saitoh; Inclined-cable Aerodynamics, Structural Design, Analysis & Testing Proceedings Structural Congress '89, ASCE/San Francisco, USA,pp 81-90, 1989

11. H.E.Langsoe & O.D.Larsen; Generating Mechanism for Cable Oscillation at Faroe Bridge, Proc. International Conference on Cable-stayed Bridges, Bangkok, pp 18-20, 1987

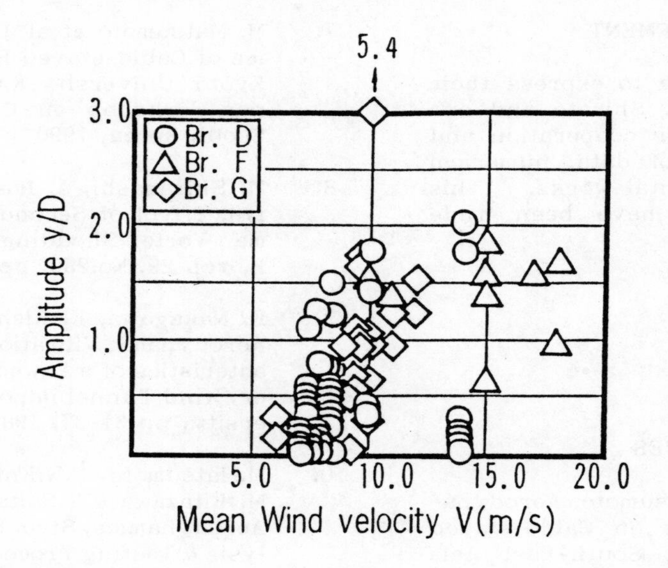

Fig. 1. Mean Wind Velocity-Amplitude Diagram of Observed Rain-Wind Induced Vibration of Prototype Cables

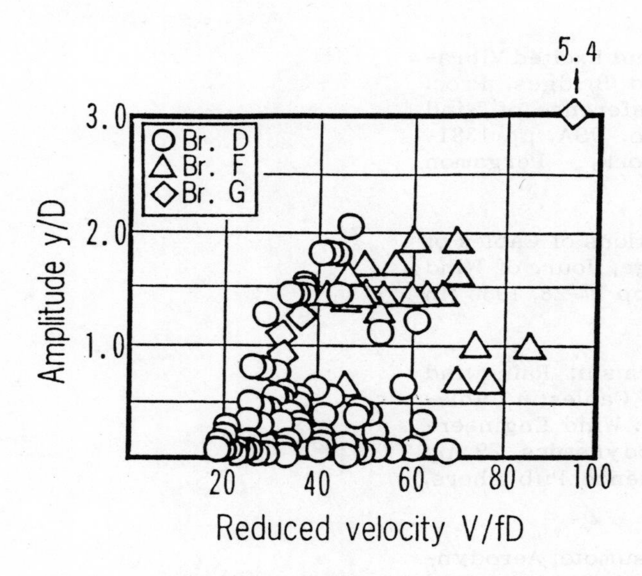

Fig. 2. Reduced Velocity-Amplitude Diagram of Observed Rain-Wind Induced Vibration of Prototype Cables

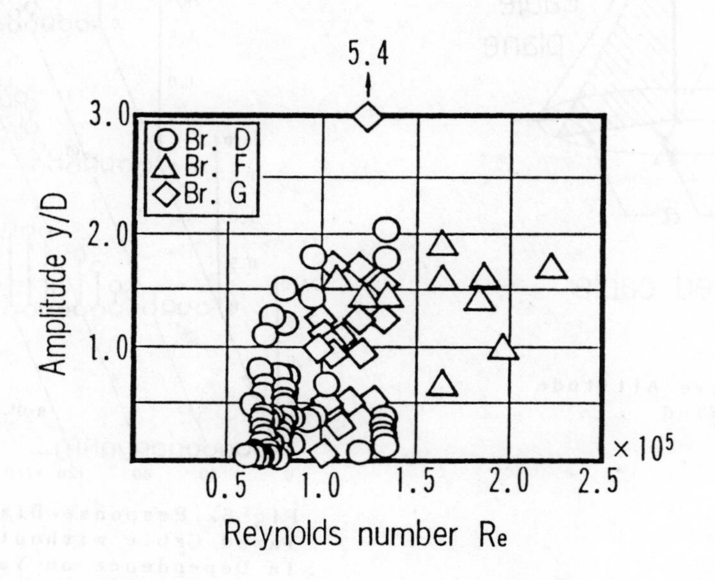

Fig. 3 Reynolds Number-Amplitude Diagram of Observed Rain-Wind Induced Vibration of Prototype Cables

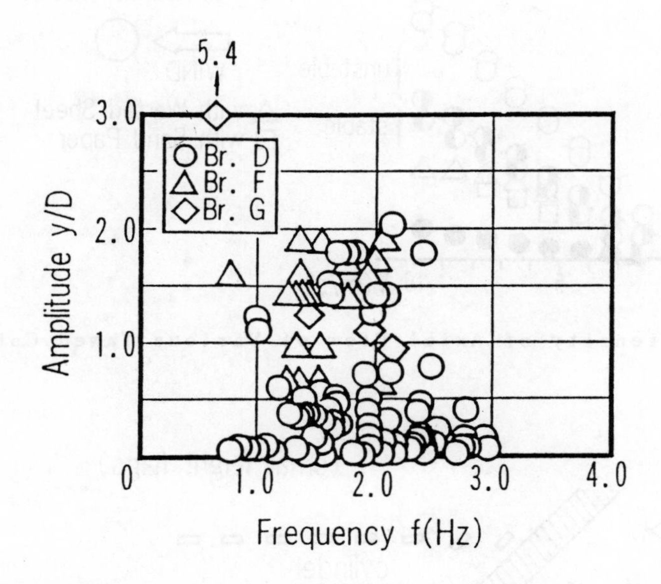

Fig. 4. Cable Frequency-Amplitude Diagram of Observed Rain-Wind Induced Vibration of Prototype Cables

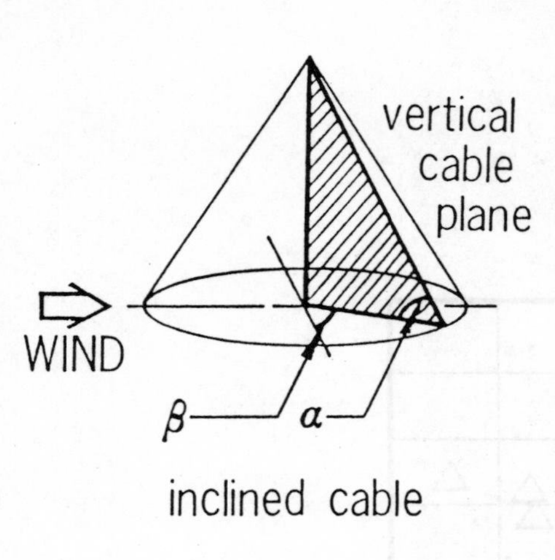

Fig. 5. Relative Altitude of Cable to Wind

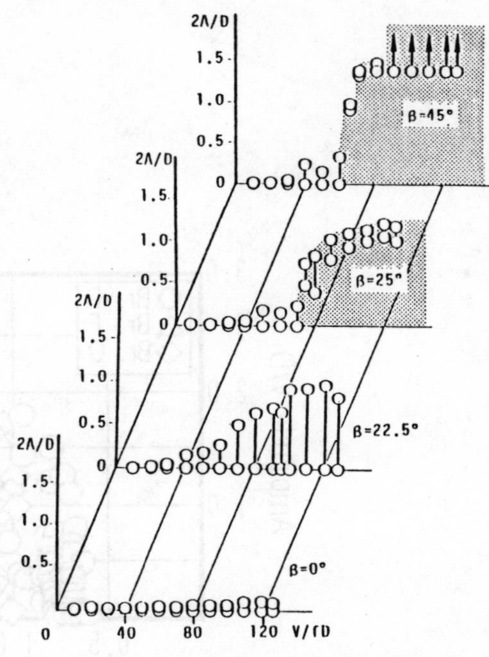

Fig. 6. Response Diagram of Yawed Cable without Rivulet in Dependence on Yawing Angle (in Smooth Flow)

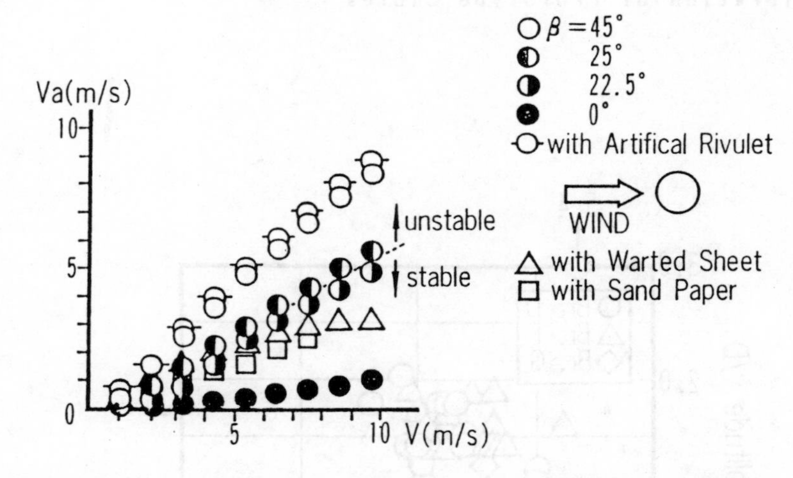

Fig. 7. Intensity of Axial Flow of Various Yawed Cables

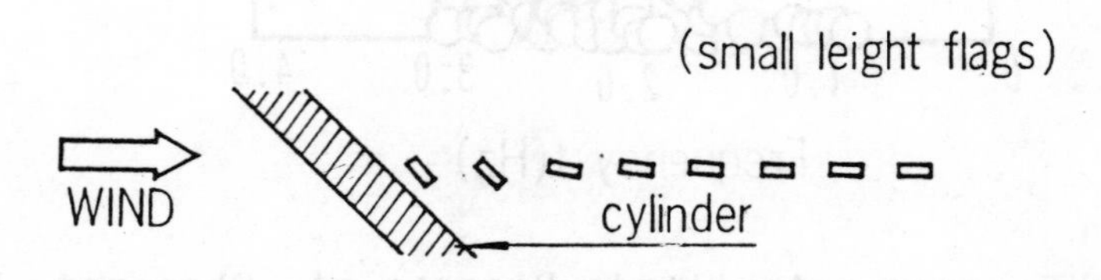

Fig. 8. Flow Pattern behind Yawed Cable (β =45 deg.)

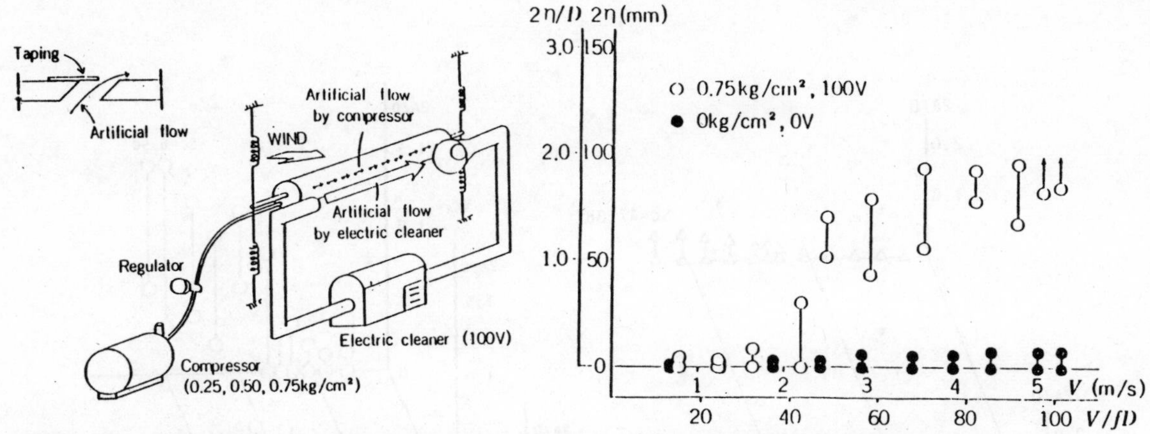

Fig. 9. Device for Generation of Artificial Axial Flow and V-A Diagram of Non-Yawed Cable with Artificial Axial Flow in Smooth Flow

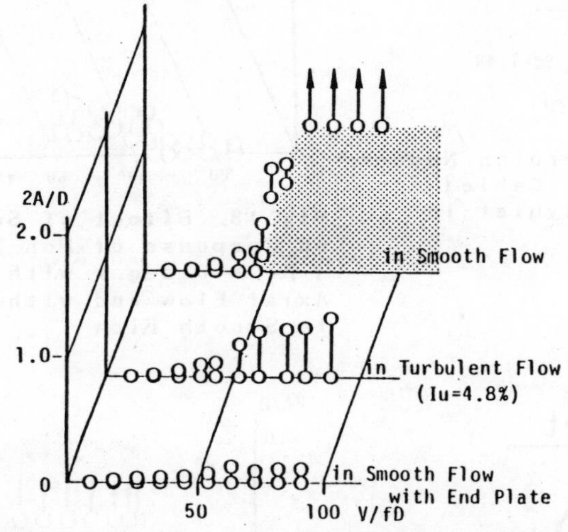

Fig. 10. V-A Diagram of Yawed Cable(β =45 deg.) without Rivulet affected by Turbulence

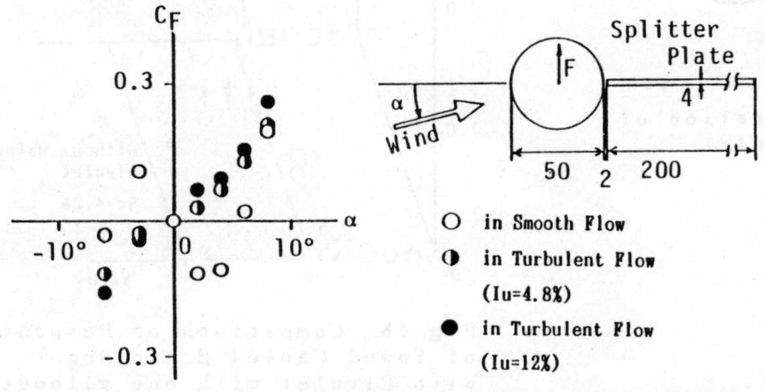

Fig. 11. Lift Force Coefficient Characteristics Affected by Turbulence of Non-Yawed Cable with Splitter Plate Installed in Near Wake

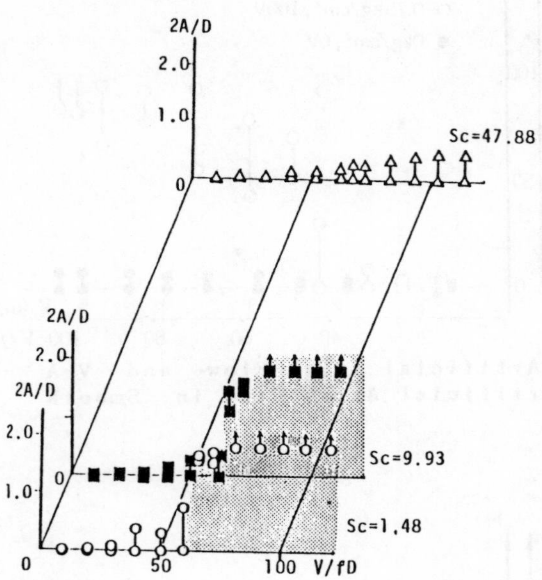

Fig. 12. Effect of Scruton Number on Response of Yawed Cable(β =45 deg.) without Rivulet in Smooth Flow

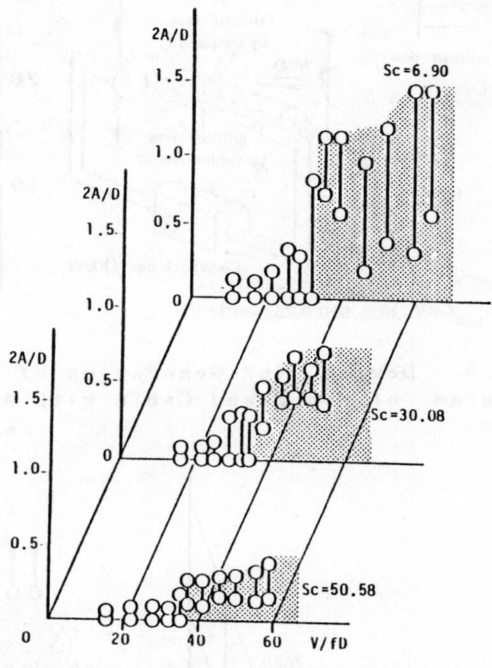

Fig. 13. Effect of Scruton Number on Response of Non-Yawed Cable (β = 0 deg.) with Artificial Axial Flow and without Rivulet in Smooth Flow

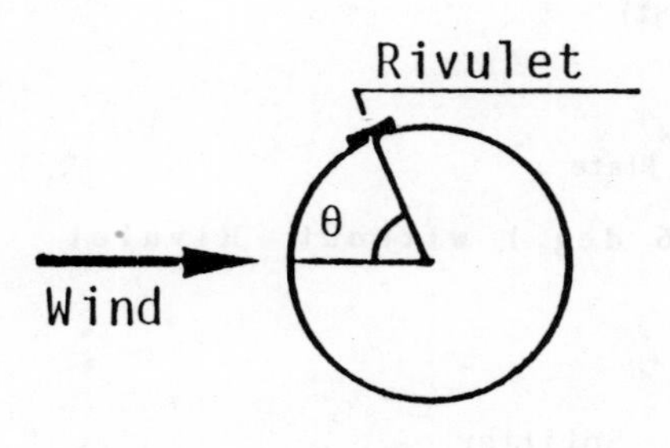

Fig. 14. Illustration of Artificial Rivulet

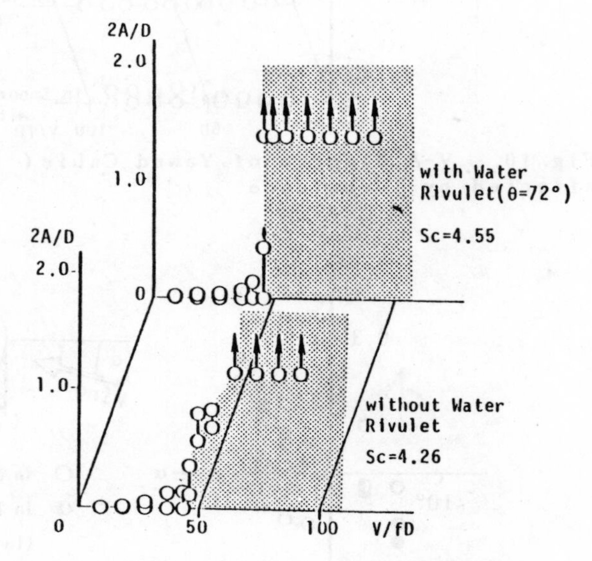

Fig. 15. Comparison of Response of Yawed Cable(β =45 deg.) with Rivulet with one without Rivulet in Smooth Flow

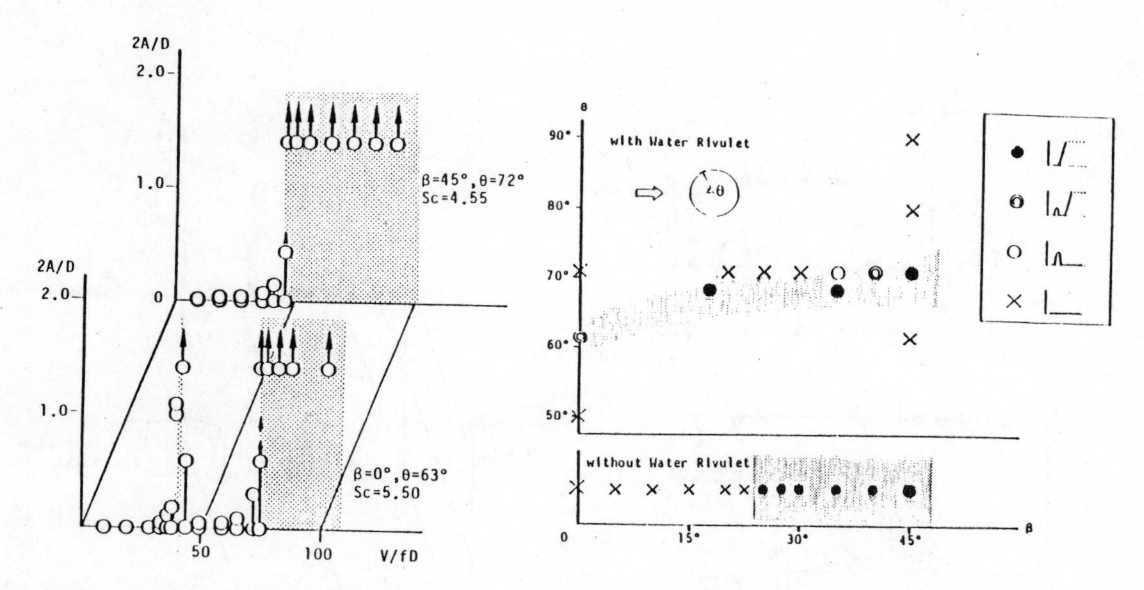

Fig. 16. Comparison of Response of Yawed and Non-Yawed Cables in Dependence on Rivulet Location

Fig. 17. Aerodyanmic Unstable Region of Rivulet Location of Various Yawed Cables (in Smooth Flow)

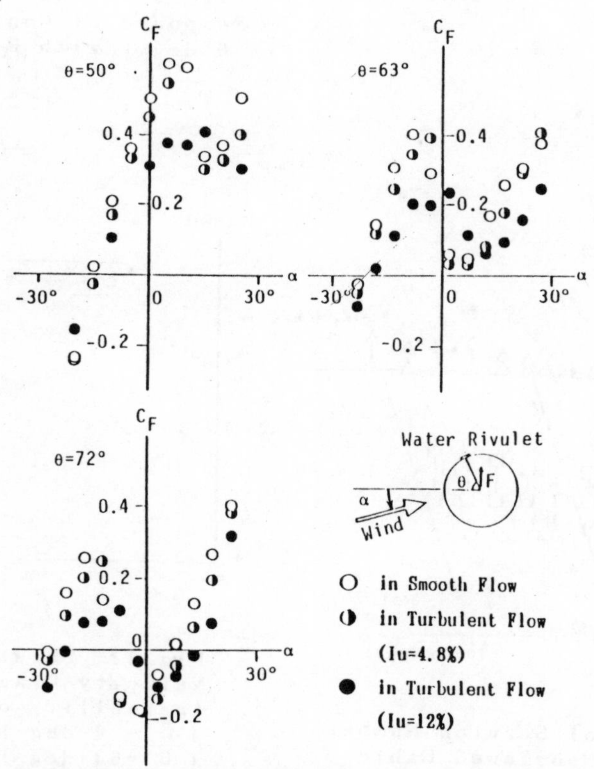

Fig. 18. Lift Force Coefficient Diagrams for Cable with various Rivulet Locations in Various Flow Conditions

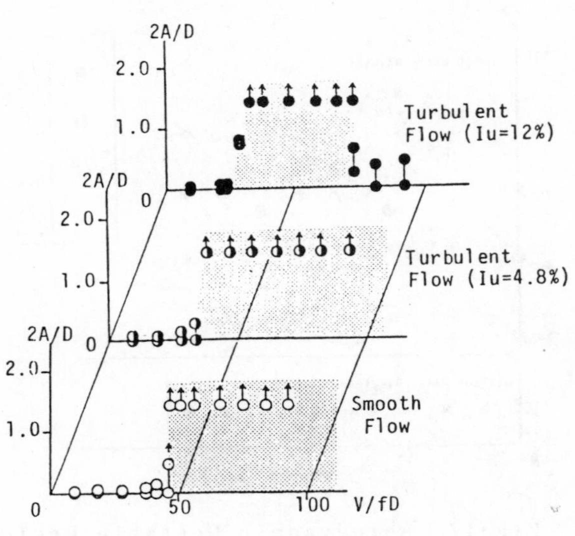

Fig. 19. Effect of Turbulence on Response of Yawed Cable(β =45 deg.) with Rivulet(θ =72 deg.)

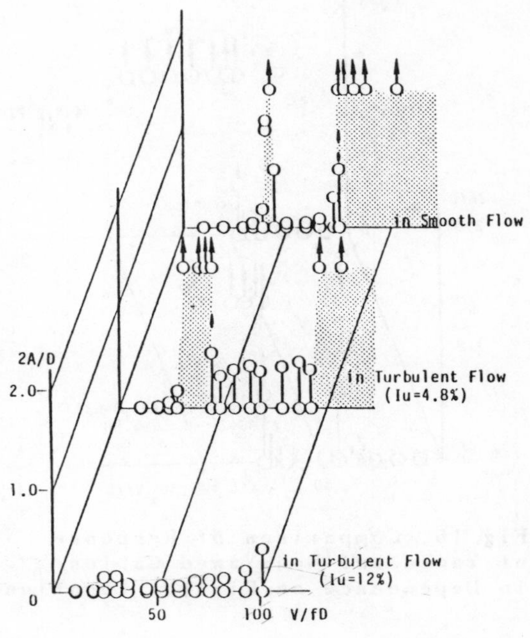

Fig. 20. Effect of Turbulence on Response of Non-Yawed Cable(β = 0 deg.) with Rivulet(θ = 63 deg.)

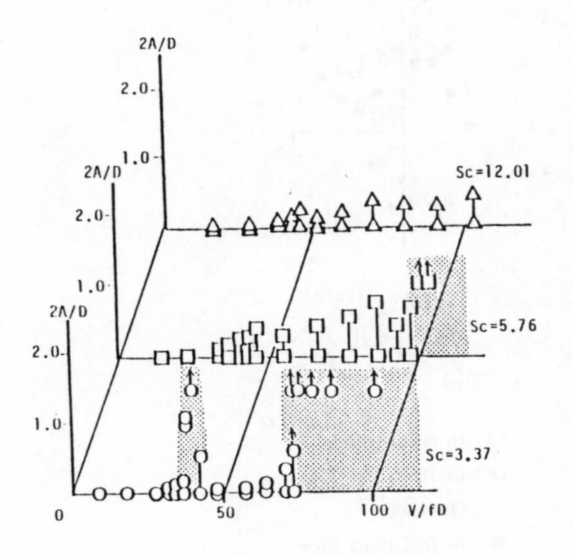

Fig. 21. Effect of Scruton Number on Response of Non-Yawed Cable (β= 0 deg.) with Rivulet(θ =63 deg.)

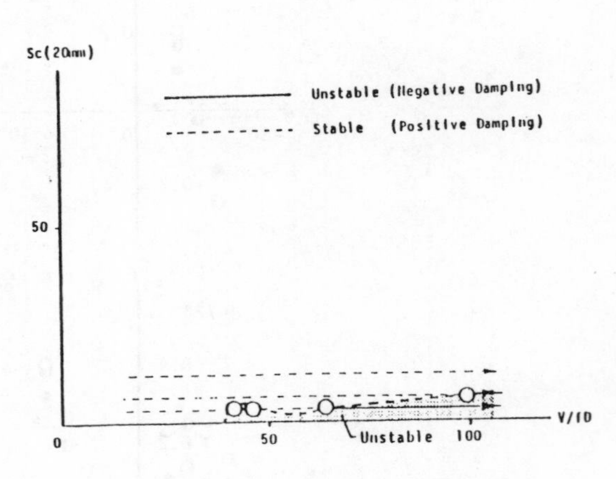

Fig. 22. Scruton Number- Reduced Velocity Diagram of Aerodynamic Instability of Non-Yawed Cable (β = 0 deg.) with Rivulet (θ =63 deg.) in Smooth Flow

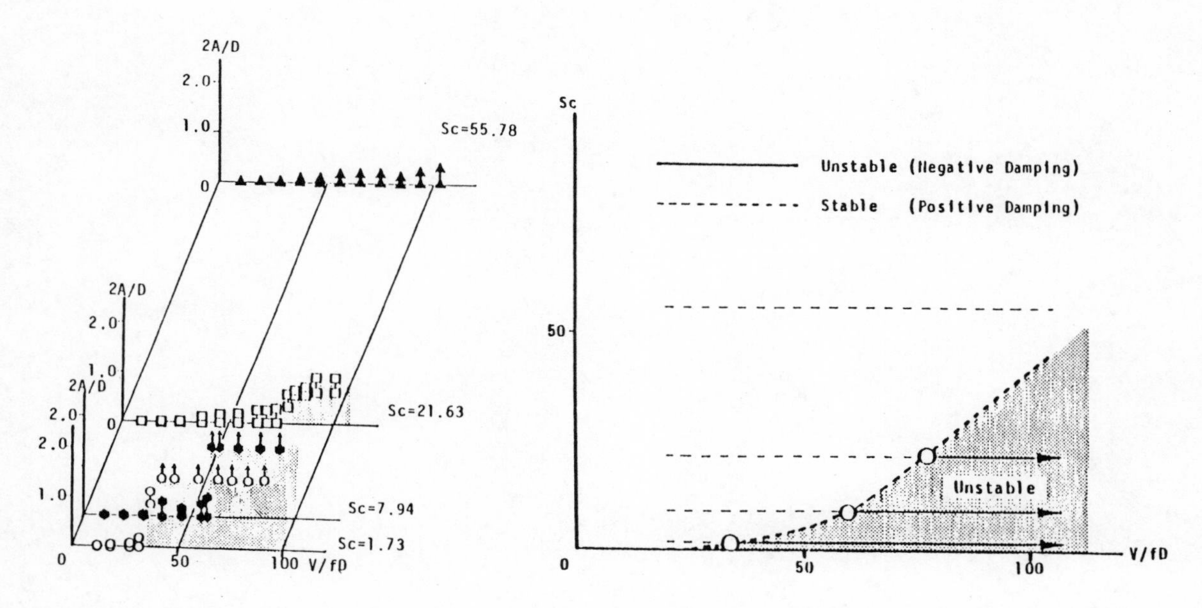

Fig. 23. Effect of Scruton Number on Response of Aerodynamic Instability of Yawed Cable (β =45 deg.) with Rivulet (θ =72 deg.) in Smooth Flow

Fig. 24. Scruton Number-Reduced Velocity Diagram of Aerodynamic Instability of Yawed Cable (β =45 deg.) with Rivulet (θ =72 deg.) in Smooth Flow

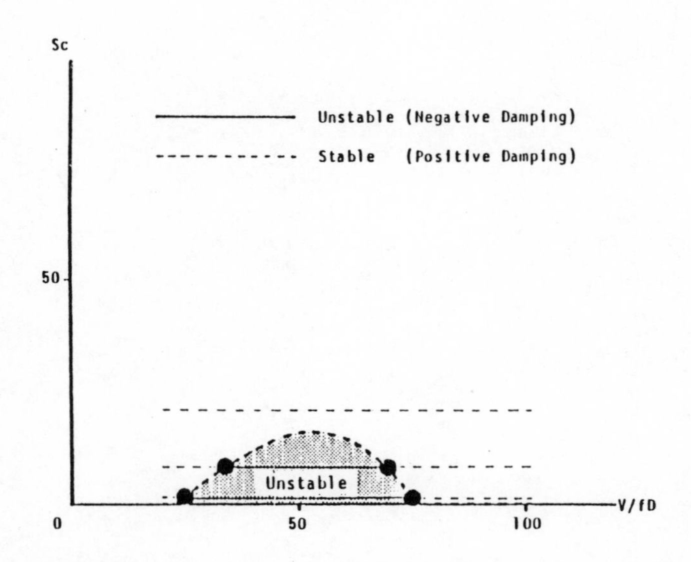

Fig. 25. Scruton Number-Reduced Velocity Diagram of Aerodynamic Instability of Yawed Cable(β =45 deg.) with Rivulet(θ =72 deg.) in Turbulent Flow(Iu=12 %)

THE USE OF DYNAMIC WIND LOAD DATA IN STRUCTURAL DESIGN

Brian E. Lee

Balfour Beatty Professor and Head

School of Civil Engineering
Portsmouth Polytechnic
Portsmouth PO1, 3QL, UNITED KINGDOM

SYNOPSIS

The last decade has seen a significant growth in both the availability of dynamic wind load design data for structural engineers as well as the need for such data in the design of increasingly wind sensitive tall buildings. The dynamic wind load data are available from wind tunnel tests using high frequency base-balance techniques as well as from a range of empirically-based, semi-theoretical calculation methods. Experience has shown that the simplifying assumptions built into both the wind-tunnel tests and the calculation procedures are approached differently by the originators of all the methods, and that inconsistencies result when direct comparisons are made. The task of adjudication between different results is compounded by the fact that very little field data are available from which an evaluation of full scale dynamic wind loads can be made.

1.0 INTRODUCTION

The design of modern high rise buildings advances as rapidly as technological evolution will permit. The dominant regulatory parameters which influence and limit building form, shape and structure are related to the three criteria which govern permissible planning, stress and loading limits.

Planning criteria determine the size and shape of a building envelope which can be accommodated on a given site. Planning criteria are related through plot-site ratios to zoning and hence environmental factors, as well as economic factors. Such criteria have changed as the economic demand for space in city centre areas has increased to the point where speculation about a "mile high building" is probably no longer unreasonable. It is to be expected that the demand for tall buildings all over the world will increase.

Stress criteria are related to the nature of materials used in the building process and to the way in which they are used. Improvements in the long term strength of steels, greater certainty in the strength of concrete, and the advent of structural glazing as a facade material has influenced the mass and stiffness of tall buildings. These factors lead to higher structures having a greater propensity to move under given loading conditions.

Load criteria in conjunction with a knowledge of the strength of construction materials determine the structural form of a building. External load criteria, whether they be for wind effects, ground bearing capacity or earthquake loads, reflect to a large extent the state of knowledge of a subject, and, thus, the degree of uncertainty with which any such specification is associated. As the forms of buildings change so too must research into the specification of loading criteria. In the past mean wind loads have been taken to be an acceptable loading specification for low-rise masonry structures; however, dynamic wind loads are now the necessary criteria when one designs tall slender buildings with low values of mass, stiffness and damping.

Thus, demands on space will maintain and increase the demand for high rise buildings. Buildings will be constructed from materials giving maximum strength with minimum cost whilst maintaining an acceptable aesthetic appearance. These factors will jointly culminate in buildings which are "wind-limited" in the sense that wind loads and wind induced motion will become the most significant factors in the structural design. This condition has already been reached in many areas of the world, notably city centres in areas subject to frequent wind storms from Texas and Florida to Hong Kong and Tokyo.

2.0 THE EVALUATION OF DYNAMIC WIND LOADS

Within the last decade, great progress has been made in the specification of dynamic wind loads on tall buildings. Such progress has been addressed by both wind tunnel experimental methods and by modified calculation procedures, based on semi-empirical methods using the aerodynamic admittance approach to the determination of the load function. However, it may be thought that progress has been made in a somewhat uncoordinated fashion and that genuine doubt should exist about the degree of uncertainty with which any individual dynamic load specification may be used. The grounds for such doubts will be discussed in this paper.

2.1 Determination of Dynamic Wind Loads by Wind Tunnel Tests

The relationship between wind characteristics and building response is a complex one involving an aerodynamic transfer function relating the gust structure to the load function, and a mechanical transfer function relating the load function to the response. Full aeroelastic modelling techniques have been evolved which enable the designer to reproduce all the elements of this process in a wind tunnel in order to assess the response of a proposed structure to its proposed local wind environment. This full modelling process, in which both transfer functions have to be represented, is considered to produce acceptably accurate predictions for both the along wind and the cross wind response cases. Comparisons with available full scale data are scarce. The cost of this full modelling procedure for any particular building has tended to be prohibitively high in all but exceptional cases. The occurrence of unacceptably large dynamic responses in modern lightweight buildings may be the penalty paid for such savings in the design process.

The other consequence of the use of the full aeroelastic modelling method is that it only enables the building response for a particular set of structural characteristics to be determined. Thus the intermediate step in the transfer process, ie. the load function, is not usually characterized. If the load function itself were capable of independent determination as a function of only the wind structure and the aerodynamic shape of the building, then this would greatly assist our understanding of the nature and origins of fluctuating wind loads.

Substantial efforts have been made to meet this objective in recent years with general success. Balances were designed for the measurement of the load function in the form of a quasi-static force spectrum which is independent of the dynamic characteristics of the model. Nonetheless, these new balance designs pose problems concerning sensitivity and frequency response. To obtain a reliable measurement of the fluctuating forces and moments on a stationary model mounted on a force balance, the resonant frequency of the entire system must be significantly higher than any expected forcing frequency of interest. At the same time the transducer should have a high sensitivity. Such systems were originally developed by Whitbread [1] for two force components and Tschanz [2] and Tschanz and Davenport [3] for five force components, and an alternative arrangement using an inverse method, in which the force balance operated at resonance was developed by Evans & Lee [4, 5]

More sophisticated dynamic base-balance instruments which have been developed over the last five years are now in current use at major consulting and research laboratories. Such balances are normally capable of yielding the modal force for up to 5 components of the total dynamic load, and may be purchased commercially,

Cook [6], or more commonly they are made in the workshops of individual laboratories. Results based on the use of such balances have been presented at recent conferences, 7th International Wind Engineering Conference, Aachen, 1987, Bluff Body Aerodynamics Colloquium, Kyoto, 1988, 8th Colloquium on Building Aerodynamics, Aachen, 1989, and the 6th US Nation Wind Engineering Research Conference, Houston, 1989. Corrections for mode shape non-linearities to be used in conjunction with high frequency base balances have been suggested by Vickery, Steckley, Isymov and Vickery [7] and Boggs and Peterka [8], and a spectrum correction for low model-balance frequencies is given by Cochran and Lee [9]. The design of a multi segment high frequency balance which make an experimental allowance for non-linear mode shape effects for both translational and torsional modes is given by Vickery and Reinhold [10], in which doubt is cast on the effectiveness of earlier analytical mode shape corrections to deal effectively with building models in highly turbulent city centre environments.

Despite the proliferation of high frequency balance systems in laboratories throughout the world, the results they yield are generally unverified with respect to full scale results, and for the most part few comparative experiments between different laboratories have been carried out. Furthermore, the relationship between the results of using these balances and results obtained from empirical calculation methods is not widely known.

2.2 Calculation Methods for the Determination of Dynamic Loads

The use of various calculation procedures for the determination of dynamic wind loads is more widespread than that of wind tunnel tests. Following the introduction of a section of dynamic wind loads in the National Building Code of Canada, NBCC [11] various adaptations of this procedure appeared in the literature, many related to codes of practice or to general usage in specific countries. The more common variants of Davenports original method [12] include the Australian code method in AS1170 [13], a method developed at the NBS [14], the Engineering Sciences Data Unit method [15] and the Swiss and Danish code procedures. The Peoples Republic of China appear to have adopted a method based on the NBCC method. The new Eurocode method for use within the European Economic Community after 1992 is likely to be based at least partially on the work of Solari [16].

A detailed study has been carried out by the author in which a comparison is presented of the results yielded by different methods when applied to the same building configurations, Lee and Ng [17].

In the study mentioned above the following codes of practice for the estimation of dynamic wind loads were examined,

(a) ESDU 76001 [15]
(b) National Building Code of Canada [11]
(c) National Bureau of Standards Method [14]
(d) NBS revised procedure [18]
(e) Australian Code AS1170 [13], and

the results obtained from them compared. All code methods were applied to three existing tall buildings for which the full scale values of all the dynamic structural parameters were known from excitation measurements.

The calculation procedures for each method and for each of the three buildings has been presented at every stage of the calculation, so that the emergence of inconsistencies may be identified. From the data presented it is clear that inconsistencies exist between all calculation methods for both the background and resonant components of the dynamic load. Such inconsistencies are reflected in the final culaculation of the buildings response. (Tables 1 and 2).

TABLE 1
Comparison of Dynamic Wind Load Calculation Methods
Peak Acceleration in m/s^2 -- In Wind Response (x-x Mode)

	ESDU 76001 Ref 15	NBCC Ref 11	NBS Ref 14	NBS Revised Ref 18	As 1170 Ref 13
Arts Tower	.047	.039	.017	.047	.056
Sutherland House	.0334	.017	.0144	.029	.017
Southampton Building	.031	.023	.011	.020	.029

TABLE 2
Comparison of Dynamic Wind Load Calculation Methods
Peak Acceleration in m/s^2 -- In Wind Response (y-y Mode)

	ESDU 76001	NBCC	NBS Revised
Arts Tower	.061	.015	.008
Sutherland House	.010	.006	.006
Southampton Building	.055	.025	.012

In summary, this study demonstrated the following points.

(i) No one calculation method used gave consistently higher or lower results than the others for all three buildings, though the ESDU method was more often high than the other methods.

(ii) The difference between the highest and lowest values of in-wind peak acceleration response ranged from 43% for the tallest building to 96% for the shortest, (Table 1). A similar comparison for the in-wind response in the orthogonal mode resulted in even greater differences, (Table 2).

(iii) For the one full scale data set readily available in a form of comparison, the John Hancock Centre, [19], the NBCC and AS1170 methods compared well, while the ESDU method predicted half an-order-of-magnitude low.

Within all of these calculation methods, a number of simplifying assumptions are built in, and, as one might expect, the treatment of these assumptions differ between different methods. The principal areas of concern are as follows:

(1) The spectrum of the horizontal component of wind velocity

A detailed discussion of the empirical forms proposed by Davenport and used in the NBCC method is given by Simiu [14], in which he notes a number of points that the users attention should be drawn to. The first of these is the absence of any allowance for height or terrain type in the expression used. Instead a characteristic wave length of 1200m is used to define the linear dimension of the reduced frequency expression. This choice was justified in an early paper by Davenport [12] on the basis that observation had shown that the spectrum showed little variation with height above ground. This observation might be rationalized by considering separately large and small eddies in the flow. For large eddies it might be expected that no appreciable variation in the fluctuating velocity component would occur over the heights of typical buildings since these are of the order of size of the large eddies themselves. In the case of small eddies, these would be expected to be highly dependent on the nature of the ground roughness; hence the eddies must vary in strength with distance from the ground. However, small eddies would be expected to have only a short duration within the flow field, and, thus, only contribute to the overall response of the building at periods close to its natural frequency. The justification for this somewhat simplified spectrum definition is that it was intended to provide a general description for a range of local wind structures at different locations based on full scale measurements. Hence an exact solution for all wind environments would be impractical for all structural engineering applications.

The compilers of the ESDU method [15] have adopted a somewhat different approach, giving a more detailed description of the spectrum based on von Karman's spectral function. Here the reduced frequency function includes a linear dimension based on the longitudinal scale of the oncoming turbulence dependent itself on the roughness length, Z_o, and an effective height above ground, equal to half the building height. The corresponding lateral and vertical turbulence length scales were used in determining the shape reduction factors for design loads.

While these two empirical forms of the longitudinal velocity spectrum only represent the turbulence characteristic approximately, the NBS calculation method [14] attempts to provide a more comprehensive definition, where the use of the friction velocity in the evaluation of this reduced frequency function marks a major distinction.

(2) The assumption of quasi-steady behaviour

The relationship between the windward face fluctuating pressure and the longitudinal velocity component assumed under the quasi-steady approach is defined as

$$p'/\overline{P} = 2\ u'/\overline{U}$$

The mean drag coefficient C_D is used in the derivation of forces on the building, and it is composed of the algebraic sum of the windward and leeward face pressure forces. As far as fluctuating pressures are concerned the use of a C_D term implies a full correlation of windward and leeward pressures. However, there are reasons to believe that the dynamic characteristics of windward and leeward pressures are different. The former is mainly generated from fluctuations in the incident wind, but the latter is governed by the mechanics of the wake structure behind the body. Thus, the linear relationship indicated in the assumption of quasi-steady behaviour is valid only when applied to low frequency fluctuations on the windward face, and, in practice, full correlation of windward and leeward pressures is unlikely to occur. As a result, an overestimation of the excitation force is to be expected.

However, the assumption of quasi-steady behaviour itself becomes invalid at high reduced frequencies as the linear transformation between the velocity and pressure terms breaks down. As a consequence of this non-linear behaviour the pressures on the surfaces could be underestimated.

Kareem [20] suggested that these two effects counteract one another and may be responsible jointly for the fact that some reasonable agreement has been obtained between estimated response and full scale observations.

(3) The coherence function

The foregoing section argues that the characteristics of the windward and leeward face fluctuating pressures are different; thus, the correlation between opposite forces will be less than unity. The same argument applies to the spatial correlation across each of the faces considered separately, since difference initiating mechanisms will result in different patterns of spatial correlation. Thus, it would be inappropriate to use the same coherence function to determine the aerodynamic admittance of both faces -- a practice common to most calculation methods. The procedure of Simiu [18] which forms the basis of the ANSI A58.1-1982 standard does include coherence functions.

The study referred to above was restricted to the calculation of along-wind dynamic wind loads. Unfortunately, the science of the prediction of cross-wind loads and responses is not as advanced as that for in-wind loads. The concept of aerodynamic admittance cannot be applied to the separated flows on building side faces with the confidence with which it can be applied to the front face using the Quasi-Steady theory and the rear face using simple pressure correlation assumptions. However, models are available for the calculation of cross-wind response based on empirical data, and two such models are outlined in the study by Ferraro, Irwin and Stone [21], who presented an evaluation of their performance in comparison with a wind tunnel model study.

2.3 Determination of Dynamic Wind Loads from Full Scale Measurements

Very little data are available from which an evaluation of full scale dynamic wind loads can be made. Reference has already been made to the observations of the response of the John Hancock Centre in Chicago [19] - though only two data points would appear to be available. The long running Canadian experiment on Commerce Court in Toronto has been extensively reported [22, 23] and, while pressure field and response data are available, it would not appear to be possible to evaluate the modal load values.

Data on the full scale wind loads acting on the Arts Tower at Sheffield University have been reported in 1981 [24] and 1982 [25]. However, these data contain the results of relatively few wind storms; thus, they have a correspondingly low statistical confidence level. More recently, however, a new data set for this building has been published [26] containing data from hundreds of hours of wind storms over an eight-year period.

The Building Research Establisment team responsible for the latest Arts Tower data set are currently running a new enlarged full scale experiment in London on a building called Hume Point. Lessons learned in the Arts Tower experiment have been applied to the Hume Point exercise and it is anticipated that this new study will yield a more reliable data set, [27]. Another full scale experiment run in Japan has been reported recently [28] from which it is hoped that modal load data may be evaluated.

2.4 Discussion

In a paper concerned with the use of field parameters in wind engineering design, Reed [29] examines the influence of meterological, structural and aero-dynamic factors. Parameters in each category were identified along with their distributions and associated levels of uncertainty, from which it was found that the two most influencial parameters in the context of structural reliability under wind loading were the structural damping ratio and the description of the turbulence intensity of the wind. The influence of these parameters is not only strong but in the case of structural damping the Coefficient of Variance (COV) range is large. Furthermore, Reed has found other discrepancies including varying descriptions of the identification and characterization of local pressure fluctuations.

It may be reasonaly concluded that different calculation methods are likely to produce different answers to the same problem, since they use different formulations of the assumptions on which they are all based. Equally, a likely parallel conclusion may be reached with respect to different wind tunnel test set-ups for the same building, where different assumptions implicit in the modelling parameters adopted will correspondingly yield different answers. No sufficiently reliable full scale data set exists by which to calibrate either approach.

3. THE PROBLEM OF CONFIDENCE LEVELS

The problem facing structural design engineers may be stated as follows.

"What degree of reliability can be placed on values of dynamic wind loads or on wind induced dynamic response of buildings given by any one individual wind tunnel test or by any one individual calculation method?"

Stated another way one might ask --

"What degree of confidence should a structural design engineer have in the results given to him or her by a wind engineering consultant?"

The use of the various calculation methods for in-wind response might imply a confidence level whose order of magnitude can be quantified by means of a comparative exercise similar to that previously carried out by the author and referred to in Section 2.2. Such an exercise would appear to indicate that the resonant component of the dynamic response was more susceptible to variation than the background component and that this might be reflected in an uncertainty level in the estimation of the peak acceleration response in a translational mode of the order of plus or minus 50%. Since the only methods of cross wind response estimation are empirically based it must follow that their uncertainty level reflects that of the wind tunnel tests on which they are based.

No such simple basis of comparison exists by which to value the confidence levels of wind tunnel tests. Notwith-standing this, all wind tunnel operations will routinely give a statement regarding the measurement accuracy of the numerical data and of its repeatability - this however is quite different from a state-ment which confirms the level of confidence with which the scaled-up values may be used in structural design calculations.

Since most clients only commission one sequence of tests, they are rarely in a position to exercise any judgement about the degree of precision contained within the data they are given. The wind engineering consultant might be in a better position to make this judgement on the clients behalf if further information on test variability from a parametric study were available.

4. PROPOSAL FOR FURTHER STUDIES

The objective of such a study would be to attempt to establish confidence levels which should be associated with wind tunnel studies of dynamic load measurement. This objective would be carried out in a parametric series of wind tunnel experiments using a dynamic balance for the determination of modal loads from which in-wind and cross-wind response values can be calculated for buildings, of known structural parameters. Comparisons with calculation methods applied over the same ranges of parameters would be completed, and comparisons with full scale data made where it was feasible.

Some specific items within this programme might include the following tasks,

(i) to determine the effect of different terrainr and changes of terrain at different fetch lengths on the dynamic loads for a given aerodynamic shape of building,

(ii) to determine the influence of overall size for a given building shape,

(iii) to determine the effect of incident wind angle for buildings of basic rectangular shape, and

(iv) to investigate the relationship of dynamic wind load and incident wind speed for simple shaped buildings.

The first task is straightforward and would additionally offer a useful basis of comparison with the various calculation procedures.

The second task would permit an examination of the scaling mismatch problem frequently encountered in wind tunnels in which buildings are modelled to a geometric scale different to that of the boundary layer simulation. An additional benefit would be to compare the results of simple scaling-up in the wind tunnel with that derived from the calculation methods.

The third task is important because it will help to clarify the validity of calculation methods as well as to establish the wind directional ranges required for tests on isolated tall buildings. Previous model students [4, 5] on buildings of simple rectangular prism forms have shown that the dynamic wind loads are at a maximum when the wind blows normal to a flat face, and, they always decrease for oblique winds. If this is generally true for a wider range of building planform shapes, then it is important since calculation methods can only be applied to the normal wind direction condition. Thus such calculation methods will produce the maximum anticipated load and not ignore potentially greater loads for oblique winds. Once this point is established, then it is possible that wind tunnel testing can be restricted to the normal wind condition.

The fourth task seeks to determine whether a general relationship between dynamic wind loads and incident mean speed can be more widely adopted. The extensive series of wind tunnel tests on three building shapes carried out by Evans and Lee [4 and 5] all indicate the universality of a cube law relationship as follows:

$$\text{Dynamic Load} \propto (\text{Mean Wind Speed})^3$$

A recent paper describing full scale results for an experiment carried out in Japan [28] gives an identical result. Th data available from the full scale measurements on the John Hancock Building gives the index as 2.75 based on two data points only.

5. CONCLUSIONS

1. A description has been given of the state-of-the-art of high frequency base balance utilization for the direct measurement of up to five components of the dynamic wind loads on wind tunnel models. No indication of the confidence level of typical results is available from comparative studies.

2. A description and comparison of some common methods of estimating in-wind dynamic wind loads for a range of accurately quantified buildings has been given. From this comparison a confidence level is suggested for the resonant component of the in-wind dynamic load. A method for the estimation of cross-wind dynamic wind loads is referenced.

3. Reference is made to the available full scale building monitoring exercises from which full scale wind loading data may be available. Few of the tests carried out so far seem capable of yielding reliable modal load data.

4. Recommendations are made for a research programme whose purpose is to help to quantify the confidence levels of wind tunnel tests for the evaluation of dynamic loads.

6. REFERENCES

1. Whitbread, R.E. The Measurement of Non Steady Wind Forces on Small Scale Building Models. *Proc. 4th Int. Conf. Wind Effects on Buildings and Structures*, Heathrow, London, 1975.

2. Tschanz, T. Measurement of Total Dynamic Loads using Elastic Models for High Natural Frequency. *Proc. Workshop on Int. Wind Tunnel Modelling for Civil Engineering Applications*, N.B.S., Gaithersburg, Md., 1982.

3. Tschanz, T. and Davenport, A.D. The Base Balance Technique for the Determinization of Dynamic Wind Loads. *6th Int. Conf. Wind Engineering*. Paper 8/1, Gold Coast, Australia, March 1983.

4. Evans, R.A. and Lee, B.E. The Measurement of Dynamic Wind Forces on Buildings Using Wind Tunnel Models. Part II: Results obtained using the modelling technique and their application to building design. Department of Building Science Report BS65, Sheffield University, 1982.

5. Evans, R.A. and Lee, B.E. The Determination of Modal Forces acting on Three Buildings, Using Wind Tunnel Methods. *6th Int. Conf. Wind Engineering*, Paper 4/1. Gold Coast, Australia, March 1983.

6. Cook, N.J. A Sensitive to Six Component High Frequency Range balance for Building Aerodynamics, *Journal of Physics*. Part E. Scientific Instruments. Vol. 1b, No. 5, pp.390-93, 1983.

7. Vickery, P., Steckley, A., Isymov, N. and Vickery, B. The Effect of Mode Shape on the Wind Induced Response of Tall Buildings. *Proc. 5th U.S. Nat. Conf. on Wind Engineering*. Lubbock, Texas, pp. 1B-41/48, Nov. 1985.

8. Boggs, D.W. and Peterka, J. Aerodynamic Model Tests on Tall Buildings. *A.S.C.E. Jnl. Engng. Mech.* Vol. 115, No. 3, pp.618-635, March 1989.

9. Cochran, L. and Lee, B.E. A Numerical Correction Procedure for High Frequency Force Balance Response Spectra. *Submitted to 8th Int. Conf. on Wind Engineering*. London, Ontario, July 1991.

10. Vickery, P. and Reinhold, T. Second Generation Force Balance. *Proc. 6th U.S. Nat. Conf. on Wind Engineering*, Houston, Texas. March 1989.

11. National Building Code of Canada, 1980 - Commentary B.

12. Davenport, A.G. The Application of Statistical Concepts to the Wind Loading of Structures. *Proceedings of the Institute of Civil Engineers*, Vol. 19, August 1961.

13. Australian Standard AS1170 - "Minimum Design Loads on Structures", Part 2 - Wind Forces, 1983.

14. Simiu, E. and Lizier, D.W. The Buffeting of Tall Structures by Strong Winds, *National Bureau of Standards Building Science Series 74*, October 1975.

15. ESDU 76001 -- Engineering Science Data Unit, 1976.

16. Solari, G. Analytical Estimation of the Alongwind Response of Structures. *Journal of Wind Engineering and Ind. Aero.*, Vol. 14, pp467-77, 1983.

17. Lee, B.E. and Ng, W.K. Comparisons of Estimated Dynamic Along Wind Responses. *7th International Conf. on Wind Engineering*, Aachen, West Germany, 1987.

18. Simiu, E. Revised procedures for Estimating Along-Wind Response, *Journal of Structural Division, ASCE*, January, 1980.

19. Davenport, A.C., Hogan, M., Vickery, B.J. An Analysis of Records of Wind-Induced Building Movement and Column Strain Taken at the John Hancock Centre (Chicago), Boundary Layer Wind Tunnel Report BLWT-10-70, University of Western Ontario, Canada, 1970.

20. Kareem, A. Synthesis of Fluctuating Along-Wind Loads on Building. *Journal of Engineering Mechanics*, Vol. 112, January 1986.

21. Ferraro, V., Irwin, P., and Stone, G. Wind Induced Building Accelerations. *Proc. 6th U.S. Nat. Conf. on Wind Engineering*, Houston, Texas, Vol. I, pp.B5-26 - B5-38, 1989.

22. Dalgliesh, W. Comparison of Model and Full Scale Tests of the Commerce Court Building in Toronto. *Proc. N.B.S. Int. Workshop on Wind Tunnel Modelling for Civil Engineering Applications*. C.U.P. 1982.

23. Dalgliesh, W.A. and Rainer, J.H. Measurements of Wind Induced Displacements and Accelerations of a 57-Storey Building in Toronto, Canada. *3rd Colloquium on Industrial Aerodynamics*, Aachen, 1978.

24. Evans, R.A. and Lee, B.E. The Determination of Dynamic Wind Loads on Buildings. A comparison of model and full scale measurements. *Proc. 4th U.S. Nat. Conf. Wind Engineering*, Seattle, Washington, July 1981.

25. Lee, B.E. Model and full scale tests on the Arts Tower at Sheffield University. *Proc. Int. Workshop on Wind Tunnel Modelling for Civil Engineering Application*. N.B.S. Gaithersburg, Md. 1982.

26. Littler, J., Ellis, B., Leary, A., and Lee, B.E. The Measurement of the Dynamic Response of Sheffield's Arts Tower to Wind Loading. *Proc. 4th Int. Conf. on Tall Buildings*, Hong Kong and Shanghai, April, 1988.

27. Littler, J.D. and Ellis, B.R. Interim findings from full-scale measurements at Hume Point. *Proc. 6th U.S. Nat. Conf. on Wind Engineering*, Houston, Texas, Vol. II, pp.B8-1 - B8-10, March 8-10 1989.

28. Ohkuma, T., Marukawa, H., Niihori, Y. and Kato, N. A Full Scale Measurement of Wind Pressures and Response Accelerations of a High Rise Building. Part 1. *8th Colloquium on Industrial Aerodynamics*, Aachen, 1989.

29. Reed, D. Use of Field Parameters in Wind Engineering Design. *A.S.C.E. Journal of Structural Engineering*. Vol. 113, No. 7, pp.1570-1585, July 1987.

CASE STUDIES

WIND TUNNEL INVESTIGATION — TYPICAL CASES

K. Seetharamulu
Professor
Civil. Engg. Deptt.
I.I.T.,
New Delhi 110 016
INDIA

B.L.P. Swami
Asstt. Professor
J.L.N. Tech. Univ
Hyderabad 500029
INDIA

K. K. Chaudhry
Professor
Applied Mech. Deptt
I.I.T.,
New Delhi 110 016
INDIA

SYNOPSIS

The innovative structures do not conform to standard shapes and the designs of such structures are not covered in the codal provisions for loading specifications. For a reliable estimation of the design wind loads these structures are modelled for tests in the boundary layer wind tunnel. At the IIT Delhi, three such typical structures, were tested in the boundary layer wind tunnel for obtaining the pressure coefficients. The details of the test results are reported herein

INTRODUCTION

There have been considerable refinements in analytical techniques for evaluation of internal stress resultants and displacements. Any such refinement is incomplete without corresponding improvement in the assessment of loading. A conservative estimate of loads need not necessarily result in a safe design. The revised Indian Standards, (IS:875 (Part 3) - 1987) specify loads for standard cases of configurations of low rise as well as high rise buildings. The innovative structures in the form of large span low rise membranes and tall towers/buildings, not conforming to standard shapes require to be modelled for testing in the boundary layer wind tunnel for reliable estimation of pressure coefficients used for computing design loads. The typical structures for wind tunnel investigations at the Indian Institute of Technology, Delhi, are the following:

(1) Shell Roof model of the prototype membrane structure covering an elliptic plan area of 120m (major axis) and 102m (minor axis).

(2) Lattice tower (63m) for interference effects by two large silos of 16m diameter and 40m height in its vicinity, and

(3) Tall twin towers of a building complex curved in plan. Two shapes were considered for arriving at an optimum configuration in respect of wind effects.

The details of the experimental set up for the test models of the low rise Shell Roof, Lattice Tower and the two cases of medium rise buildings are reported herein. Because these belong to the category of rigid structures, aero-elastic investigations are not necessary.

DETAILS OF EXPERIMENTAL INVESTIGATIONS

Shell Roof Model

In the case of thin shells the unsymmetrical wind loading aggravates the transverse bending effect for which thin shells are most ill suited. It is, therefore, important to predict the wind loads with accuracy.

The proposed roof was meant to cover a large span swimming pool [1,2]. The form is obtained by the intersection of a hollow cone with an inclined plane. The truncated cone with an elliptic base is suitably tilted. For the case investigated, major and minor axes of the ellipse were 120m and 102m respectively. The vertex angle of the triangle for obtaining a surface of revolution is 105 degrees. The generated shell surface is symmetrical about the longitudinal major axis only. The shell has a circular opening of 10m diameter at the vertex. The geometrical details are shown in Figs. 1 and 2. This innovative form has the advantage of providing natural light over the stage and activity zone at one end of the covered area and the large covered areas is meant to be used for spectator gallery. In the case of reinforced concrete shells, it is possible to stiffen the weak zones by V-grooves while casting. The presence of groove meant for stiffening cum strengthening is investigated herein for its effect on wind pressures.

The motivation for the present investigation was an attempt to fill the gaps in the design information for such shells. It may be shown that extending the provisions of Indian Standards for such structures results in the values of pressures which are grossly different from the true values. Hence, wind tunnel model study is a requirement for estimating wind loads on such shell surfaces.

It may be observed from the shell configuration that the stiffness along the longest slanting portion OB, is insufficient because of the small curvature in the hoop direction [2] and relatively long length. Therefore, a second model for the same covered area and overall configuration was tested but this time with a V-groove in the form of a folded stiffner provided along OB. The stiffner is expected to modify the wind pressures while stiffening the shell in the weak zone. The pressure coefficients were found for changed surface conditions based on the test in the boundary layer wind tunnel.

Details of shell roof model

The shell model was cast to a scale of 1:200 with a thickness of 1cm. For casting the model, glass fibres with general purpose epoxy resin were used. The finished model demensions were 600mm x 510mm with a height of 150mm.

As many as 41 Nos. of 3mm size brass pressure taps were inserted on the surface of the model to obtain clear pressure distribution. After mounting the model on the turn-table in the test section of the wind tunnel, the pressure taps were connected from inside to a multitube manometer with the help of PVC tubes. A low speed blower type wind tunnel with a test section of 750mm x 450mm was used. To create the conditions of turbulent boundary layer in the wind tunnel while testing the model, a triangular wooden strip of 25mm x 25mm with the toe pointing towards the flow direction was fixed to the floor of the wind tunnel at a distance of 3m from the edge of the model. This produced 1/7th power law velocity profile.

Discussions, shell roof model:

1. Practically for all orientations of the model, suction (positive pressures) are predominant in both the cases, with and without stiffner. In certain orientations, there are some negative pressures occuring on the windward side of the model without stiffner, whereas with stiffner almost the entire surface is under suction. Therefore, when both dead and live loads are considered the net loading on the shell surface is reduced. This is a useful design information. Conservative designs based on the code would have involved high cost of structure.

2. For 0 degree orientation, with the introduction of the stiffner, there is considerable reduction (20% to 30%) in the value of surface pressures for the shell model, Figs. 3 and 4. However for 180 degree orientation, there is no significant change in the pressures.

3. With the introductuion of a single stiffner, Figs. 5 and 6 in the form of V-groove over the longest slanting portion of the shell, there is significant change in the pressure gradients. High pressures are observed at the crown.

4. It is inferred that with the introduction of more stiffners in the form of grooves, randomly located at large number of locations on the surface, the overall surface pressure distribution is expected to be considerably modified.

Interference Effects on Lattice Tower due to Two Large Silos

The case study pertained to the estimation of wind pressure acting on the 11.5m x 10m in plan and 63m high transfer tower of the belt conveyer system in the vicinity of two large silos each of 16m diameter and 40m high [3]. The entire complex is located in the vicinity of sea coast. Except for some hills situated at fairly remote distance, the surrounding environment is closest to the "open country environment". The surface pressures, with and without the silos on the scale model at various orientations have been obtained to quantify the interefe-rence effects of silos on the transfer tower. Secondly, the dependence of surface pressures due to the variation of the spacing between the towers was also studied.

Details of lattice tower model

A wooden model of the main transfer tower, conveyer tower and two large silos was made to a scale of 1:200. Based on the solidity ratio of the tower 25 percent permeability was provided in the main tower. The plan of the model arrangement is shown in Fig.7. The main transfer tower was provided with 60 Nos. brass pressure taps at selected locations.

To simulate open country environment in the wind tunnel, small wooden cubical blocks of 1 cm side were arranged at equal spacing of 4 cm over the floor of the wind tunnel. The power law coefficient of the velocity profile obtained with this arrangement was 0.16.

Discussion, lattice tower model:

1. From the results of pressures on the tower, Table 1, it was noticed that at 0 degree orientation, the windward face was under negative pressures (thrust maximum 1.4) and leeward face was under positive pressures (maximum suction 1.25). When the silos were present both windward and leeward faces are under suction, Table 2.

Table-1: Pressure coefficients on the tower without silos. (For 0 deg. orientation)

Ht. (m)	Elevation	Face A	Face B	Face C	Face D
54	+58.00	-1.06	+1.40	+1.25	+1.39
45	+49.00	-1.04	+1.33	+1.21	+1.29
36	+40.00	-1.04	+1.29	+1.19	+1.19
27	+31.00	-1.02	+1.19	+1.07	+1.10
18	+22.00	-1.01	+1.13	+1.00	+1.00
9	+13.00	-1.01	+1.13	+0.85	+1.00

Table-2: Pressure coefficient on the tower with silos (For 0 deg. orientation)

Ht. (m)	Elevation	Face A	Face B	Face C	Face D
54	+58.00	+1.20	+1.50	+1.45	+1.45
45	+49.00	+0.96	+1.48	+1.38	+1.40
36	+40.00	+0.73	+1.42	+1.30	+1.40
27	+31.00	+0.48	+1.40	+1.30	+1.40
18	+22.00	+0.38	+1.40	+1.30	+1.40
9	+13.00	+0.20	+1.35	+1.27	+1.40

Thus, the resultant design pressures for the tower with silos is reduced. This is be attributed to the shielding effect of the silos on the towers under consideration.

Further, at 180 degrees orientation, the silos being in the downstream location their windward effect is not perceptible. Thus, there is no change in the nature of pressure, Table 3.

Table-3: Pressure coefficients on the tower with silos. (For 0 deg. orientation)

Ht. (m)	Elevation	Face A	Face B	Face C	Face D
54	58.00	1.10	0.85	1.10	0.80
45	49.00	0.96	0.73	0.98	0.71
36	40.00	0.55	0.66	0.98	0.60
27	31.00	0.35	0.66	0.98	0.48
18	22.00	0.20	0.38	0.98	0.48
9	13.00	0.20	0.17	0.96	0.29

2. When the spacing between silos was decreased from 17 cm to 12.5 cm, an increase of 20% to 30% in the magnitude of surface pressures was noticed at 0 degree orientation, Table 4. From the design considerations, this indeed is a significant increase in pressures.

Table-4: Variation of pressure coefficients with spacing of silos (for 0 degree orientation)

Height in mts.	Elevation	Face A			Face C		
		17cm	15cm	12.5cm	17cm	15cm	12.5cm
54	+58.00	+1.20	+1.43	+1.62	+1.45	+1.78	+2.35
45	+49.00	+0.96	+1.10	+1.15	+1.38	+1.75	+2.22
36	+40.00	+0.73	+0.88	+1.02	+1.30	+1.62	+1.98
27	+31.00	+0.48	+0.62	+0.76	+1.30	+1.55	+1.82
18	+22.00	+0.38	+0.48	+0.58	+1.30	+1.48	+1.68
9	+13.00	+0.20	+0.26	+0.32	+1.27	+1.39	+1.61

Twin Tower Building Complex

The case study pertains to the study of wind pressure distribution on two alternative configurations of a proposed twin tower building complex, curved in plan. The first alternative configuration [4] was with symmetrical twin tower blocks 83m high each. It was in the form of segments of circle, with convex sides facing each other forming a narrow venturi with a throat width of 18m, Figs. 8 and 9. The two tower blocks had a common base up to plaza level (+6.20m). These were connected by bridges at 32.7m and 66.2m levels in the throat of the venturi shaped region between the blocks. The inner and outer faces of each block were generated by circular area of radii 22.4m and 40.85m giving a total width of 18.45m. Each block subtended an angle of 158.3 degrees at the centre. The connecting bridges are glazed externally.

Wind pressures may sometimes become critical in the case of tall buildings due to shape effects. For such a configuration the codal provisions are not directly applicable. It was, therefore, tested in the wind tunnel for the nature and magnitude of pressure distribution on the various surfaces of the twin tower building complex. The effect of the curvature and mutual interference effect of the blocks was studied. The study was particularly directed to the pressures in the venturi shaped regions between two tower blocks.

In the modified configuration, two towers of unequal heights shown in perspective view, Fig.10, were tested in the wind tunnel. Again both the blocks were curved in plan but in this case the concave sides faced each other. The heights of the towers were 83.4m and 67m, Figs. 11 and 12. Each block had an external radius of 48m and internal radius of 28m, giving a total width of 20m. Each block substended an angle of 140.3 degrees at the centre.

Details of twin towers Models

Both the models of the alternativeswere prepared to a scale of 1:30 in wood. The rigid models were made hollow from inside to facilitate pressure tappings from within. The models had pressure tapping points equal to 101 and 112 for the two cases for obtaining pressure distributions at various surfaces of the tower blocks. The urban environment was simulated in the wind tunnel by means of cubical wooden blocks of 2 cm side placed at a spacing of 6 cm on the floor of the wind tunnel. The power law coefficient of the velocity profile obtained with this arrangement was 0.36.

Discussions, twin tower building complex:

1. Table 5, summarises the results of the two cases of configurations. It may be noted that the magnitude of pressures are lesser in the second alternative configuration than in the case of the first alternative.

2. The connecting bridges between the tower blocks situated in the venturi region experience high pressures. The critical pressure coefficients were +2.8 and -0.5 respectively at 45 degrees orientation on the opposite sides, giving the net effective pressure of 3.3. Therefore, the venturi type region has been avoided in the revised configuration.

3. The maximum velocity of flow in the narrow portion was

Table-5: Comparison of Pressure Coefficients in the Original and Modified Configurations of Twin Tower Building Complex.

Sl. No.	Orientation to the flow (Deg.)	For Twin Towers(Original Case)				Bridges		For Twin Towers(Original Case)			
		Pressure on 'A'	Pressure on 'C'	Pressure on 'E'	Pressure on 'G'	I	II	Pressure on 'A'	Pressure on 'C'	Pressure on 'E'	Pressure on 'G'
1.	0	+1.1	+1.1	+1.4	+1.1	+1.8	+1.0	+1.14	+0.46	+0.98	+0.46
2.	45	-0.6	+1.6	+1.3	-0.6	+2.8	-1.0	+1.04	-0.44	-0.98	-0.39
3.	90	-0.7	+1.6	+1.2	+1.6	-	-	+0.65	+0.31	-1.07	+0.47
4.	135	-0.6	+0.8	+1.3	-0.2	-0.5	+2.8	+1.14	-0.21	-1.04	+0.38
5.	180	+1.4	+1.1	+1.4	+1.1	-1.0	+1.8	+1.14	+0.52	+1.04	+0.46
6.	225	+1.3	-0.6	-0.6	+1.6	-0.5	+2.8	+1.77	+0.62	+0.92	+0.00
7.	270	+1.2	+1.6	-0.7	+1.6	-	-	-1.10	+0.75	+0.64	+0.04
8.	315	+1.3	-0.2	-0.6	+0.8	+2.8	+0.5	+1.42	+0.69	-1.00	-0.06

found to be 1.6 times the free stream velocity in the wind tunnel at 0 degree orientation.

4. Table 6 summarises the comparison of the net effective pressures for the original and revised configuration. It may be seen that in the case of second configuration, there is considerable advantage in respect of reduction in net effective pressures. The maximum net effective pressures in the modified configuration are -1.85 and -1.54 for the taller and shorter blocks respectively as compared to -2.3 for the first configuration. Thus, the reduction of maximum net effective pressures is 20% and 33% for the taller and shorter blocks respectively as compared to the first configuration.

CONCLUDING REMARKS

The wind tunnel tests in the present study gave quantitative data of wind pressures on the Shell Roof Model, Lattice Tower and two configurations of the Twin Tower Building Complex. Such

Table-6: Comparison of net effective pressure coefficients on the original and modified configurations of twin tower building complex.

Orientation to the flow (degrees)	Effective pressures for twin towers (Original Case)		Effective presures for twin towers (Modified case)	
	A-C	E-G	A-C	E-G
0	+0.3	+0.3	+0.68	+0.52
45	-2.2	+1.9	+1.48	-1.37
90	-2.3	-0.4	+0.96	-1.54
135	-1.4	+1.5	+1.35	-1.39
180	+0.3	+0.3	+0.62	+0.58
225	-1.9	-2.2	+1.15	+0.92
270	-0.4	-2.3	-1.85	+0.60
315	+1.5	-1.4	+0.82	+1.06

shapes and configurations are not included in the codal provisions, IS-875. Therefore, in the absence of wind tunnel testing the available codal provisions would have been grossly inadequate.

In the three cases discussed herein, the study being comparative in nature, that is, (a) with and without groove in the slanting portion of the Shell Roof. (b) effect of spacing and location of silos with respect to Lattice Tower. (c) with and without venturi effects caused by the orientation of surface curvature, the blockage effect in the wind tunnel may not be a serious limitation. However, with the modern wind tunnels with larger cross-sectional area and test section lengths, the errors due to blockage effects are practically eliminated.

REFERNENCES

1. Swami B.L.P., K.K. Chaudhry and K. Seetharamulu, Wind Effects on a Shell with Stiffners - Asia Pacific Symposium on Wind Engineering, Dec. 5-7, 1985, Univ. of Roorkee, India, pp. 181-187.

2. Swami B.L.P., K. Seetharamulu and K.K. Chaudhry, Wind Pressures on a Shell Roof Model - International Symposium on Innovative Applications of Shell and Special Forms, Nov. 21-25, 1988, Bangalore, India, pp. 845-854.

3. Swami B.L.P., K. Seetharamulu and K.K. Chaudhry, Interference Effects on Wind Pressures - A Case Study, Asia Pacific Symposium on Wind Engineering, Dec., 5-7, 1985, Univ. of Roorkee, India, pp. 79-85.

4. Swami B.L.P., K. Seetharamulu and K.K. Chaudhry, Wind Tunnel Study of Twin Towers Curved in Plan, Journal of Structural Engg. Vol., 13, No.1, April 1986, pp. 43-49.

5. Swami B.L.P., K. Seetharamulu and K.K. Chaudhry, Wind Pressures on Twin Tower Building Complex, Journal of Structural Engg., Vol., 15, No.3, Oct. 1988, pp. 81-86.

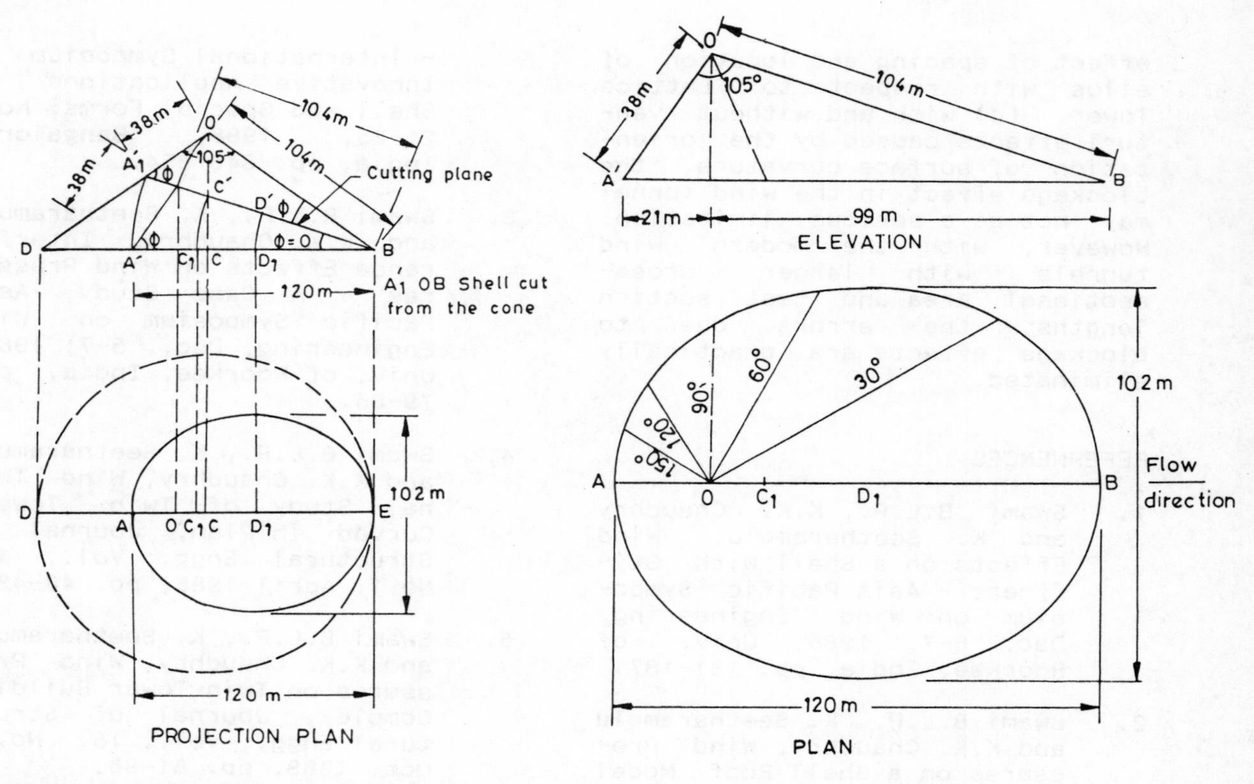

FIG·1 GENERATION OF SHELL SURFACE FROM A CONE

FIG.2 CONFIGURATION OF THE SHELL ROOF

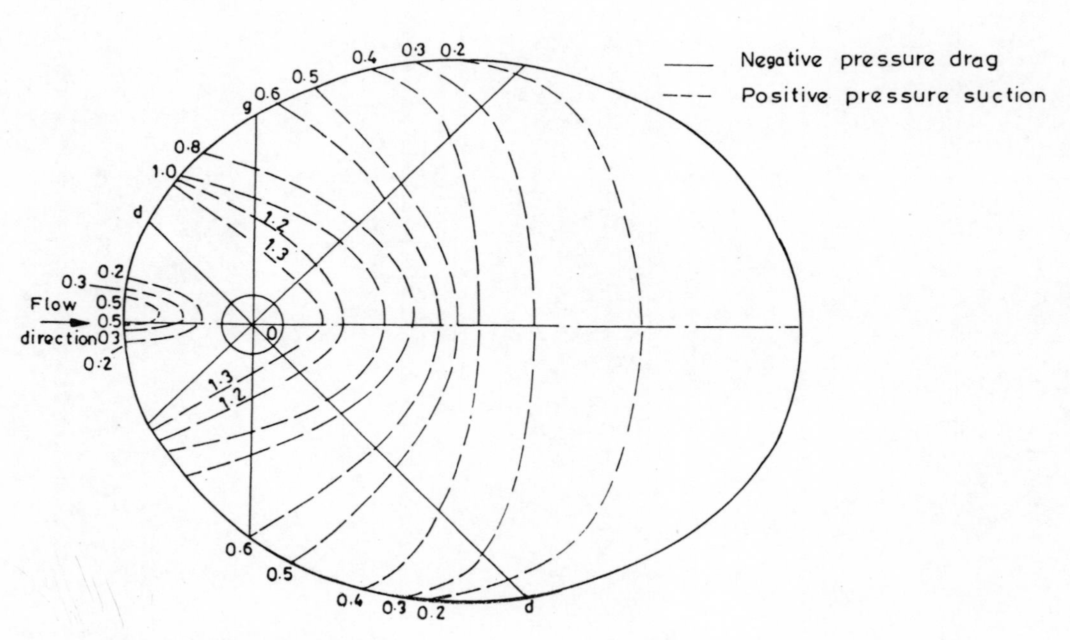

FIG.3 PRESSURE CONTOURS MODEL AT 0 DEGREE ORIENTATION (WITHOUT GROOVE)

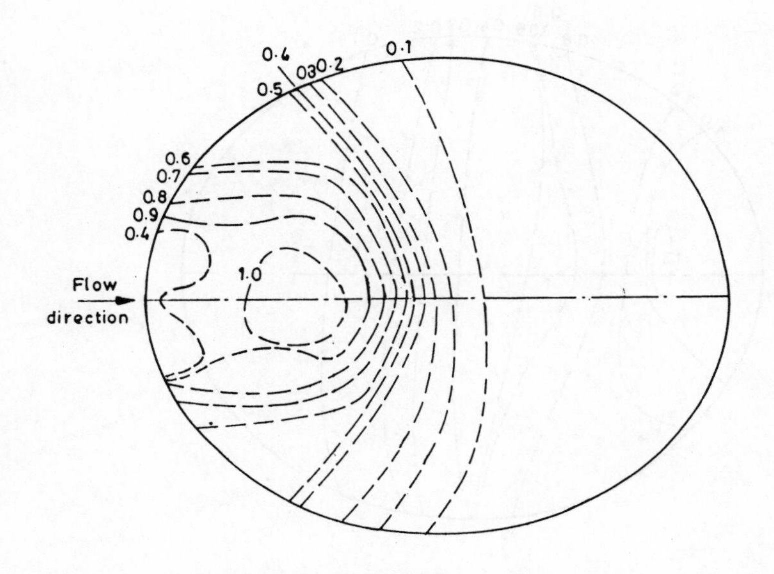

FIG.4 PRESSURE CONTOURS FOR MODEL AT 0 DEGREE ORIENTATION (WITH GROOVE)

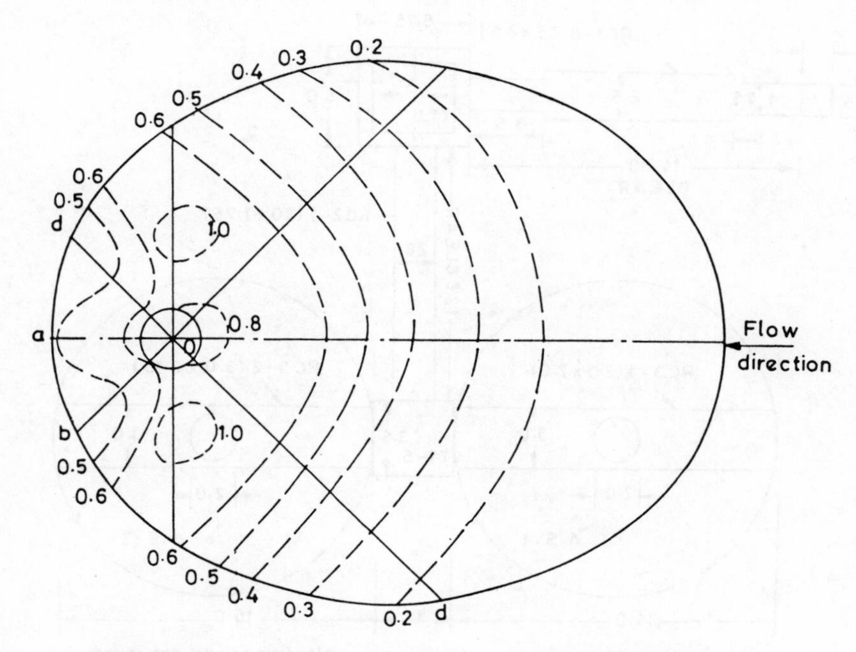

FIG.5 PRESSURE CONTOURS FOR MODEL AT 180 DEGREE ORIENTATION (WITHOUT GROOVE)

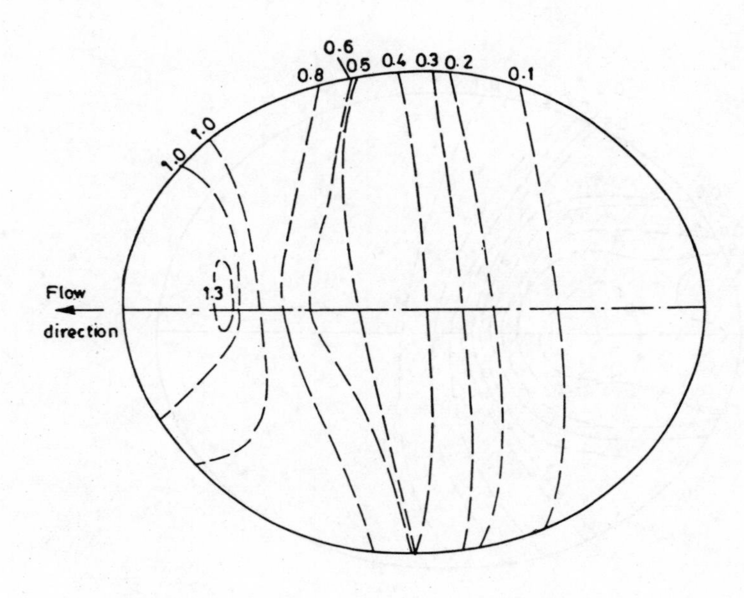

FIG.6 PRESSURE CONTOURS FOR MODEL AT 180 DEGREE ORIENTATION (WITH GROOVE)

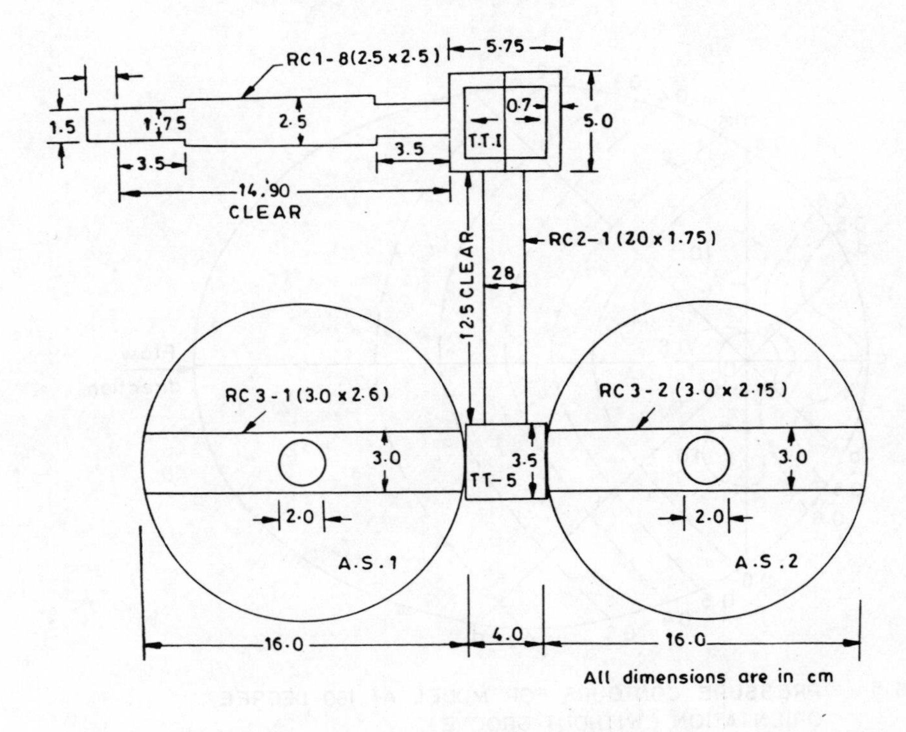

FIG.7 PLAN OF THE MODEL ARRANGEMENT OF THE LATTICE TOWER

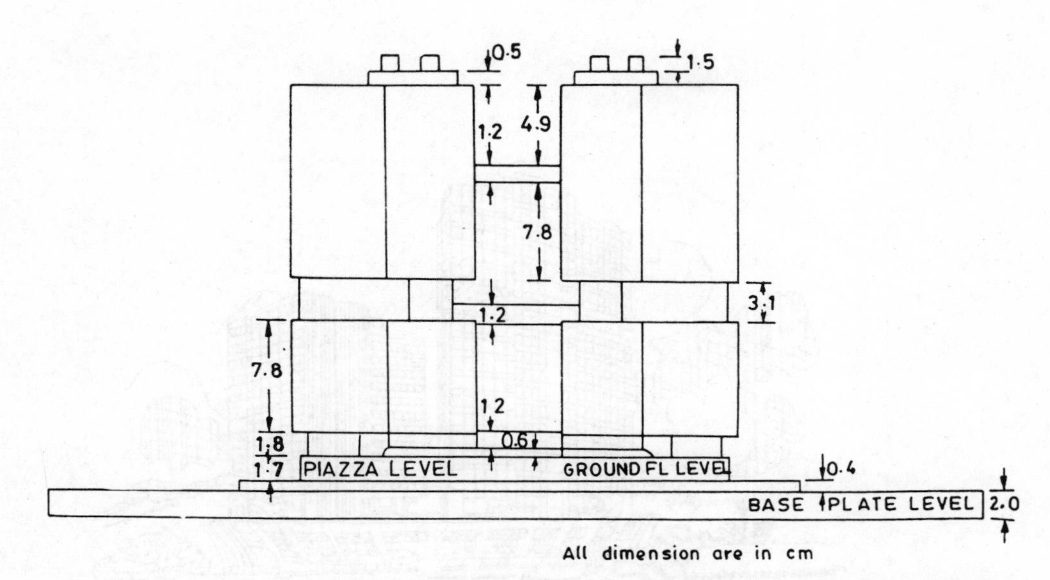

FIG.8 MODEL SECTION AT CENTRE OF TWIN TOWER BUILDING COMPLEX (FIRST ALTERNATIVE)

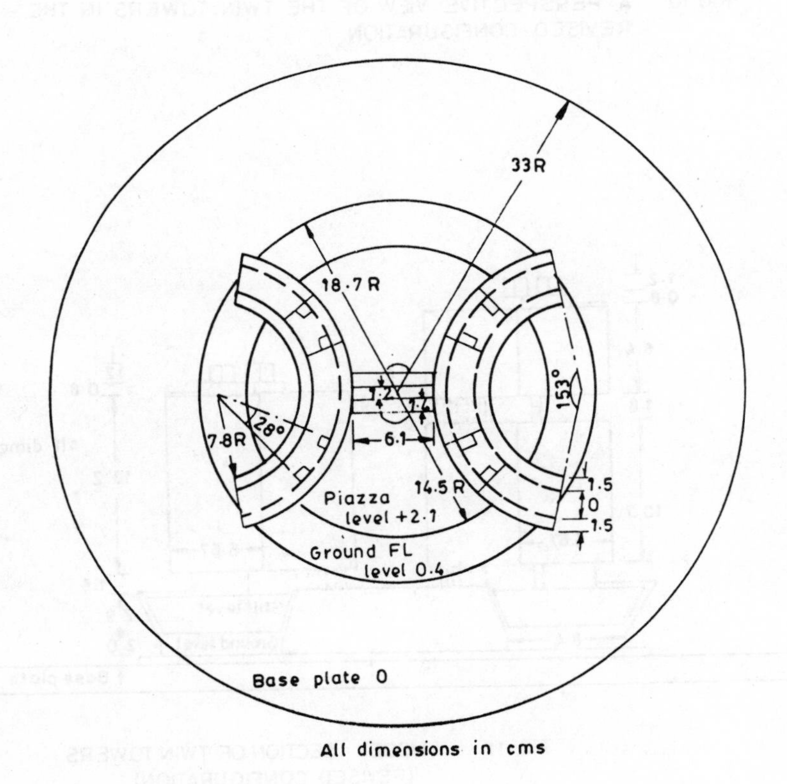

FIG.9 MODEL PLAN OF TWIN TOWER BUILDING COMPLEX (FIRST ALTERNATIVE)

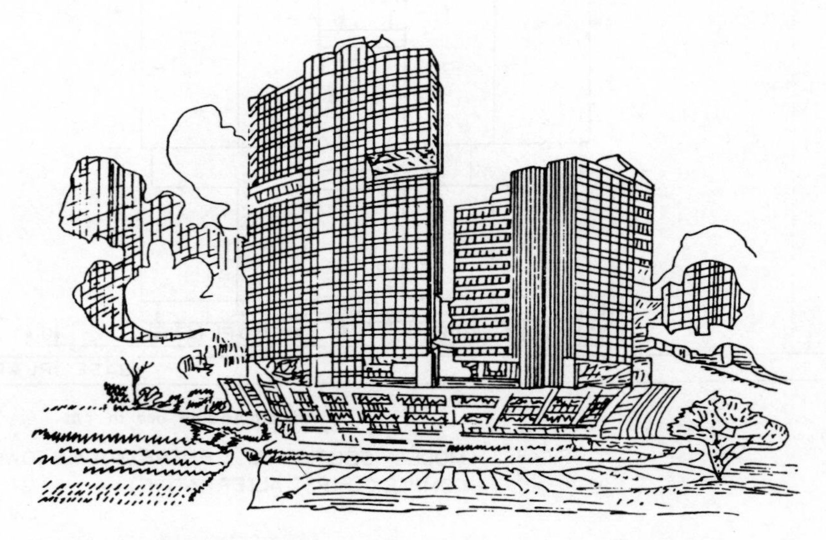

FIG·10 A PERSPECTIVE VIEW OF THE TWIN TOWERS IN THE REVISED CONFIGURATION

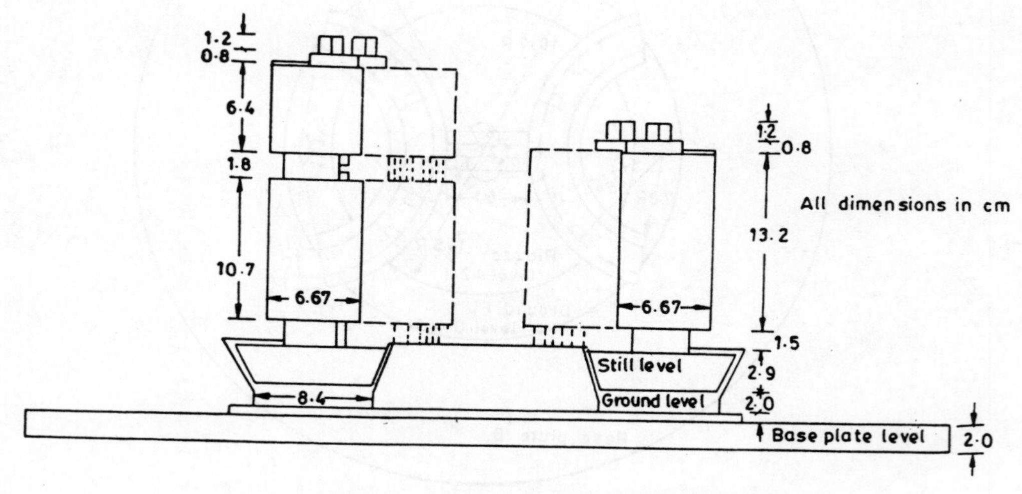

FIG.11 MODEL SECTION OF TWIN TOWERS (REVISED CONFIGURATION)

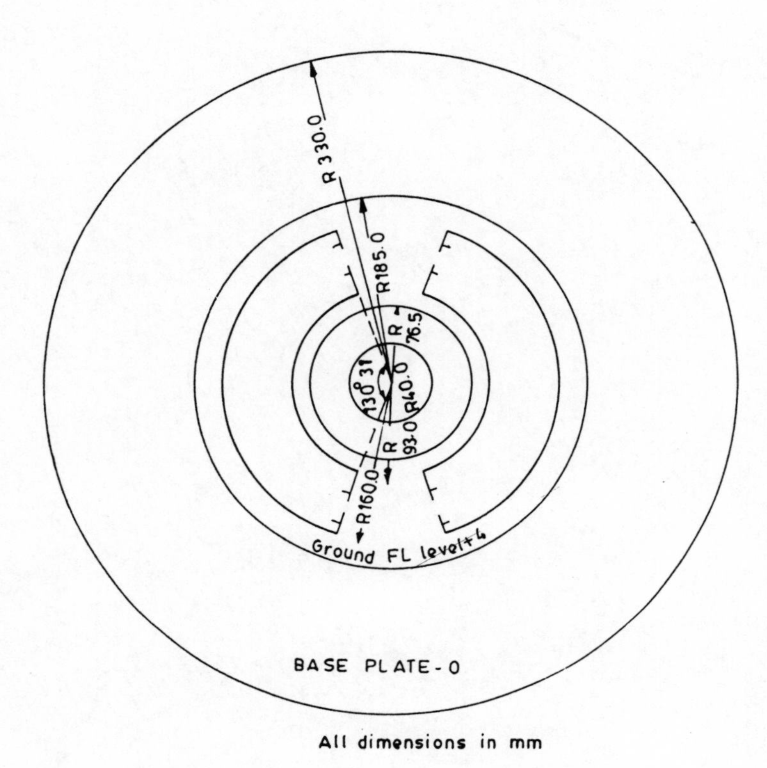

FIG.12 MODEL PLAN OF TWIN TOWERS
(REVISED CONFIGURATION)

WIND LOADS ON CYLINDRICAL ROOFS

A.K. Ahuja *Prem Krishna* *P.K. Pande*

Civil Engineering Department
University of Roorkee
Roorkee, INDIA

SYNOPSIS

Wind loads on buildings are affected by structural characteristics as well as flow characteristics of wind. It is well recognised that the shape and size of the structure affect pressure coefficients, and so do the flow characteristics, though the knowledge of the quantitative measure of the latter is comparatively limited. This paper deals with an experimental study to determine the effect of permeability on the wind pressure distribution on cylindrical roofs. Variations of maximum pressure coefficients and total force on the roof due to variation in the geometrical proportions of the roof are also reported.

INTRODUCTION

Evaluation of design wind loads on the roof of a building requires information about pressure coefficients and design wind speed, which are generally obtained from the relevant code of practice. Whereas design wind speed is affected primarily by parameters such as the geographical location, type of terrain, local topography, size and design life of the building, wind pressure coefficients are mainly affected by the shape of the building, its length to width ratio, direction of wind and the scale and intensity of turbulence. Variations of wind pressure on flat and pitched roofs with the change in wind velocity, wind angle, building shape and dimensions have been studied in detail and such information is available in National Codes of a number of countries and in books on Wind Effects on Buildings and Structures [3,5,9,10,12]. However, similar information for curved roofs is rather limited [7,11,12].

The subject of wind loading on curved roof surfaces has drawn attention only since the last few decades. The state-of-the-art is well depicted by stating that a suspended roof (which is in a category of roofs, inherently flexible and thus affected by the pressure distribution over the surface as well as by the gustiness in wind) was designed in the 50s for a uniform static wind load over its surface [4]. Much work has since been put in to create the necessary information on wind loading and its effects on curved roofs to enable more rational (and perhaps safer) designs. The information is nevertheless not yet comprehensive by any means. The current state-of-the-art vis-a-vis the types of curved roofs used, the problem of wind loading on them, the relevant information available, and the future needs has been reported in ref. 8.

The various geometrical forms of curved roofs can be classified as (i) Domes, (ii) Hyperbolic Paraboloids and (iii) Cylinders —— concave or convex. While domes have a double positive curvature, hyperbolic paraboloids have a double negative Gaussian curvature, and cylindrical surfaces a single curvature. Roofs in all these forms (Fig. 1) can either be rigid shells in concrete, timber, reinforced glass or plastic, ferrocement, or, these can fall under the more flexible type, supported on cable systems or air supported membranes.

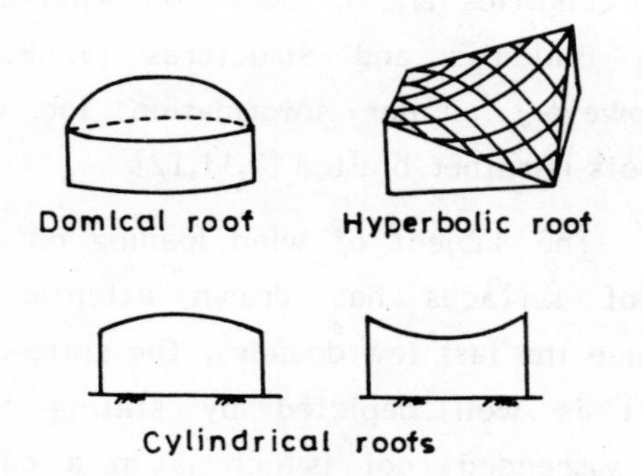

Fig. 1. Buildings with Different Forms of Curved Roofs

Wind pressures on these various roof forms can be obtained by placing appropriately scaled rigid models in a wind tunnel and obtaining the pressure coefficients therefrom. The oft needed result is the mean pressure distribution over the surface. This is adequate for the design of rigid roofs. Whereas this is also very valuable information for the more flexible roof forms, it is obviously not possible to determine the dynamic effects from such a test. If, however, simultaneous pressure fluctuation records are taken from a rigid model, dynamic effects can be obtained therefrom using analytical techniques.

National Codes of Australia, Canada, Czechoslovakia, Germany, New Zealand, Switzerland, USSR [12] and India [6] include the values of wind pressure coefficients on cylindrical roofs with convex surfaces for wind normal or parallel to eaves. However, no codal information is available regarding wind pressure coefficients on cylindrical roofs with concave surfaces.

A review of recent wind tunnel tests indicates that there is only one case reported by Ishizaki and Yoshikawa [7] regarding measurement of wind pressure coefficients on rigid models of cylindrical roof with concave surface. The report gives the values of mean pressure coefficient and gust pressure spectra on a cylindrical model roof of different sag to span ratios and height to span ratios, laid-across the total width of a wind tunnel.

A detailed study on this subject has been made by the first author to obtain comprehensive information about the effects of various structural and flow parameters on wind pressure distribution [1]. Results dealing with wind pressure distribution on cylindrical roofs with concave surfaces, as influenced by the roof geometry and the angle of incidence of wind and its velocity, have been reported earlier [2]. It was observed that pressure coefficients (i) vary with sag to span ratio of the roof for all the angles except when wind is parallel to eaves, (ii) change significantly with change in angle of incidence of wind and (iii) do not vary significantly with height of the buildings and the velocity of wind.

As mentioned earlier, curved roofs are quite often supported on cable trusses or networks. The knowledge of both overall load as well as the disposition of load on the roof is needed to analyse such roofs for obtaining the complete design information. This paper presents the variation of (i) the maximum value of the pressure coefficient on the roof, (ii) the overall roof load, with a change in its geometry, and (iii) the wind pressure coefficients when the walls in the building are permeable.

EXPERIMENTAL STUDY

Models and Measurements

For the work presented herein, rigid models of timber were prepared to represent a building with a cylindrical roof, on a scale of 1:240 (Fig. 2). The prototype chosen for study is a rectangular building of size 120m x 60m and 18m height, with a roof sagging in the 60m direction. Three models, one with a flat roof and the other two with sag to span ratio of 0.06 and 0.10 respectively, were prepared to study the effect of sag. Twenty five pressure points were provided on the roof surface of each model (Fig. 3). Photo. 1 shows one of the models of cylindrical roof.

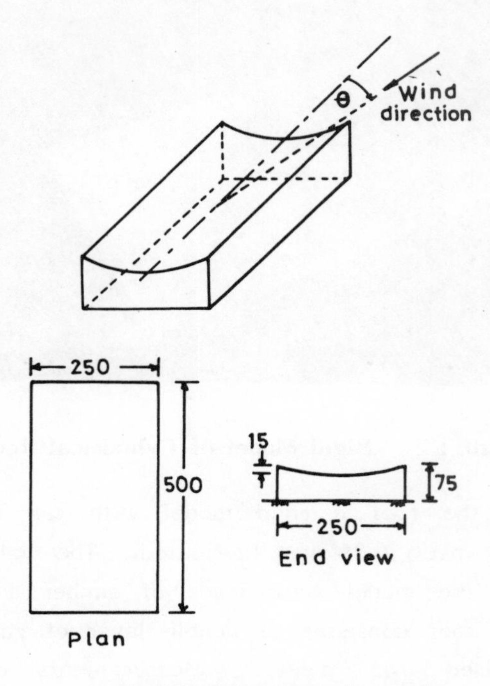

Fig. 2. Model of Cylindrical Roof with Sag to Span Ratio = 0.06

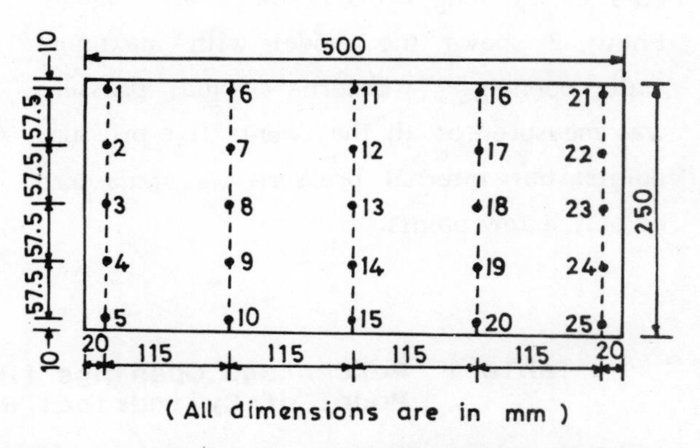

Fig. 3. Plan of A Rigid Model Showing Locations of Pressure Points on Roof Surface

To study the effect of openings (windows and doors) on wind pressure distribution

Photo. 1 - Rigid Model of Cylindrical Roof

on the roof, a rigid model with sag to span ratio 0.06 was fabricated. The walls for the model were made of timber and the roof consisted of double layer of galvanised iron sheet. Measurements on the model were made for 90^{0} angle of incidence of wind. On the same model, the openings were varied as 3.20, 12.67 and 30.00 percent of wall area by making cuts in the long walls (Table 1 and Fig. 4). Photo. 2 shows the model with maximum wall opening. External wind pressure was measured at all the twenty five pressure points, but internal pressure was measured at only a few points.

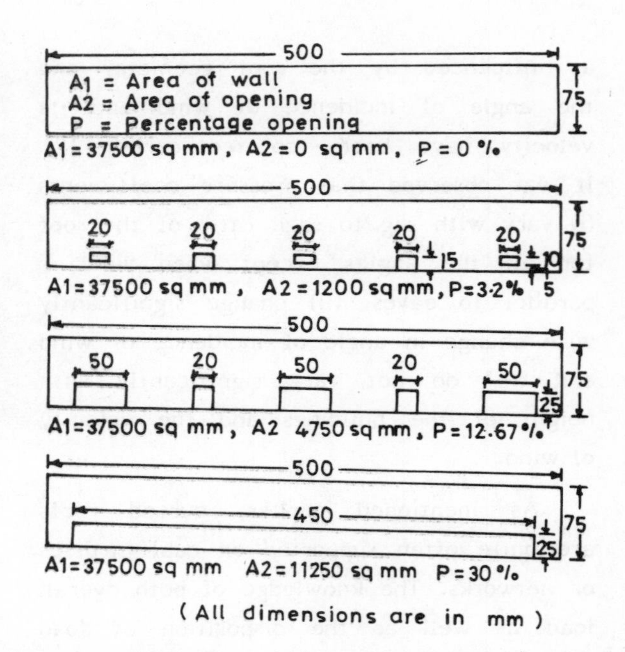

Fig. 4. Elevation of Longitudinal Wall with Openings

Flow Conditions

All the above mentioned models were tested in a closed circuit wind tunnel with a 80 cm x 110 cm cross-section, in which shear flow is generated with the help of a grid at the entrance. The wind velocity profile obtained conforms to a power law index equal to one-seventh. This is a flow with low turbulence intensity

Table 1 Percentage Openings (Permeability) in Walls of the Model of Cylindrical Roof with Sag to Span Ratio=0.06

Case No.	Percentage openings	
	Windward wall	Leeward wall
1	0.00 %	0.00 %
2	3.20 %	3.20 %
3	12.67 %	12.67 %
4	30.00 %	12.67 %
5	30.00 %	30.00 %

Photo. 2 - Rigid Model of Cylindrical Roof with Openings in Walls

of 1% at gradient height and 4% (max. value) near the floor.

RESULTS AND DISCUSSIONS

Mean Pressure Coefficients

The results are discussed in terms of the mean pressure coefficients defined as -

$$C_{pm} = \frac{p - p_o}{\frac{1}{2}\rho V_m^2}$$

where

- p = the absolute pressure at any point on the model
- p_o = the ambient or reference pressure
- ρ = density of air
- V_m = mean value of wind velocity at mid height of building model.

From the measurements already reported [2], it is observed that the whole surface of the roof is subjected to suction for wind angles 0°, 15°, 30° and 90°. At other angles, namely 45°, 60° and 75°, a few spots are subjected to pressure though of very small magnitude. Table 2 indicates such values for 60° wind incidence angle.

Maximum Pressure Coefficients

Study of the maximum values of negative pressure coefficients ($C_{pm\ max}$) on three rigid models show that maximum pressure coefficients occurs close to the windward corner(s) of the model. Variation of $C_{pm\ max}$ with sag to span ratio for different wind incidence angles is summerised in Table 3 and also shown in Fig. 5. It is observed from this figure and the table that the maximum value of negative pressure coefficient reduces with increase in sag to span ratio at most of the wind incidence angles (except 60°).

Ishizaki and Yoshikawa [7] also observed similar trend of variation of maximum suction coefficient. They measured a maximum value of suction as -1.70 on flat roof at the central section for wind incidence angle of 90° (this being a case of two dimensional flow) and found that curved roofs have values less than this. In the study being reported here, for a comparable case, the coefficient of suction at the central section was found to be -1.37 and that at the extreme section -1.73. Further, as reported above in ref. 7, suctional forces are lower for curved roofs as compared to a flat roof, except for the angle of incidence of 60° studied by the authors.

Table 2 Comparison of Mean Wind Pressure Coefficients (C_{pm}) on Different Roofs (wind incidence angle=60°)

Pressure Point No.	Sag to span ratio		
	0.00	0.06	0.10
1	-0.738	-4.384	-4.100
2	-0.358	-0.777	-1.052
3	-0.964	-0.738	-0.476
4	-0.975	-0.487	-0.262
5	-0.856	-0.649	-0.572
6	-1.507	-1.442	-1.442
7	-0.750	-1.342	-1.751
8	-0.299	+0.013	+0.097
9	-0.341	-0.118	+0.137
10	-0.501	-0.485	-0.508
11	-1.173	-0.956	-0.833
12	-1.111	-1.185	-1.178
13	-0.579	-0.810	-0.870
14	-0.304	-0.041	+0.076
15	-0.356	-0.321	-0.145
16	-1.044	-0.738	-0.585
17	-0.921	-0.741	-0.723
18	-0.652	-0.798	-0.833
19	-0.446	-0.597	-0.372
20	-0.326	-0.343	-0.211
21	-1.005	-0.637	-0.478
22	-0.719	-0.508	-0.525
23	-0.607	-0.577	-0.560
24	-0.510	-0.592	-0.562
25	-0.319	-0.487	-0.495

Table 3 Comparison of Maximum Wind Pressure Coefficients ($C_{pm\,max}$) on Different Roofs

Wind Incidence Angle	Sag to span ratio		
	0.00	0.06	0.10
0°	-1.349	-1.246	-1.209
15°	-1.511	-1.426	-1.326
30°	-2.377	-1.934	-1.460
45°	-1.722	-1.789	-1.723
60°	-1.507	-4.384	-4.100
75°	-2.941	-2.791	-2.167
90°	-1.726	-1.249	-1.133

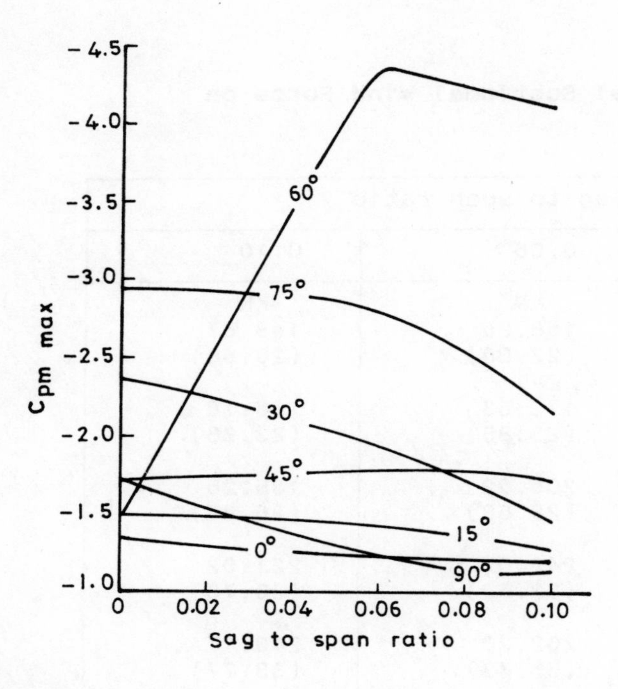

Fig. 5. Variation of Maximum Pressure Coefficient with Sag to Span Ratio

Total Wind Force

The total suctional force on the prototype roof due to wind is calculated as -

$$W_L = \sum_{n=1}^{25} A_n P_n$$

$$= \sum_{n=1}^{25} A_n \left(\frac{1}{2} \rho V_m^2 C_{pn}\right)$$

$$= \frac{1}{2} \rho V_m^2 \sum_{n=1}^{25} A_n C_{pn}$$

where

A_n = the area of roof on which average pressure coefficient is C_{pn}

V_m = mean velocity of flow at mid height of prototype building

ρ = density of air.

For the results reported herein, ρ was taken as 1.204 N-sec^2/m^4 and V_m was taken as 9.85 m/sec (= 35.46 Km/hour) with a gradient velocity as 15 m/sec (= 54 Km/hour). Plan area of the roof of the prototype building is 7200 square meter. Table 4 compares the total wind force on different roofs for various wind incidence angles. It is observed from Fig. 6 and Table 4 that the total suctional force on the roof decreases with increase in slope at most of the wind incidence angles. However, at 90° wind incidence angle, there is no significant change in values of total suctional force with change in sag to span ratio. Further it is seen that total wind load is minimum at 0° wind incidence angle on any roof. It increases with increase in angle of wind direction and attains a maximum value at 90° wind incidence angle. Average suctional pressure varies between 20 N/m^2 and 50 N/m^2 on an average.

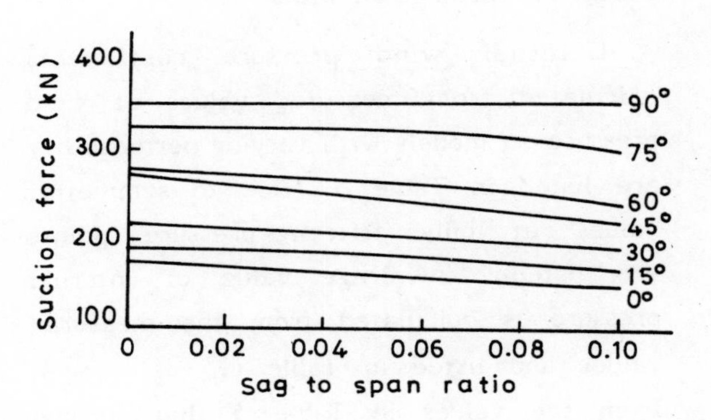

Fig. 6. Variation of Total Wind Load on Roof with Sag to Span Ratio

Table 4 Comparison of Total Suctional Wind Force on Different Roofs

Wind Incidence Angle	Sag to span ratio		
	0.00	0.06	0.10
	KN	KN	KN
0°	175.37 (24.36)	158.85 (22.06)	148.87 (20.68)
15°	163.19 (22.66)	172.53 (23.96)	166.78 (23.96)
30°	215.29 (29.90)	206.52 (28.68)	189.26 (26.29)
45°	273.80 (38.03)	242.07 (33.62)	221.02 (30.70)
60°	277.75 (38.58)	262.30 (36.43)	243.17 (33.77)
75°	327.17 (45.44)	325.04 (45.14)	302.54 (42.02)
90°	353.28 (49.07)	353.37 (49.08)	357.13 (49.60)

Note : Values in parentheses are the average values of suction in N/m^2.

Effect of Openings on Walls

External wind pressure coefficients calculated from measured values of wind pressure on models with varying permeability are listed in Table 5. (Due to symmetry, values at only fifteen pressure points are listed). Average value of internal pressure is calculated from the measured values and listed in Table 6. It is seen from the values in Table 5 that suction on the roof decreases with increase in openings in the walls, as more and more air escapes below the roof. Further, as is evident from Table 6, internal pressure increases with increase in openings. Resultant force on the roof (suction) increases with the increase in openings (Table 7).

CONCLUSIONS

Following conclusions are drawn from the study presented in this paper.

1. Maximum value of negative pressure coefficient occurs close to the windward corners of the roof which reduces with increase in sag to span ratio at most of the wind incidence angles. But at 60° angle, windward corners are subjected to very high suction

Table 5 Variation of External Pressure Coefficients with Openings in the Walls (sag to span ratio = 0.06, wind incidence angle = 90°)

Pressure Point No.	Percentage opening				
	WW=0.00 % LW=0.00 %	WW=3.20 % LW=3.20 %	WW=12.67 % LW=12.67 %	WW=30.00 % LW=12.67 %	WW=30.00 % LW=30.00 %
1	-1.150	-1.136	-1.130	-1.128	-1.118
2	-1.224	-1.208	-1.198	-1.185	-1.151
3	-0.855	-0.844	-0.787	-0.782	-0.629
4	-0.404	-0.395	-0.320	-0.314	-0.146
5	-0.296	-0.257	-0.237	-0.218	-0.154
6	-1.024	-0.989	-0.988	-0.981	-0.929
7	-1.078	-1.058	-1.057	-1.031	-1.001
8	-0.932	-1.035	-1.026	-1.022	-0.943
9	-0.819	-0.795	-0.786	-0.734	-0.644
10	-0.489	-0.460	-0.428	-0.412	-0.347
11	-0.951	-0.921	-0.894	-0.890	-0.845
12	-0.998	-0.990	-0.962	-0.952	-0.905
13	-0.997	-0.996	-0.964	-0.956	-0.892
14	-0.889	-0.857	-0.829	-0.815	-0.714
15	-0.587	-0.560	-0.533	-0.522	-0.399

in case of curved roofs, whereas suction is of moderate value in case of flat roof.

2. Total wind load (suction force) is minimum at 0° wind incidence angle on any roof. It increases with the increase in angle of wind direction and attains a maximum value at an angle of 90°.

3. Total force of suction on the roof decreases with the increase in sag to span ratio at most of the wind incidence angles. However at 90° angle there is no significant change.

4. Suction pressure on flat and cylindrical roofs with sag to span ratio 0.06 and 0.10 varies between 20 N/m^2 and 50 N/m^2 on an average.

Table 6 Variation of Average Pressure Coefficients on the Inner Surface of the roof, with the Openings in Walls (sag to span ratio = 0.06, wind incidence angle = 90°)

Percentage opening	ww = 0.00 % Lw = 0.00 %	ww = 3.20 % Lw = 3.20 %	ww = 30.00 % Lw = 12.67 %
Average pressure coefficient	Zero	+0.466	+0.921

Table 7 Variation of Resultant Pressure on Roof with the Openings in the Walls (sag to span ratio = 0.06, wind incidence angle = 90°)

Percentage opening	ww = 0.00 % Lw = 0.00 %	ww = 3.20 % Lw = 3.20 %	ww = 30.00 % Lw = 12.67 %
Coefficient at pressure point No.13	-0.997*	-1.462*	-1.877*

* Negative sign implies suction on the roof.

5. Suction on roof decreases with increase in openings on walls. Similarly, internal suction also decreases or internal pressure increases with increase in openings. Resultant force on roof is increase of suction with the increase in permeability.

REFERENCES

1. Ahuja, A.K., "Wind Effects on Cylindrical Cable Roofs", Ph.D. Thesis, Deptt. of Civil Engineering, University of Roorkee, Roorkee, India, Nov. 1989.

2. Ahuja, A.K., Krishna, P., Pande, P.K., and Pathak, S.K., "Wind Loads on Curved Roofs", Proc. of the Asia-Pacific Symp. on Wind Engg., Univ. of Roorkee, Roorkee, India, Dec. 5-7, 1985, pp. 145-154.

3. Blessmann, J., "Wind Pressure on Roofs with Negative Pitch", Jour. of Wind Engg. and Industrial Aerodynamics", Vol. 10, No. 2, 1982, pp. 213-230.

4. Esquillan, N., and Saillad, Y. (ed.), "Hanging Roofs", Proc. of the IASS Colloquium on Hanging Roofs, Continuous Metallic Shell Roofs, and Superficial Lattice Roofs (Paris, 1962), North-Holand Publishing Co., Amsterdam, 1963.

5. Houghton, E.L. and Carruthers, N.B., "Wind Forces on Buildings and Structures-An Introduction", Edward Arnold (Publishers) Ltd., London, 1976.

6. IS:875 (Part-3)-1987, "Indian Standard Code of Practice for Design Loads (other than Earthquake) for Buildings and Structures - Part 3 Wind Loads", Bureau of Indian Standards, New Delhi, India.

7. Ishizaki, H. and Yoshikawa, Y., "A Wind Tunnel Model Experiment of Wind Loading on Curved Roofs", The Bulletin of the Disaster Prevention Research Institute, Kyoto University, Vol. 21, March 1972.

8. Krishna, P., "Wind Effects on Curved Roofs", Proc. of the Second Asia-Pacific Symp. on Wind Engg., Beijing, China, June 1989, pp. 39-53.

9. Sachs, P., "Wind Forces in Engineering", Pergamon Press, New York, 1972.

10. Simiu, E. and Scanlan, R.H., "Wind Effects on Structures-An Introduction to Wind Engineering", John Wiley & Sons, New York, 1978.

11. Vaessen, F., "Wind Channel Tests to Investigate the Wind Pressure on a Hypershell Roof", Proc. of the IASS Colloquium on Hanging Roofs, Continuous Metallic Shell Roofs, and Superficial Lattice Roofs, Paris, 9-11 July, 1962, pp. 87-92.

12. "Wind Resistant Design Regulations - A World List - 1975", Published by Gaknjutsu Bunker Fukyn-Kai, Ch-Okayama, Meguroku, Tokyo, Japan, 1975.

INTERFERENCE EFFECTS ON TALL RECTANGULAR BUILDINGS

M. Yahyai *P.K. Pande* ***Prem** Krishna* *Krishen Kumar*

Civil Engineering Department
University of Roorkee
Roorkee, INDIA

ABSTRACT

With the increase in construction of tall buildings in cities, there is a growing concern about the importance of the effects of interference among such structures under the action of wind. Prediction of the mean and fluctuating forces and moments on a building which is partially or completely sheilded by an interfering structure is practically impossible without model tests. An extensive wind tunnel study of the wind effect on a tall rectangular building model in isolation and with an interfering building on upstream as well as downstream side of the principal building was carried out. Both rigid and aeroelastic models were tested, and the results in terms of the buffeting factors due to interference obtained.

This paper however presents only the results of the rigid model study. The remaining part of the investigation is under finalisation and will be reported subsequently.

INTRODUCTION

Tall buindings are prone to wind induced excitation, both in the along-wind and the across-wind directions. This has been the subject of many analytical and experimental investigations. However, the construction of a new building leads to a modification of the wind induced response of other buildings in proximity, while the response of the new building itself is also affected by the existing buildings. This interference may lead to a reduction of the wind effects due to sheltering or to larger wind-induced forces on structural frame and/or claddings.

Within the last two decades attention has increasingly been given to the buffeting effects of a building on neighbouring tall buildings as a result of interference and proximity, but no general solutions have emerged so far. Most of these studies are situation specific and of necessity experimental, as the problem is far too complex to be handled with the analytical procedures available. It is worth mentioning that till today no code of practice has considered the effect of interference for designing a tall building.

The first study on interference effects is apparently made by Harris [1]. He showed that significant changes in pressure distributions could be developed on a building model due to the presence of an adjacent structure. Systematic studies were made by Melbourne [2] and Saunders [3] highlighting the interference of a medium size building due to the presence of an upstream structure. The interference effects were also investigated by Kareem [4], Bailey and Kwok [5,6], Sykes [7], Sakamoto and Haniu [8] and Blessmann and Riera [9]. However, as mentioned earlier, the information available on the effects of interference is too meagre to enable generalisation which could be useful for design purposes. Keeping this in mind, a programme was undertaken at the University of Roorkee to generate more data on the effects of interference between tall buildings. The first study under this programme was made by Mir [10] who conducted rigid model tests on two identical prismatic buildings of square base, with an aspect ratio of 8. These tests-conducted in the BLWT at the University of Roorkee - showed a three-fold increase in the torsional moment coefficient when the interfering building model was close to the instrumented model. The mean lift and drag forces were found to decrease in general - as would be expected for a close spacing in rigid model tests.

A subsequent study under the programme comprised of testing both rigid and aeroelastic models of tall rectangular buildings in the tunnel, with and without interference. The present paper reports some of the results, primarily on the rigid model tests. The aeroelastic model test results are under finalisation and will be reported later.

EXPERIMENTAL PROCEDURES

Rigid models of a prototype whose dimensions were 150 m height (h); 30 m breadth (b) and 25 m depth (d) were tested in the open-circuit boundary layer wind tunnel at the University of Roorkee. The rigid models were made out of plywood of 6 mm thickness. A geometric scale of 1/250 was adopted for the fabrication of the models. The wind tunnel velocity was kept at about 10 ms^{-1} throughout the experiment which corresponds to a reduced frequency (Strouhal number) equal to 0.16, expected to be a critical value. The building model was first kept in the tunnel without any interfering model in a simulated open country terrain with a power law coefficient of $\alpha = 0.10$. The profiles of velocity and turbulence intensity of the wind tunnel are shown in Fig. 1. The principal model (downstream model) was firmly fixed on a three component load cell which can sense the two shear forces along and across the wind - and a bending moment about an axis perpendicular to the wind directions. Figure 3 shows a photograph of the models in the wind tunnel.

EXPERIMENTAL RESULTS AND DISCUSSION

The upstream interfering building was placed at a number of locations in the tendem as well as offset arrangement as shown in Fig. 2. The distance between the

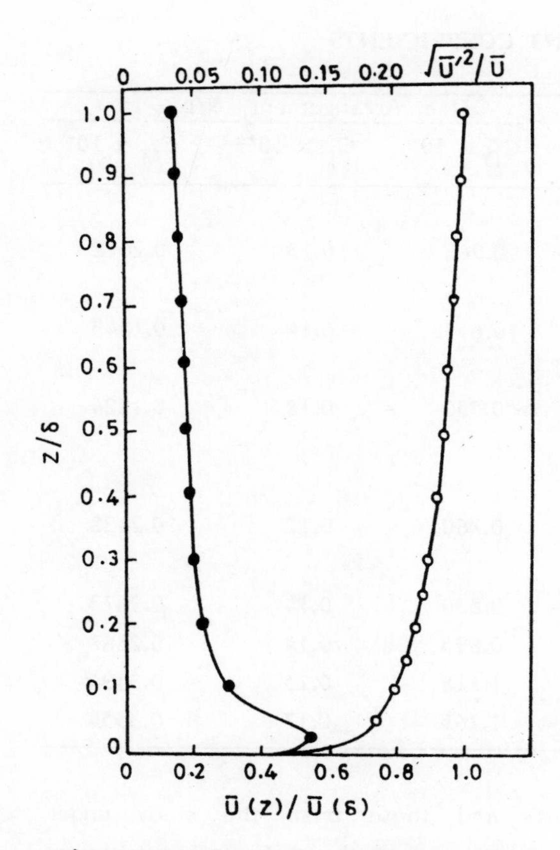

Fig.1. Profiles of the wind Model

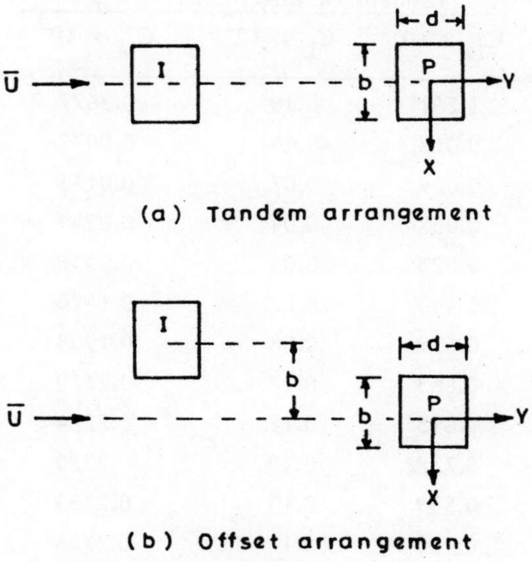

two models y, was varied from 2d to 24d the direction of the wind, and the offset x, used was equal to b. The results are presented in the form of buffeting factors defined as,

$$B.F = \frac{\text{force on a building with interference}}{\text{force on an isolated building}}$$

Table 1 depicts the values of drag, lift and along-wind bending moment coefficients for both the tandem and offset arrangements. These coefficients are defined by the following equations :

$$C_D = \frac{F_D}{1/2\,\rho\,\bar{u}^2\,hb}$$

(a) Tandem arrangement

(b) Offset arrangement

Fig.2. Building arrangement

$$C_L = \frac{F_L}{1/2\,\rho\,\bar{u}^2\,hb}$$

$$C_M = \frac{M}{1/2\,\rho\,\bar{u}^2\,h^2 b}$$

in which F_D, F_L, M and ρ are the total drag and lift forces, total bending moment and the air density respectively. Figures 4 through 6 show the variation of these coefficients with spacing y/d on upstream side for both the cases. in the tandem arrangement, the mean forces and moments are very low upto a distance of about 4d as is clear from Table 1 as well as Figs. 4, 5 and 6. Similar findings are reported by Sakamoto [8]. Flow observations by him

TABLE : 1 FORCE AND MOMENT COEFFICIENTS

	Tandem Arrangement, x = 0				Offset Arrangement, x/b = 1		
$\frac{y}{d}$	$C_D \times 10$	$C_L \times 10^2$	$C_M \times 10^3$	$\frac{y}{d}$	$C_D \times 10$	$C_L \times 10^2$	$C_M \times 10^3$
-	1.261	0.19	0.3677				
2	0.018	-0.04	0.0077	2	0.963	0.13	0.2612
2.5	0.029	0.07	0.0171				
3	0.038	0.04	0.0249	3	0.655	0.14	0.1843
4	0.025	0.04	0.0550				
5	0.440	0.07	0.1420	5	0.550	0.18	0.1824
6	0.639	0.08	0.1903				
7	0.683	0.17	0.2179				
8	0.693	0.18	0.2274	8	0.760	0.12	0.2438
9	0.738	0.15	0.2320				
11	0.839	0.15	0.2733	11	0.830	0.14	0.2673
13	0.900	0.17	0.2726	13	0.895	0.14	0.2867
21	1.137	0.18	0.3292	21	1.118	0.15	0.3495
24	1.209	0.19	0.3461	24	1.165	0.17	0.3554

have revealed that negative pressures on the rear face are smaller compared to that on the front face. He has explained this on the basis of the observed' behaviour of the shear layer which gets separated at the leading edges on the two sides of the upstream model and reattaches itself on the side faces of the downstream model after which it rolls up weakly into the region to the rear. When the spacing becomes larger than 4d, the buffeting factors increase monotonically with increase in spacing. However, the buffeting factors for the downstream building exhibit values near to 1 at a spacing of about 24d.

As mentioned earlier, Mir [10] carried out an intensive study on the interference effect between two square buildings. No direct comparison is possible between his results and those from the study under report, because of the different geometrical proportions of the buildings tested as well as the flow profiles used. However, for such cases where it was possible, a comparison has been attempted in Figs. 3, 4 and 5. The correlation is fair. In Fig. 3 the buffeting factors of the mean drag coefficients for a square building model reported by Sakamoto [8] are also marked. His values are once again comparable with those measured for rectangular building.

In the case of the interfering building placed at an offset of one width, the buffeting factors are higher as compared to the values for the tandem arrangement for values of y/d upto about 5. This is seen in Figs. 4 through 6 and depicted in Table 1. As observed by Sakamoto [8], the increase in

Fig. 3. Building Models Placed in BLWT

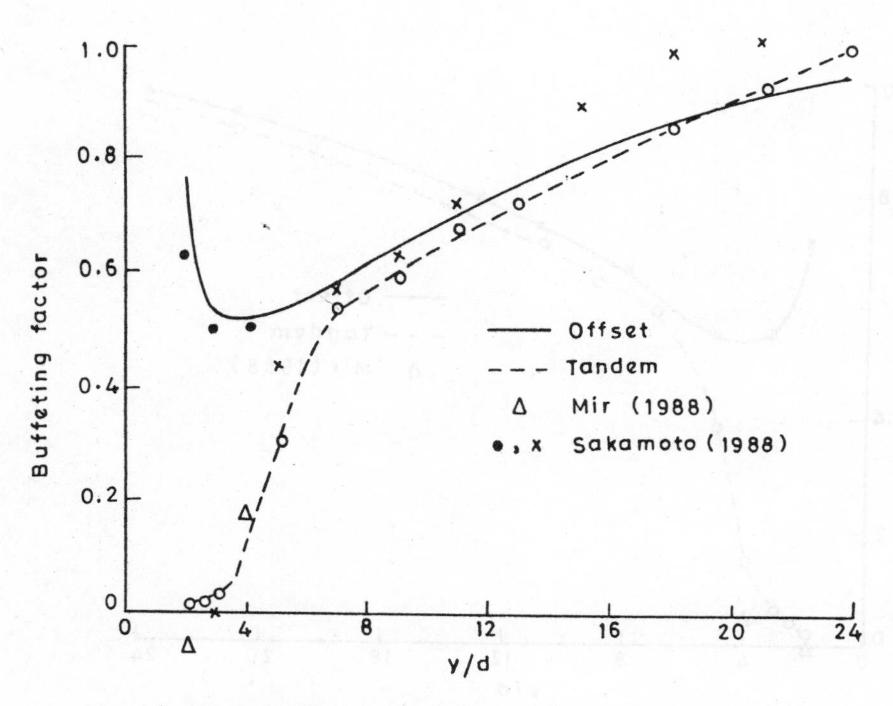

Fig. 4. Variation of B.Fs for C_D with spacing

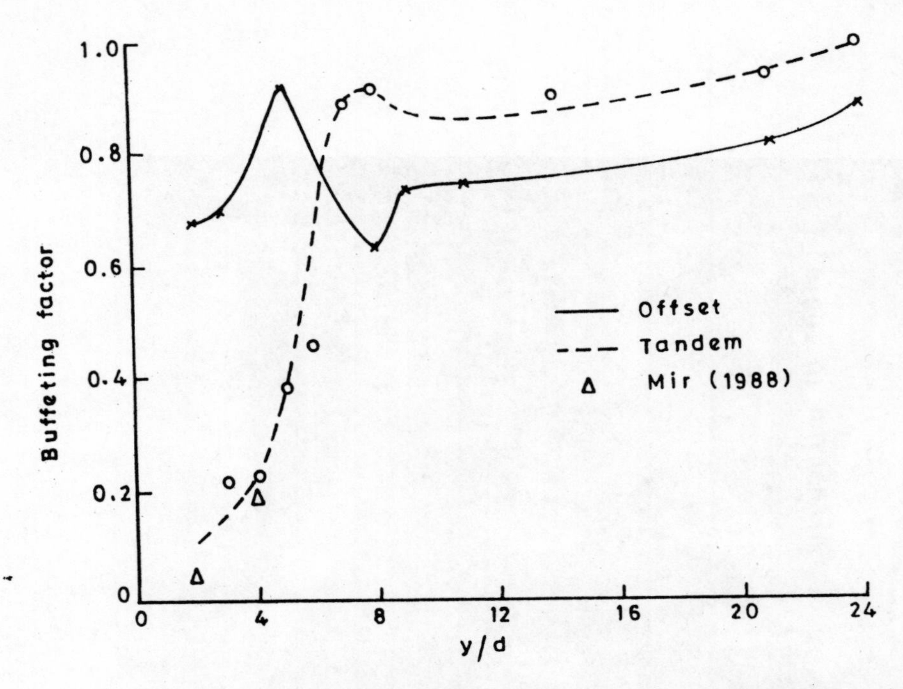

Fig.5. Variation of B.Fs for C_L with spacing

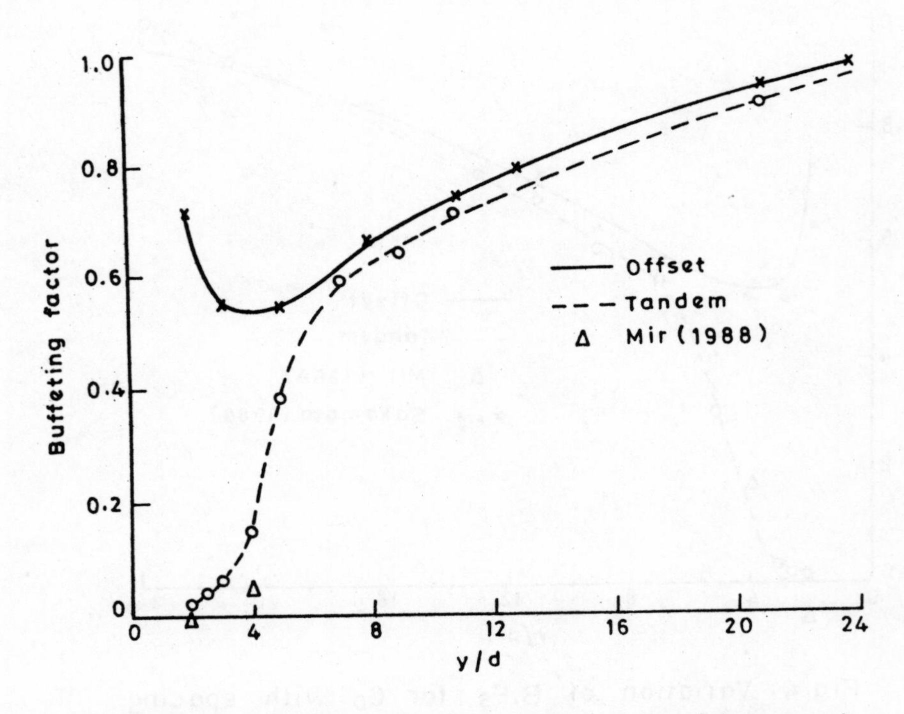

Fig.6. Variation of B.Fs for C_M with spacing

buffeting factors may be due to the fact that the shear layers generated from upstream building collide with the front surface of the downstream model and hence one gets a circulating flow around the model causing increase in the negative pressure. When the interfering model moves farther than y/d = 5, the buffeting factors increase gradually and once again at larger values of y/d (equal to about 24) they approach unity.

CONCLUSIONS

The study presented in this paper showed that an upstream interfering building causes shielding effect on a downstream building. This effect reduces the force and moment coefficients of the downstream building. This beneficial effect i.e. shielding effect will be more in the case of tandem arrangement. The interference effect between two buildings of the same size is extended upwind of a downstream model for a distance of about 24 times the depth of the building d. It needs to be emphasised, however, that the results pertain only to the rigid model studies and the fluid-structure interaction may considerably modify the results. This interaction has been studied on an aeroelastic model and the preliminary analysis indicates that an interfering building causes generally enhanced dynamic motion of a principal building in both along-wind and across-wind directions.

ACKNOWLEDGEMENTS

This paper is based on a study carried out for his doctoral thesis by the first author under the supervision of the other three authors. Help received from Mr. R.P. Gupta and other staff in the Wind Tunnel in carrying out the experimental work is gratefully acknowledged.

REFERENCES

1. C.L. Harris, "Influence of Beighbouring Structures on the Wind Pressure on Tall Buildings", Bureau of Standards, Journal of Research, Vol. 12, 1934.

2. W.H. Melbourne and D.B. Sharp, "Effects of Upward Buildings on the Response of Tall Buildings", Proc. Regional Conf. Tall Building, Hongkong, 1976.

3. J.W. Saunders and W.H. Melbourne, "Buffeting Effects of Upstream Buildings", Proc. 5th Int. Conf. Wind Engg., Fort Collins, Col. 1979.

4. Ahsan Kareem, "The Effect of Aerodynamic Interference on the Dynamic Response of Prismatic Structures", J. Wind Engg., Vol.25, 1987, pp. 365-372.

5. P.A. Bailey and K.C.S. Kwok, "Interference Excitation of Twin Tall Buildings", J. Wind Engg., Vol.21, 1985, pp. 323-338.

6. K.C.S. Kwok, "Interference Effects on Tall Buildings", 2nd Asia Pasific Symposium on Wind Engg., Peking, China, 1989.

7. D.M. Sykes, "Interference Effects on the Response of a Tall Building Models", J. Wind Engg., Vol.11, 1983, pp. 365-380.

8. H. Sakamoto and H. Haniu, "Aerodynamic Forces Acting on Two Square Prisms Placed Vertically in a Turbulent Boundary Layer", J. Wind Engg., Vol. 31, 1988, pp. 41-66.

9. J. Blessmann and J.D. Riera, "Wind Excitation of Neighbouring Tall Buildings", J. Wind Engg., Vol.18, 1985, pp. 91-103.

10. S.A. Mir, "Wind Interference Effects Between Two Prismatic Buildings", M.E. Dissertation, Deptt. of Civil Engg., University of Roorkee, 1988.

AERODYNAMIC STUDY OF A CHIMNEY

Krishen Kumar *Prem Krishna* *P.K. Pande* *P.N. Godbole*

Department of Civil Engineering
University of Roorkee
Roorkee, INDIA

SYNOPSIS

A 218 m tall reinforced concrete chimney has been studied for its behaviour under wind forces and the findings presented. Both experimental as well as theoretical studies have been carried out. As a first step, a rigid model with a 1/250 scale was tested in a boundary layer wind tunnel to predict the along-wind base forces and moments duly accounting for the gustiness of wind. This was followed by tests on an aeroelastic model. The necessary corrections for distortions, sometimes introduced intentionally, have been derived and applied for estimating the prototype along-wind as well as across-wind responses on the basis of results obtained from the aeroelastic model.

The chimney is of constant section in approximately the top half of its height and has a "large" taper in the lower portion, which makes its analysis more involved. The theoretical results obtained by using the Davenport's approach for along-wind analysis and Vickery's approach with some modifications for the across-wind analysis, are compared with the experimental results. These are seen to be on the higher side for the across-wind response, but compare well with the experimental values for the along-wind response.

INTRODUCTION

Wind response of tall chimneys is affected by a large number of flow and structural parameters with a complex interaction between the two. The subject of chimney aerodynamics has therefore attracted much attention of designers and research workers over the last several decades. Studies carried out range from theoretical work to measurements of chimney response on models in wind tunnels and on prototypes in the field. A special feature of chimney aerodynamics is that it has significant responses in the along-wind as well as the across-wind directions. Despite considerable research effort, the across-wind response of a chimney still defies an accurate theoretical estimate due to its sensitivity to a number of parameters that are involved. Variations in the parametric functions used in the procedures given by Vickery [1], Scruton [2] and Rumman [3] are indicative of the empirical nature of the theories.

Vickery and Basu [4] have reported levels of accuracy at 80% confidence as ±25% for lift coefficient C_L, ±10% for Strouhal number, S, ±25% for mass dampings δ, and ±20% for correlation length L, yielding the resultant level of reliability on the prediction of across-wind response of the order of ±40%. The chimney under study, 218 m high, has a constant diameter in its top 90 m and is tapered in the rest of its height, throwing in an additional factor of complexity.

Experimental studies present their own limitations. Whereas such studies are carried out on the assumption that the model results can be extrapolated to the full scale values, the model response will be an exact scaled down version of the prototype response only if all the similarity requirements are satisfied. These include similarity of flow as well as similarity of certain structure parameters. It is well known that it is impossible to achieve complete similarity. One has therefore to identify the parameters which are of importance to any phenomenon and then try to simulate them to the extent possible and include corrections for the distortions that can not be avoided. It will be seen that corrections to account for distortions need to be applied invariably when trying to predict prototype values from model measurements.

This underlines the need to carry out both experimental as well as theoretical studies and correlate the two. This will enable the design of a chimney under wind loading to be carried out with greater confidence and sense of certainty. The paper describes such a study for a 218 m high chimney designed for a thermal power station.

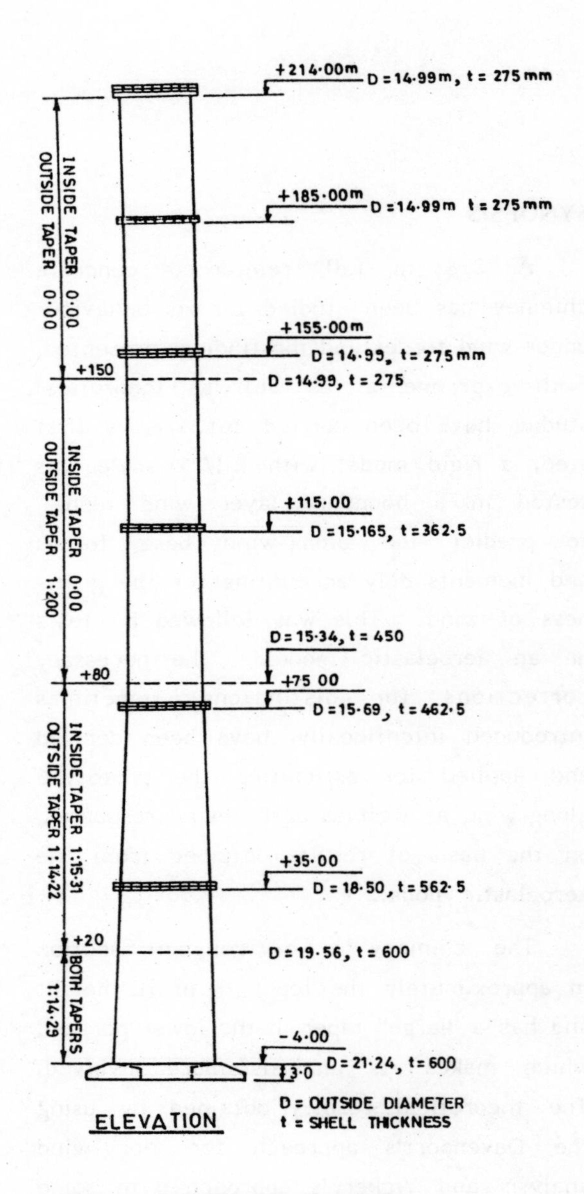

FIG. 1 : PROTOTYPE CHIMNEY

The dimensions of the chimney are given in Fig. 1. Its shell is built in grade M-25 concrete. The study comprises of the following aspects :

(1) Estimation of the mean bending momnent and shear force at the base of the chimney from a rigid model wind tunnel study.

(2) Measurement of the response on a suitably scaled aeroelastic model, prediction of prototype values therefrom and comparison with theoretical results.

(3) Estimation of the along-wind and across-wind response of the chimney to wind based on analytical procedures.

In pursuance of the above objectives it was decided to build first a rigid model, and then an aeroelastic model of the chimney. For the purposes of the objectives (2) and (3) above the chimney shell alone was modelled. Although the chimney consists of the shell, platforms as well as the lining, the analysis as well as aeroelastic modelling was done for the shell alone since it would be the simplest, and would serve the objective of validating the analytical procedure adequately.

THE RIGID MODEL STUDY : SALIENT FEATURES

A solid timber model which represented the outside dimensions of the chimney to a geometric scale of 1/250 was tested in the boundary layer wind tunnel with a mean-wind velocity profile conforming to a power law index of 0.09. A scale of 1/250 was chosen, primarily to limit the model height to remain within the thickness of the boundary layer in the wind tunnel which is about 1 m in the 2 m high tunnel cross section. The terrain surrounding the chimney was also modelled and placed on the tunnel's turn-table such that it could be rotated independently of the chimney model. This enabled simulation of the condition of wind approaching the chimney from different directions.

The maximum base shear coefficient and base bending moment coefficient, C_s and C_m respectively, were obtained from this test as 0.857 and 0.425. Using 1% of critical damping and funcamental frequency of vibration of the chimney as 0.2407 Hz, the gust factor was obtained as 2.28 by the procedure laid down in the CICIND Code [5]. Somewhat smaller values of 1.96 and 2.13 were obtained by the ACI Draft Code formula [6] and the Davenport's method [7,8] respectively.

Adopting a gust factor value of 2.3 for design, the peak values of base shear and base bending moment for the prototype chimney worked out to be 4.7×10^6 N and 5.1×10^6 N-m respectively for a mean hourly speed of 112 kmph at 10 m height, consistent with the design wind velocity. Considering the inadequacy of a rigid model to represent the aeroelastic effects and to account for possible higher mode contributions these estimates were increased by 10%, to become, 5.17×10^6 N and 5.61×10^8 N-m respectively. It must also be recognised that a rigid model could not be expected to show up any across-wind aerodynamics.

AEROELASTIC MODEL STUDY

Modelling Considerations

Based on dimensional analysis certain important parameters for an aeroelastic model study can be identified resulting in the following similarity requirements [9, 10].

(a) Flow parameters : These include,

(i) Reynolds number

$$(\rho_a VL/\mu)m = (\rho_a VL/\mu)p \quad (1)$$

where ρ_a – mass density of air
V – wind velocity
L – characteristic length scale
μ – dynamic viscosity of air

Subscripts m and p refer to the model and the prototype respectively.

(ii) Froude number

$$(V/\sqrt{gL})m = (V/\sqrt{gL})p \quad (2)$$

(iii) Atmospheric boundary layer characteristics

This includes mean velocity profiles, turbulence intensity and scale, etc.

(b) Structure parameters : These include,

(i) Geometric similarity

$$Lm = Lp/N1 \quad (3)$$

N1 being the model scale

(ii) Density ratio

$$(\rho_s/\rho_a)m = (\rho_s/\rho_a)p \quad (4)$$

which can also be written as

$$(m/\rho_a{}^2)m = (m/\rho_a L^2)p \quad (5)$$

where m is the mass per unit length of the structure.

(iii) Structural damping

$$(\delta_s)m = (\delta_s)p \quad (6)$$

(iv) Stiffness distribution along the height

$$(EI/\rho_a V^2 L^4)m = (EI/\rho_a V^2 L^4)p \quad (7)$$

This can also be written for thin walled structures in the form

$$(Et/\rho_a V^2 L)m = (Et/\rho_a V^2 L)p \quad (8)$$

Where E is the elastic modulus and t the thickness of the structure.

As has been pointed out earlier, it is rarely possible to meet all the above similarity requirements and one has therefore to identify the influence of each parameter in order to get meaningful results from model studies. A brief discussion of these is given in the following paragraphs.

The simulation of Reynolds number is impossible in view of the large scales and velocities required for the same. For rounded shapes, the effect of Reynolds number is quite complex and hence great care is required in carrying out model tests. A study of the variation of drag and oscillatory coefficients for a cylinder suggests that the tests should preferably by carried out in the range of Reynolds number 10^4-10^5 and the values obtained be corrected for the prototype Reynolds numbers which are in the range of 10^6-10^8. This can be done if the drag values obtained from model tests are miltiplied by 0.8/1.2, 0.8 and 1.2 being typical C_d values in these Reynolds number ranges.

The Froude number effects are known to be insignificant for structures like chimneys which oscillate essentially in a horizontal plane.

The simulation of atmospheric boundary layer characteristics also can be obtained only to a limited extent. A wind tunnel with a long test section and appropriate roughness, vortex generators, etc., in the upstream fetch gives a reasonable simulation.

The simulation of density ratio, stiffness distribution and mass distribution can be achieved to a fair degree of accuracy. The simultaneous simulation of density ratio and stiffness distribution, however, requires a model material with specified values of density and elastic modulus which may always not be possible to obtain. The practice in this respect has been to simulate the stiffness distribution and then simulate the density ratio by adding discrete masses to the model. It may also be mentioned that such a procedure results in model thicknesses which are very small, varying along the height, and difficult to achieve even with precision machining.

Model Design and Fabrication

The structure of the chimney consists of the concrete shell, the plateforms at different levels along the height and the lining. It was realised that whereas the along-wind response will not be significantly affected by the addition, or otherwise, of the platforms and the lining ro the chimney shell, the 'shell alone' situation will yield the maximum across-wind response and the addition of the platforms and the lining will decrease this response. Thus, it was the shell alone that was modelled for the aeroelastic testing of the chimney to try and measure the maximum possible response. The geometric scale for the model was kept the same as for the rigid model, i.e. 1/250.

To determine the thickness of the model chimney, the principle of scaling the inertia forces and the restoring forces by the same ratio was followed. That is,

$$\left[\frac{E\ t}{\rho_a V^2 L}\right]_m = \left[\frac{E\ t}{\rho_a V^2 L}\right]_p$$

or

$$\frac{Er\ tr}{\rho r\ Vr^2\ Lr} = 1 \qquad (8)$$

where the subscript r denotes the ratio of prototype to model value. In this study $Lr = 250$, $\rho r = 1$, and the modular ratio is given by equation 9.

$$Er = Ep^*/Em = 3.6\text{x}10^4 / 7.2\text{x}10^4 = \frac{1}{2} \qquad (9)$$

However, in view of $Er = 1/2$, the thickness ratio was chosen as 2Lr. Thus, using eq 8, and substituting $\rho r = 1$,

$$Vr^2 = (1/2)\,(2Lr)/Lr = 1$$

$$\text{i.e. } Vr = V_p/V_m = 1 \qquad (10)$$

It may be noted that choosing tr = 2Lr will introduce a distortion in mass scaling. This is so because the density

* The dynamic Young's modulus for M-25 concrete has been taken as $3.6\text{x}10^4$ N/mm^2 [11].

of aluminium and concrete are very nearly equal - being 27 kN/m^3 and 25 kN/m^3 respectively. The reduction of thickness by a factor of 2 will thus give a model whose mass is nearly half of the desired value. The distortion in mass scaling can be corrected by adding discrete masses to the model.

The model was fabricated by mechining a solid bar of aluminium. A 16 mm thick x 19.5 mm wide base was integrally cast with the model for fixing it in position. The scaled dimensions for the model with length scale as 1/250 and thickness scale as 1/500 are shown in Fig. 2. Despite the use of precision machining, difficulty was faced in achieving accurately the small thicknesses involved in the model.

It was noted that the average thickness in the model was 25% higher than the required value. The shell thickness, t_m therefore actually became 5/8 of $t_p/250$ instead of being 1/2 of $t_p/250$.

Thus, $t_p/t_m = 2Lr/1.25 = 1.6\ Lr$,

and, $Vr^2 = (1/2)(1.6\ Lr^4)/Lr^4 = 0.8$

so that

$$Vr = V_p/V_m = 0.9 \quad (11)$$

The first mode natural frequency of vibration of the model without any added mass was measured to be 120 Hz. The desired natural frequency of the model to maintain the appropriate scaling ratio, i.e., Lr should be 98.8 Hz. This can be obtained by adding discrete masses.

A priliminary prototype analysis showed that across-wind excitation occurs at an approximate wind speed of 26 m/s. Since $V_p/V_m = 0.9$ (eq 11) the model will experience accross-wind excitation at a speed of

$$26/0.9 = 28.8 \text{ m/s}$$

This estimate was required to be modified for the Strouhal number dependence on the Reynolds number. Thus while the Strouhal number for the prototype chimney

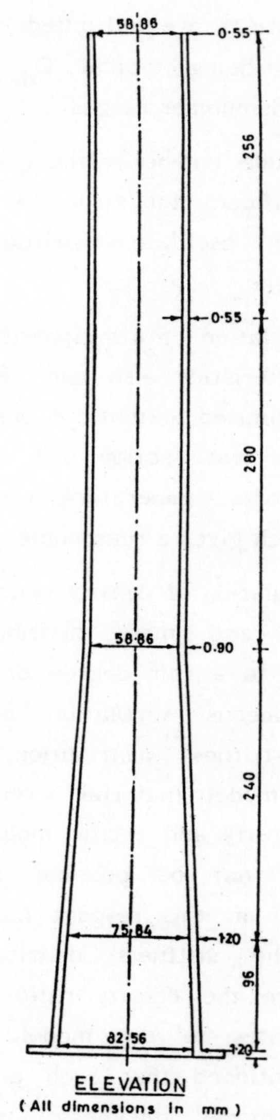

FIG.2 : AEROELASTIC MODEL : Scale - 1/250

may be taken as 0.22, a value of 0.17 for the model will be more realistic [9], since the model is being tested at a smaller Reynolds number. The estimated excitation velocity will thus become,

$$28.8 \times 0.22/0.17 = 37.3 \text{ m/s}$$

However, since the maximum speed attainable normally in the low speed wind tunnel used was only about 22 m/s, a two-pronged solution was used :

(i) the excitation velocity was lowered by adding enough mass to the model to reduce its frequency to 75 Hz (the estimated excitation velocity thus became 37.3 x 75/98.8 = 28.3 m/s), and,

(ii) constricting the tunnel cross section to raise the attainable speed.

Wind Simulation and Instrumentation

The aeroelastic model was tested in a boundary layer wind tunnel with a 2m x 2m cross section in the same manner as the rigid model and a mean wind velocity profile having a power law index of 0.09.

Turbulence intensity was also measured at two free flow velocities of 10 m/s and 20 m/s (at 1m height) using a Constant Temperature Anemometer (CTA) System of Dantec Electonik, Denmark. The values were found to be 13.5% and 11.5% respectively and were almost uniform over the height of the model except in the bottom 15 cm portion, where the intensities were higher, reaching values of 17% and 14% respectively at 5 cm height. The values achieved experimentally are close to those expected in the atmosphere. The base bending moments were measured by using resistance strain gauges fixed to the shell near the base, and calibrating the strains therein against known bending moments.

Figure 3 shows the rigid model mounted in the wind tunnel with terrain and Fig. 4 show the aeroelastic model of the chimney.

Fig. 3: Rigid Model with Terrain

Fig. 4: Aeroelastic Model

Model Results and Prediction of Prototype Values

The model as fixed in the tunnel had a first mode natural frequency of vibration equal to 75 Hz and damping 2.3%. The bending moments obtained are as follows :

Along-wind

The model was tested at a wind velocity of 20.33 m/s at 1.0 m height, and following base bending moments were obtained :

Mean base bending moment	4590 N-mm
Fluctuating peak of base moment	4475 N-mm
Maximum (total static+dynamic) value	9065 N-mm

Gust factor = 9065/4590 = 1.98

Using the established power law for the mean velocity profile, velocity at the tip of model is obtained as 20.08 m/s corresponding to the observed velocity of 20.33 m/s at 1.0 m height. The corresponding prototype velocity V_p at the tip is obtained as

$$V_p = 0.9 \times 20.08 = 18.07 \text{ m/s}$$

$$V_{ref} = V_{10}(\text{mean hourly}) = 112 \text{ kmph or } 31.11 \text{ m/s.}$$

Velocity (mean hourly) at the tip of chimney

$$= 31.11 \times (218/10)^{0.09} = 40 \text{ m/s}$$

The prototype bending moment will be scaled in the ratio of the cube of the length scale Lr, so that the total base bending moment in prototype would be as follows :

$$M_p = M_m \, Lr^3 \, Vr^2$$

$$= 9065 \times (250)^3 \times (40/18.07)^2$$

$$= 6.94 \times 10^{11} \text{ N-m}$$

$$= 6.94 \times 10^{8} \text{ N-m}$$

The corresponds to a damping value of 2.3%. Though a lower value of damping is expected in the prototype no correction is made here since the along-wind response is largely insensitive to damping.

The design base bending moment for the prototype when corrected for the Reynolds number is obtained as

$$(0.8/1.2) \times 6.94 \times 10^8 = 4.63 \times 10^8 \text{ N-m}$$

Across-wind

The peak across-wind response was found to occur at a velocity of 27 m/s measured at 1m above the base of model.

Peak bending moment observed at this critical velocity

= 17500 N-mm

The velocity at the tip of the model from the observed value of 27 m/s at 1 m height= $27.0 \times (0.872/1.0)^{0.09} = 26.67$ m/s. This is quite close to the estimated value of 28.3 m/s.

The observed value of the peak base bending moment of 17500 N-mm should first be modified for the change in the critical velocity that would occur due to the deviation of the natural frequency of vibration from the desired value of 98.9 Hz to the measured value of 75 Hz. The change in velocity will be proportional to the change in natural frequency. Hence the pressures and moments will vary as the square of the natural frequency.

Therefore base moment corrected on this account

$= 17500 \times (98.8/75)^2 = 30369$ N-mm

Two more corrections need to be applied to this value before the corresponding value can be predicted for the prototype.

1. The net damping for the test can be taken as

   $= 2.3\% - 0.4\% = 1.9\%$

   0.4% corresponds to the negative aeroelastic damping. If, therefore, the value is to be reduced to 1% structural damping estimated for the prototype, the corresponding value for the model will be

   $= 1.0\% - 0.4\% = 0.6\%$

   At resonance, the excitation and hence the bending moment is related to damping in the ratio of square-root of the damping.

   Thus base moment corrected for damping

   $= 30369\sqrt{(1.9/0.6)} = 54042$ N-mm

2. The coefficient of lift, C_L is affected by Reynold's number. R_e for prototype is 2.6×10^7 and for model 1.08×10^5. A study of literature (2, 12, 13) shows that it will be reasonable to assume the values of C_L as 0.27 [2] and 0.20 [12] for the model and prototype respectively.

   Hence prototype design base bending moment

   $= 54042 \times (0.2/0.27) \times (250)^3 = 6.25 \times 10^{11}$ N-mm

   $= 6.25 \times 10^8$ N-mm

ANALYTICAL STUDIES

The along-wind response can be seen to consist of a quasi-static component due to the mean wind force and a dynamic component on account of the gustiness of wind. The mathematical formulations for the latter are based on Davenport's approach [7,8]. The across-wind response is essentially dynamic and a result of vortex-shedding. Unlike the gusts, vortex-shedding produces forces which originate in the wake behind the structure, act mainly in the across-wind direction and the resulting oscillation could be resonant in character. In the present study, the across-wind response has been obtained based on Vickery's formulation [1,12] with the values of lift coefficient. Strouhal number and correlation length as proposed by Ahmad [14].

The alongwind analysis starts with some specified value of wind speed which is then increased in steps till the maximum wind speed is obtained. The mean response and the resonant and non-resonant components due to gusts are computed separately, from which the total alongwind response is worked out. For across-wind analysis, the velocity is incremented in steps and for each velocity the point of resonance is traced, whereafter the across-wind response is computed.

One of the important factors in the analysis of across-wind excitation is the assessment of the lift coefficient, C_L, Strouhal number, S, and the correlation length, L. While Vickery treated these as constants in his earlier work [12], he as well as some other workers [1,13,14] advocate some modifications in these values for 2nd and 4th modes. Based on this and Ahmad's work [14], the following values were adopted for these coefficients.

For 1st mode : for all values of R_e;
C_L = 0.20, S = 0.22, L = $D(Z_o)$, Z_o being the height at which resonance is occuring.

For 2nd mode : for $R_e < 3 \times 10^6$;
C_L = 0.20, S = 0.22, L = $D(Z_o)$

For $R_e \gg 3 \times 10^6$; $C_L = 0.15$, $S = 0.25 + 0.04 (R_e - 3 \times 10^6)/(1.65 \times 10^7)$,
L = $D(Z_o)$

Analysis was carried out assuming the following data :

Terrain coefficient, $C_T = 0.005$
Constant B' related to the width of the spectral peak = 0.30
Background turbulence factor, B = 0.65
% of steel in the chimney cross section = 3.0

COMPARISON OF RESULTS

The peak alongwind and acrosswind base bending moments obtained from the various studies are as given below, at 1% damping.

	Alongwind BM	Across-wind BM
Rigid model	5.61×10^8 Nm	
Aeroelastic model	4.63×10^8 Nm	6.25×10^8 Nm
Analysis	5.11×10^8 Nm	8.1×10^8 Nm

The comparison is reasonably good for the along-wind values but the analytical value is rather high for the across-wind BM.

It is well known that both the aero-elastic model studies as well as the theory have their own limitations. While the aeroelastic model testing suffers from the limitations of proper simulation of the wind environment and the Reynold's number, the theory for across-wind excitation has a number of parameters whose uncertainities may cause the predicted values to change. This is underlined by the present study too and the comparison made above. It may thus be prudent to adopt the values on a conservative side.

Some changes were made in the final design of the chimney and instead of resorting back to the model studies, it was considered expedient as well as reliable, to arrive at the design forces by the analysis.

ACKNOWLEDGEMENTS

Useful discussions were held with the team of concerned Engineers from Bharat Heavy Electricals Ltd., India and Central Electricity Authority, India, during the course of this study.

Assistance received from Sri R.P. Gupta and his staff at the wind tunnel in carrying out the experimental work is acknowledged.

REFERENCES

1. Vickery, B.J., "Wind Induced Loads on Reinforced Concrete Chimneys", National Seminar on Tall Reinforced Concrete Chimneys, New Delhi, pp. 1-18, April 1985.

2. Scruton, C., "On the Wind Excited Oscillations of Stacks, Towers and Masts", Proc. Symp. on Wind Effects on Bldgs. and Strs., National Physical Laboratory, Teddington, U.K., Vol. 2, pp. 798-832 and discussions pp. 833-836, 1963.

3. Rumman, W.S., "Reinforced Concrete Chimneys", Handbook of Concrete Engineering by M. Fintel, 2nd Edn., Von Nostrand, pp. 565-586, 1986.

4. Vickery, B.J. and Basu, R., "Simplified Approaches to the Evaluation of the Across-wind Response of Chimneys", Jr. of Wind Engg. and Ind. Aerodyn., Vol. 14, pp. 153-166, 1983.

5. CICIND, "Model Code for Concrete Chimneys", Part A The Shell, International Committee on Industrial Chimneys, Dusseldorf, Oct. 1984.

6. Draft ACI Standard, "Design and Construction of Reinforced Concrete Chimneys", ACI 307-84, Detroit.

7. Davenport, A.G., "The Response of Line-like Structures to a Gusty Wind", Proc. Inst. Civil Engrs., London, Vol. 23, 1962.

8. Devenport, A.G., "Gust Loading Factors", Proc. ASCE, Jr. Strl. Div., Vol. 93, No. ST3, June 1967.

9. Vickery, B.J., "The Aeroelastic Modelling of Chimneys and Towers", Int. Workshop on Wind Tunnel Modelling for Civil Engg. Applications, Gaithersburg, Maryland, USA, pp. 408-428, April 1982.

10. Melborne, W.H., "Comparison of Model and Full-Scale Tests of a Bridge and a Chimney Stack", Proc. Int. Workshop on Wind Tunnel Modelling for Civil Engg. Applications, Gaithersburg, Maryland USA, pp. 637-653, April 1982.

11. Draft IS:4998, BDC 38 (3972), "Criteria for Design of Reinforced Concrete Chimneys Part I : Assessment of Loads", Bureau of Indian Standard, New Delhi, 1987.

12. Vickery, B.J. and Clark, A.W., "Lift or Across-wind Response of Tapered Stacks", Jr. Strl. Div., ASCE, Vol. 98, ST1, pp. 1-20, Jan. 1972.

13. Cincotta, J.J. and Jones, G.W., "Experimental Investigation of Wind Induced Oscillation Effects on Cylinders in Two Dimensional Flow at High Reynold's Numbers", Meeting on Ground Wind Load Problems in Relation to Launch Vehicles, NASA, Langly Research Centre, June 1966.

14. Ahmad, M.B., Pande, P.K. and Krishna, P., "Self-Supporting Towers Under Wind Loads", Jr. Strl. Div., ASCE, Vol. 110, No. 2, pp. 370-384, Feb. 1984.

WIND LOADS ON TRANSMISSION LINE TOWERS

A.R. Santhakumar

Anna University
Madras, INDIA

SYNOPSIS

A 230 KV transmission line on the east cost of Tamil Nadu is instrumented to provide conductor loads due to severe climatic events. The details of the line, instrumentation used and methodology are briefly described. Such field data will be useful for statistically evaluating the loads.

1. INTRODUCTION

A tower has to withstand the loadings ranging from straight runs upto varying angles and dead ends. To simplify the the designs and ensure an overall economy, tower designs are confined to a few standard types of tangent, angle and dead ends. Wind constitues the major load on Transmission Line Towers and dictates the final weights.

2. WEIGHTS OF TYPICAL TOWERS USED IN INDIA

The weights of various types of towers used on transmission lines 66 KV to 400 KV together with the spans and sizes of conductor and ground wire used on lines are given in Table 1. Assuming .that 80 percent are tangent towers 15 percent are 30° angle tower and the remaining 5 percent are 60° and dead-end towers and allowing for 15 percent extra weight for stubs and extensions of towers for a 10 Km line are also given in the Table.

3. LOAD COMBINATIONS

The various loads coming on the tower under the normal and the broken wire conditions (BWC) and their appropriate combination considered for design is shown in Table 2 for a typical 400 KV Transmission Tower (Figure 1). In this table the following notations have been used

Tension at 32°C without wind = T

Maximum tension = MT

Wind on Conductor = WC

Wind on insulator = WI

Angle of deviation = ϕ

4. VARIATION OF LOADING

Generally, there are two kinds of loads imposed on the tower - fixed loads and fluctuating loads. All actual loads vary according to statistical law but the variation of wind loads and resulting conductor tension are large.

Again the effects of these two types of loads on the design of a tower are not quite identical or all the loads are not resisted by different components of the tower in equal proportions.

(Metric Tonnes)

	400 kV Single circuit	220 kV Double Circuit	220 kV Single Circuit	132 kV Double Circuit	132 kV Single Circuit	66 kV Double Circuit	66 kV Single Circuit
Span (m)	400	320	320	320	320	245	245
Conductor	Moose 54/3.53 mm Al + 7/3.53 mm St.	Zebra 54/3.18 mm Al.+7/3.18 mm St.	Zebra 54/3.18 mm Al+7/3.18 mm St.	Panther 30/3 mm Al +7/3 mm St.	Panther 30/3 mm Al. +7/3 mm St.	Dog 6/4.72 mm Al+7/1.57 mm St.	Dog 6/4.72 Al. +7/1.57 mm St.
Groundwire	7/4 mm 110kgf/mm² quality	7/3.15 mm 110kgf mm² quality	7/3.15 mm 110kgf/mm² quality	7/3.15 mm 110kgf/mm² quality	7/3.15 mm 110kgf/mm² quality	7/2.5 mm 110kgf/mm² quality	7/2.5 110kgf/mm² quality
Tangent Tower	7.7	4.5	3.0	2.8	1.7	1.2	0.8
30° Tower	15.8	9.3	6.2	5.9	3.5	2.3	1.5
60° and Dead-end Tower	23.16	13.4	9.2	8.3	4.9	3.2	2.0
Weight of towers for a 10-km line	279	202	135	126	76	72	48

Note: Recent designs have shown 10 to 20% reduction in weights.

Table 1 Weights of towers used on various voltage categories in India

Tower type	Longitudinal loads		Transverse loads	
	Normal condition	Broken-wire condition	Normal condition	Broken-wire condition
A	0.0	0.5xTxCos ϕ/2	WC+WI+D	0.6WC+WI+0.5DA
B	0.0	1.0xMTxCos ϕ/2	WC+WI+D	0.6WC+WI+0.5D
B Section Tower	0.0	1.0xMT	WC+WI+0.0	0.6WC+WI+0.0
C	0.0	1.0xMTxCos ϕ/2	WC+WI+D	0.6WC+WI+0.5D
C Section Tower	0.0	1.0xMT	WC+WI+0.0	0.6WC+WI+0.0
D 60°	0.0	1.0xMTxCos ϕ/2	WC+WI+D	0.6WC+WI+0.5D
Dead-end with slack span (slack span side broken)	0.7MT	1.0xMT	0.65WC+WI	0.6WC+WI
Dead-end with slack span line side broken	0.7MT	0.3MTxCos 15°	0.65WC+WI+ 0.3MTxSin 15°	0.25WC+WI+0.3 MTxSin 15°
Dead-end	MT	Nil	0.5WC+WI+0.0	0.1WC+WI+0.0

Table 2 Various load combinations under the normal and broken-wire conditions for a typical 400 kV line

Vertical loads are primarily resisted by the leg members whereas transverse and longitudiual loads are resisted jointly by the legs and diagonal bracing members.

The design method incorporating the appropriate degree of structural reliability can be adopted by specifying a particular reliability class for the tower considered. The recent codes [1,2] include reliability based design considering the effect of random loads based on statistical methods. The difficulty in implementing this type of design is the lack of statistical data for the various loading conditions at particular tower sites.

5. OBJECTIVES OF TEST PROGRAMME

Load and resistance factor design approach can be successfully implemented if additional and improved wind load data is available. Towards this end a typical 220 KV double circuit tangent tower has been instrumented. The objectives of this study are:

(i) to obtain realistic conductor loads

(ii) evaluate the drag coefficient, height factor, gust factor and span effect.

(iii) review the procedure proposed in IS 802.

This study is in progress and at present the salient features and details of this study is presented. The project is undertaken by Anna University as a research project.

6. THE DETAILS OF THE LINE

The 230 KV transmission line is on east coast of Tamil Nadu and has a normal span of 320 m and the height of the tower is approximately 33m. The line belongs to Tamil Nadu Electricity Board. The details of the test tower is shown in Figure 1. The tower is in the plains (altitude 50 m above MSL). The sorrounding is mainly paddy fields with no trees or obstructions which may produce eratic wind patterns. The prevailing wind direction is NW in one season and SE in the other.

7. WEATHER STATION

The wind direction and velocity is recorded at the site using an anemometer. This is installed on the tower itself at 10m height and provides reference wind speed, wind direction and atmospheric temperature. The weather station produces analog signal which is converted and stored as digital data. The data is analysed using time series techniques.

8. FORCE MEASUREMENTS

Wind forces at conductor support and ground wire support points are monitored by load cells installed as shown in Figure 2. The load cell has a built-in swing angle indicator (Inclinometer) in two directions. The resolved force components in vertical and horizontal directions transverse to the line are as indicated in Figure 2.

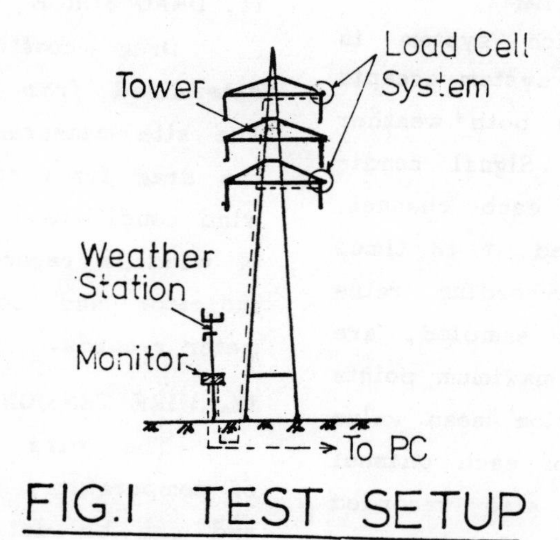

FIG.1 TEST SETUP

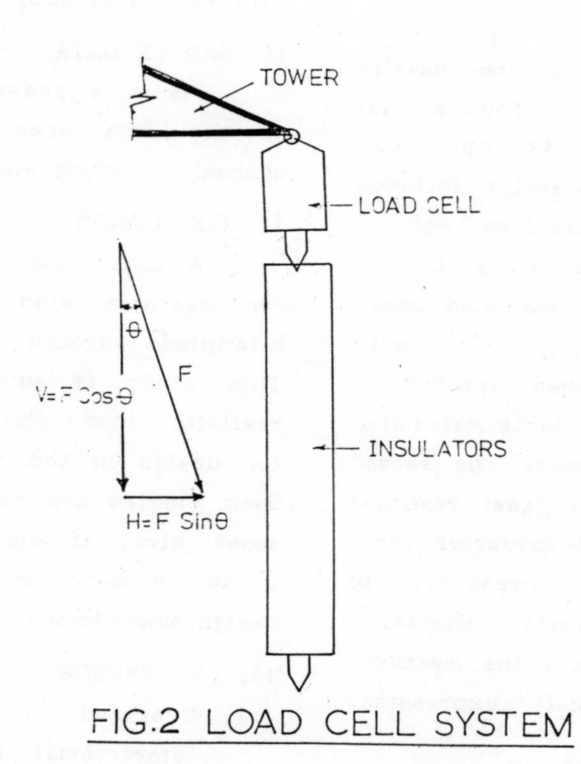

FIG.2 LOAD CELL SYSTEM

9. DATA ACQUISITION SYSTEM

The data acquisition system is shown in Figure 3. The system accepts 6 channels of data from both weather and force transducers. Signal conditioning is provided for each channel. All input data is sampled at 12 times per second. Three recording rates are set. Not all data sampled, are recorded. Only data at maximum points are recorded. In addition mean value and standard deviation of each channel during observation are also recorded on magnetic tapes of 5 megabyte capacity. The recorded data is brought back and analysed by the Personal Computer which as a link with main frame computer also.

10. DYNAMIC RESPONSE

The size of gusts are usually small compared to span. Thus spatial distribution of gusts is important. Two approaches are generally followed to estimate gust response of a conductor. One approach is to use a "gust factor" to the mean wind speed to find the peak gust. A "span reduction factor" is then applied to arrive at the conductor force calculated based on peak gust speed. The second approach is to apply a "gust response factor" directly to the conductor force calculated from wind pressure, to account for all additional effects. It is intended to compare the measured values with analytical expressions based on Davenport [3]

11. DRAG FORCE

Drag coefficients are generally determined from wind tunnel studies. The site measurements will also reveal the drag force coefficient under normal wind conditions. The angle of incidence of wind is recorded by wind direction indicator and co-related with inclinometer records.

12. WIRE TENSION

The wire tension is a function of temperature, sag and wind velocity and can be determined by Sag-tension calulations [4]. The tension is maximum at the attachement points where the slope is maximum. The site measurements is intended to be co-related with analytical computer studies.

13. SAMPLE DATA

Figure 4 shows a typical sample of the data acquired from the six channel recording station.

14. CONCLUSION

A data base for conductor loads for maximum wind exposure is being attempted through this field study. This study is intended to result in realistic load statistics and spectra for design in the maximum wind zone. Such studies are needed in other wind zones also, if significant advancement is to be made on reliability - based design methodology.

15. REFERENCES

1. Overhead Line Support Loadings, International Part 2 : Wind and Temperature Loadings Publication

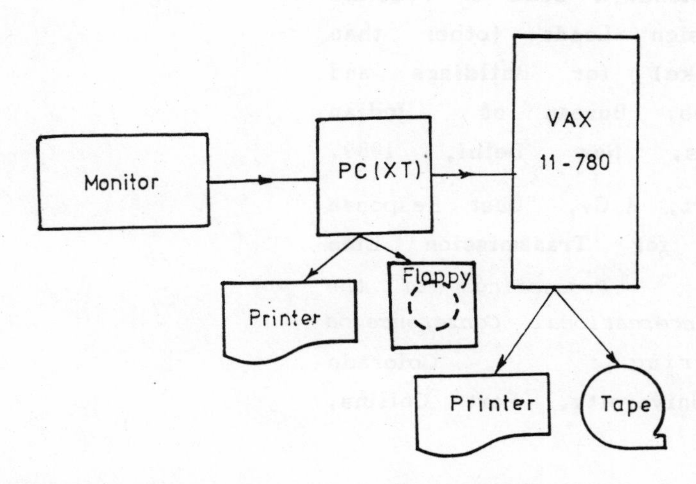

FIG.3 DATA ACQUISITION

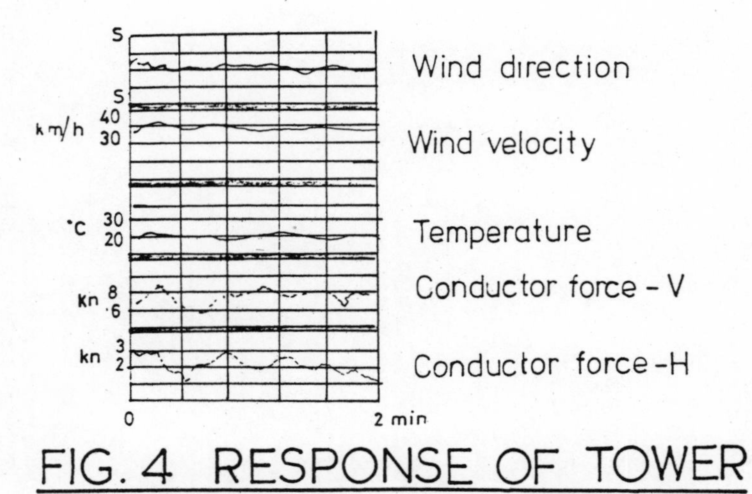

FIG. 4 RESPONSE OF TOWER

826-2, International Electrotechnical Commission (IEC), 1985.

2. Indian Standard Code of Practice for Design Loads (other than Earthquake) for Buildings and Structures, Bureau of Indian Standards, New Delhi, 1989.

3. Devenport, A.G., "Gust Response Factors for Transmission Line Loading, " *Proceedings of the Fifth International Conference on Engineering,* , Colorado State University, Fort Collins, 1979.

4. Murthy,S.S. and Santhakumar,A.R., *Transmission Line Structures,* McGraw-Hill Book Co., Singapore, pp.92-104, 1990.

WIND EFFECTS ON HYPERBOLIC COOLING TOWERS

N. Prabhakar

Technical Manager
Gammon India Limited
Prabhadevi
Bombay 400 025, INDIA

SYNOPSIS

Wind force forms the major external applied loading in the design of hyperbolic cooling towers. The paper reviews assessment of wind pressures acting on the towers with reference to Indian codes, and influence of meridional ribs on cooling tower shell with case studies. It also deals with sensitiveness of shell and steel reinforcement due to wind induced tension, and briefly on the dynamic effect in large cooling towers.

1. INTRODUCTION

Many hyperbolic cooling towers have been built in the country at several thermal and nuclear power stations. In view of their large size with very small shell thickness, they are very sensitive to horizontal loads such as wind. In this paper an attempt has been made to review applied wind loadings with reference to Indian codes, influence of meridional ribs on circumferential wind pressure distribution and its effect on the shell with case studies, and current design methods and specialised problems associated with wind effects on hyperbolic cooling towers.

2. WIND

2.1 Wind Pressure

Till the recent publication of the Indian Standard Code of Practice IS:875 (Part 3)-1987 [1] in February 1989, the design wind pressures on large number of cooling towers built since mid-1960s were calculated on the basis of the earlier code of practice IS:875-1964 [2] which adopted wind pressure as static loads, the intensity of which varying with height and the zone at which the structure is located. The new code IS:875 (Part 3)-1987 determines wind pressures based on peak wind speed of 3 second-gust with a return period of 50 years. The zones of basic wind speed at 10m above ground at speeds of 33, 39, 44, 47, 50 and 55 m/sec. are shown in the code on a wind map of the country. The design wind speed is calculated by considering factors related to probable life of structure, terrain, local topography and size of structure separately, and their combined effect is determined by multiplying the factors. Fig. 1 shows comparison of design wind pressures as per the old and new IS codes for cooling tower type structure in an open terrain. It is seen that the pressures at 39, 47 and 55 m/sec. wind speed more or less, tally with the pressures of the earlier code, and new wind zones at wind speed of 33, 44 and 50 m/sec. are introduced in the new code.

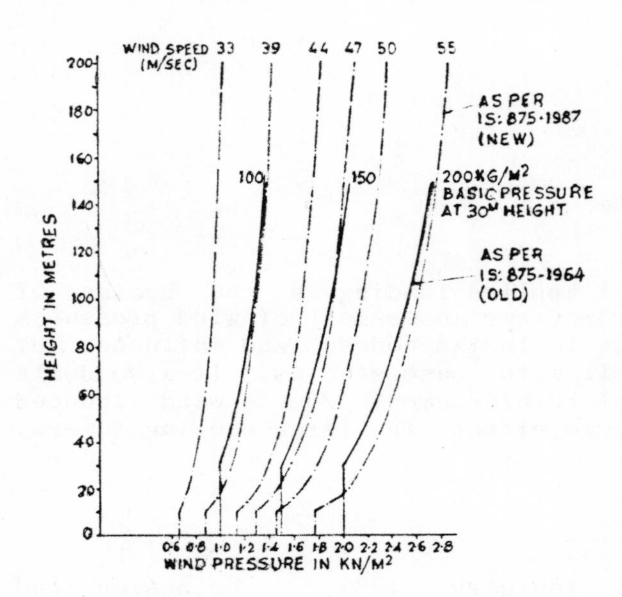

Figure 1: Design wind Pressures Old and New IS Codes for Open Terrain, Class C Structures.

The Indian cooling towers built so far have been designed for peak-wind pressures of short duration by static method. It is very well established now that wind effects on the tower are characterised by the presence of a large steady-state component and a significant random component due to air turbulence. The response of the random component can be calculated in the frequency domain by spectral analysis. This component contributes strongly to the total response peaks at a rate of atleast 50%. Although this theory is well established in principle, it is not used for practical design of cooling towers as Niemann [3] has found that large amount of computations are to be made involving several factors in both meridional and circumferential directions at different elevations of the tower, for separate cases of tensile, compressive, shear forces and bending moments in the shell. The objective approach as adopted in many codes, has been to translate the loading and structural response into a quasi-static method by applying a factor, often called as the 'Gust-Factor', in the static analysis of the tower. It must be said however, that deficiencies if any, of the equivalent quasi-static load concept are balanced by a set of provisions such as minimum shell thickness and reinforcement, high buckling safety, etc. which are observed in the practical design.

The gust factor depends on the natural frequency in the fundamantal mode, wind speed, terrain and size of structure. In view of large size of the structure, the peak response occuring in a time interval of 1 hour duration is considered appropriate for the design of cooling towers. The gust factor method given in the new IS code IS:875 (Part 3)-1987 is shown for regular shaped slender structures such as cubes, cylinders with hardly any taper. For hyperbolic shape, the diameter at the throat level is considered as the breadth of the structure, on a conservative approach. The gust factor is calculated by the following equation as per the code :

$$G = 1 + g_f r \sqrt{B(1 + \phi)^2 + SE/\beta}$$

where '$g_f r$' is a function of terrain and height of the structure, 'B' is background turbulence factor depending on terrain and size of the structure, 'ϕ' is usually zero as cooling towers are over 75 m height, and 'SE/β' is related to wind fluctuations near the natural frequency of the structure. It is found that for cooling towers, the 'G' value is governed by terms '$g_f r$' and 'B' only, and the other factors are very small and of little significance. The 'G' value usually varies between 1.6 and 2.2, the value increasing with smaller tower height and rough terrain. The gust factor given in IASS recommendations for cooling tower [4] for 'VH/(f a)' value of 0.8, 1.6 and 2.0 are 1.85, 2.0 and 2.15 respectively, where 'VH' is the mean-hourly wind at the top of tower, 'f' is the lowest mode frequency and 'a' is the throat radius. The 'G' values of IS Code and IASS recommendations are more or less close to each other. The gust factor given in German VGB guidelines [5] varies between 1.0 and 1.15, and in ACI-ASCE Report 334 [6] it is considered as 1.0, but these are for peak wind pressures instead of the mean-hourly wind pressure considered earlier. It is found that design wind pressures as calculated by the gust-factor method are more than those due to the peak-wind method by about 15 to 20%.

2.2 Distribution of Wind Around the Shell

The circumferential distribution of wind around the shell at any height is usually defined by normalising values of equal angle increments from the windward direction, and is represented by a Fourier series, $H = \sum An \cos N\Theta$. Table I shows the wind pressure cofficients 'An' which have been extensively used for the Indian towers. The IS code IS:11504-1985 [7] for natural draught cooling towers specifies the same cofficients as in BS:4485 [8].

TABLE - I

Fourier Coefficient 'An'

Harmo-nic	BS 4485 1975	Niemann 1971	Zerna 68
0	-0.00071	-0.3923	0.128056
1	0.24611	0.2602	0.435430
2	0.62296	0.6024	0.511731
3	0.48833	0.5046	0.372272
4	0.10756	0.1064	0.104642
5	-0.09579	-0.0948	-0.045549
6	-0.01142	-0.0186	-0.027082
7	0.04551	0.0468	0.018113

Note :

1. BS:4485 includes 0.4 internal pressure.
2. Niemann excludes internal pressure.
3. Zerna includes 0.5 internal pressure.

The coefficients by Zerna are based on measurements on full-scale hyperbolic cooling tower having small meridional ribs on shell. These ribs create surface roughness of the tower, and the effect of this is to reduce the suction on the sides of the tower. The roughness parameter is characterised by the ratio of K/S as shown in Fig. 2. The projection 'K' is usually 50 mm to 100 mm, and the spacing 'S' is about 2 m to 6 m. The width of the rib is taken between 2K and 5K, and this has no influence on pressure coefficients.

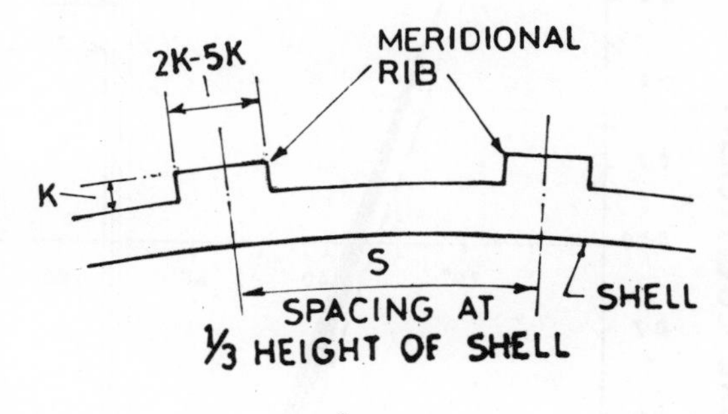

Figure 2: Roughness Parameters Ref. IASS [4] and VGB [5]

The IASS recommendations and the VGB guidelines give equations for the pressure coefficients around circumference of the shell for four different cases of rib projection, and these are given in Table.II. Graphical presentation of these coefficients are shown in Fig. 3.

TABLE - II

Coefficients for Circumferential Distribution of Wind Around Shell as per IASS Recommendations [4] and VGB Guidelines [5].

Roughness Parameter K/S	Curve	Coefficient		Zone III
		Zone I	Zone II	
0.025 to 0.100	K 1.0	$1-2.0(\sin \frac{90}{70}\theta)^{2.267}$	$-1.0+0.5(\sin(\frac{90}{21}(\theta-70)))^{2.395}$	-0.5
0.016 to 0.025	K 1.1	$1-2.1(\sin \frac{90}{71}\theta)^{2.239}$	$-1.1+0.6(\sin(\frac{90}{22}(\theta-71)))^{2.395}$	-0.5
0.010 to 0.016	K 1.2	$1-2.2(\sin \frac{90}{72}\theta)^{2.205}$	$-1.2+0.7(\sin(\frac{90}{23}(\theta-72)))^{2.395}$	-0.5
0.006 to 0.010	K 1.3	$1-2.3(\sin \frac{90}{73}\theta)^{2.166}$	$-1.3+0.8(\sin(\frac{90}{24}(\theta-73)))^{2.395}$	-0.5

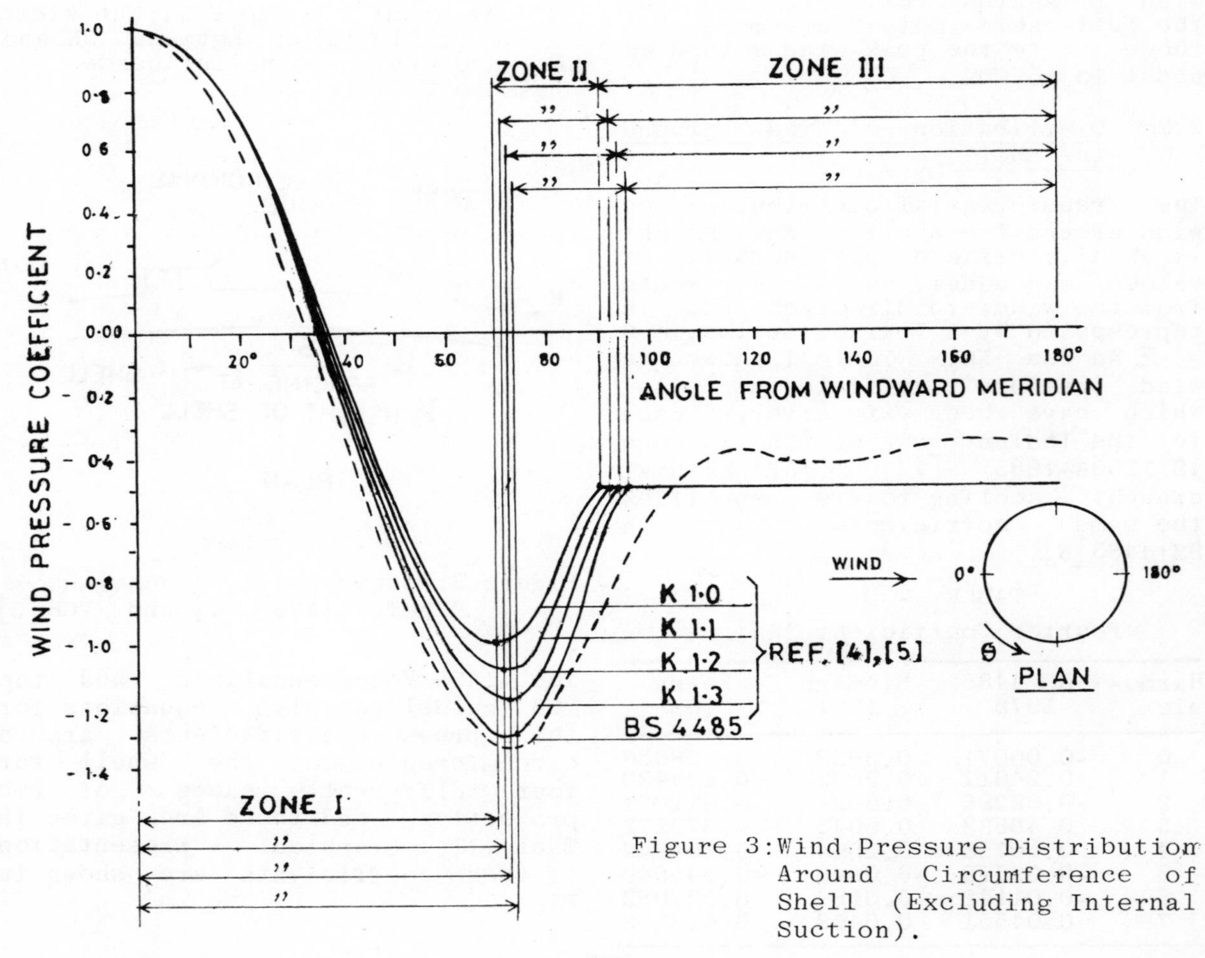

Figure 3: Wind Pressure Distribution Around Circumference of Shell (Excluding Internal Suction).

For the purpose of comparison of the effect of surface roughness, stress resultants for a cooling tower of the size given in Fig. 4 are worked out for the cases of smooth shell and four grades of roughness with meridional ribs. The shell thickness in each case is based on a minimum factor of safety of 5 against local buckling as per the IASS recommendations.

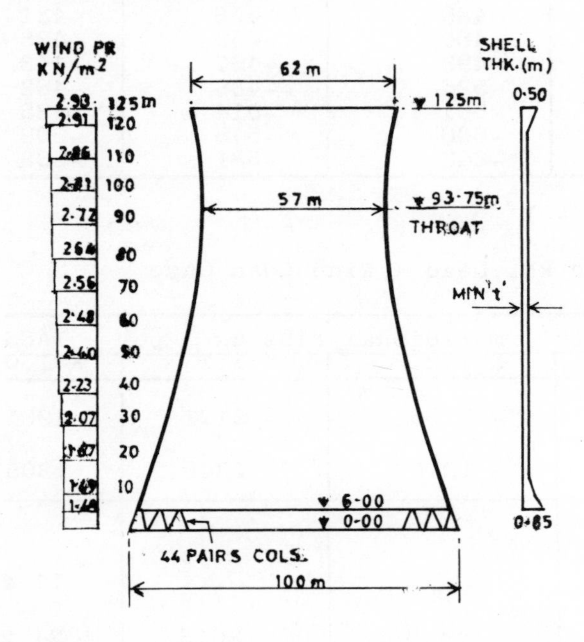

Figure 4: Cooling Tower for the Purpose of Comparison.

The results of meridional stress resultants in shell due to wind for the cases of maximum tension and compression are given in Tables III and IV. It is seen that the values of stress resultants in both the Tables are considerbly reduced at lower levels as the surface roughness increases. The significance of this reduction in the stress resultants has much effect on the requirement of shell thickness, quantity of shell concrete and reinforcement, and also on loads on raker columns and tower foundation, and these are given in Tables V, VI and VII. The maximum percentage of reductions are given below:

(i) Shell concrete quantity 4.7%

(ii) Shell reinforcement 16.4%

(iii) Raker column load compn. 11.4%

- do - Tension 24.2%

(iv) Foundation load compn. 19.5%

- do - Tension 68.0%

Undoubtedly, cooling tower shells with meridional ribs offer an economical solution, particularly for towers located in zones of high wind pressures.

TABLE - III

Meridional Stress Resultants due to Wind in kN/m, Tension at $\theta = 0$ Deg.

Level (m)	Smooth Shell as per BS:4485	Shell with meridional ribs as per IASS			
		K 1.3	K 1.2	K 1.1	K 1.0
125	0	0	0	0	0
115	40	33	31	28	28
105	164	133	127	114	114
95	342	289	278	258	256
85	508	463	450	433	426
75	650	634	620	607	590
65	740	738	716	697	671
55	800	794	763	732	702
45	834	815	779	737	706
35	851	819	780	732	702
25	863	817	778	727	701
15	875	817	780	723	705
6	894	825	789	737	717

TABLE - IV

Meridional Stress Resultants due to Wind in kN/m, Compression θ=63-72 Deg.

Level (m)	Smooth Shell as per BS:3385	Shell with meridional ribs as per IASS K 1.3	K 1.2	K 1.1	K 1.0
125	0	0	0	0	0
115	- 43	- 41	- 36	- 32	- 29
105	-180	-169	-151	-135	-121
95	-358	-345	-312	-281	-253
85	-491	-486	-447	-409	-361
75	-558	-523	-488	-453	-417
65	-563	-514	-489	-458	-431
55	-558	-502	-480	-459	-438
45	-577	-516	-493	-467	-443
35	-615	-557	-526	-488	-462
25	-664	-601	-561	-514	-485
15	-708	-648	-600	-545	-509
6	-757	-700	-643	-581	-538

TABLE - V

Axial Load on Raker Column in kN, Dead + Wind Load Case

Case	Smooth Shell as per BS:4485	Shell with meridional ribs as per IASS K 1.3	K 1.2	K 1.1	K 1.0
Max. Compn.	- 5548	- 5554	- 5356	- 5171	- 4981
Max. Tension	1621	1442	1370	1348	1305
% reduction over BS :4485					
Compression	-	-	3.6	7.3	11.4
Tension	-	12.4	18.3	20.3	24.2

TABLE - VI

Meridional Load on Tower Foundation in kN, Dead + Wind Case

Case	Smooth Shell as per BS:4485	Shell with meridional ribs as per IASS K 1.3	K 1.2	K 1.1	K 1.0
Max. Compn.	- 9665	- 9212	- 8765	- 8184	- 8091
Max. Tension	2859	2362	2115	1800	1702
% reduction over BS:4485					
Compression	-	4.9	10.3	18.1	19.5
Tension	-	21.0	35.2	58.8	68.0

TABLE - VII

Quantities of Concrete and Steel Reinforcement in Shell

Material	Smooth Shell as per BS:4485	Shell with meridional ribs as per IASS			
		K 1.3	K 1.2	K 1.1	K 1.0
Min. shell Thickness (mm)	205	200	200	195	190
Concrete (cu.m)	6414	6306	6292	6184	6129
Reinforcement (M.T.)	510	490	472	452	438
% reduction over BS:4485					
Concrete	-	1.7	1.9	3.7	4.7
Reinforcement	-	4.1	8.1	12.8	16.4

2.3 Internal Suction

The draught and flow of air through the cooling tower creates an internal negative pressure or suction, and a value of 0.4 to 0.5 is usually considered in the design. The effect of the negative internal pressure results an increase in circumferential compressive forces to the extent of 40-50% of the forces due to wind, and corresponding reduction in the values of circumferential tensile force in the shell. The stress resultants in meridional direction are least affected. It may be prudent to consider the negative pressure for the purpose of calculating buckling safety, and ignore it for calculation of circumferential reinforcement in the shell.

2.4 Cooling Towers in Group

Where hyperbolic cooling towers are located in a group, the values of design wind pressures and pressure coefficients around circumference are much affected due to aero-dynamic interference effect depending on the spacing of towers or other structures of significant dimensions in the vicinity, and the angle of wind direction in relation to the axis of alignment of the towers. For such cases, in view of not many measured data being available on full-size towers, aero-elastic model testing in wind tunnel including all adjacent local topographical features, building and other structures is necessary although the test is valid for values of Reynolds number (Re) upto about 3×10^5 for laminar airflow as against Re of more than 10^8 in actual condition under turbulent wind flow.

Generally, a clear spacing of 0.5 times the base diameter is provided between the towers, and the wind pressures are enhanced between 10 and 40 percent when designing cooling towers in groups. For some of the Indian towers built in recent years, the design wind pressures are based on wind tunnel model test carried out at the Indian Institute of Science, Bangalore. The enhancement factors considered in some of the Indian Towers in groups are given in Table VIII.

TABLE - VIII

Sr. No.	Location	Basic wind pressure (kN/m^2) height	Enhance-ment Factor
1.	Wanakbori	1.5	1.33
2.	Neyveli Stage I	2.0	1.43
3.	Raichur	1.0	1.60
4	Kutch	1.5	1.35
5.	Panipat Stage III	1.5	1.50
6.	Kawas	1.47*	1.573

Note : * at 10 m height

3. DESIGN ASPECTS

3.1 Shell Thickness

The behaviour of hyperbolic cooling tower is quite different from that of a cantilever structure such as a chimney, in that maximum meridional tension in shell occurs at azimuth 0 deg. on the windward side and the maximum meridional compression occurs at azimuth 65-75 deg. from wind direction, following the same pattern as circumferential wind pressure distribution. Circumferentially, wind load produces compression and tension, and wind moments throughout. The magnitude of wind moments both in meridional and circumferential direction are quite small and are of little significance in the design.

The concrete shell thickness is generally governed by buckling consideration resulted by self weight and wind load, and a factor of safety of 5 is provided under service load condition. The buckling safety is calculated either by using equation derived by Der and Fidler for overall safety, based on wind tunnel tests, or alternatively by the inter-active formula developed as a result of experimental studies on local buckling by Kratzig, Zerna and Mungan at the University of Bochum, Germany.

The shell thickness is also governed by its tensile strength to avoid propagation of cracks in the tension zone, and for this reason, the tensile stress in concrete is limited to about 3.0 N/mm2. There is a close relation between a high wind load factor causing tensile failure and buckling safety factor of 5 as the latter leads to the choice of a reasonable wall thickness against tensile failure.

3.2 Shell Reinforcement

The shell reinforcement is usually governed by direct tension and bending moment acting on the section arising out of dead load + wind + temperature. The reinforcement is calculated on the basis of either factored loading of 1.4 for wind and 1.0 for dead weight at steel stresses limited to 87% of the yield stress of steel as per BS:4485, or in accordance with IS:456-1978 [9] by working stress method, but without considering 33% increase in permissible stresses in concrete and reinforcement, normally permitted under wind load case. It is found that the quantity of meridional reinforcement calculated by BS:4485 is generally greater than those by IS:456 by about 10%.

The shell reinforcement is very sensitive to wind loads, and Table IX shows how wind load factor drops rapidly with the increase in wind speed. For example, if a tower is designed for an under-estimated wind speed of 39 m/sec. and the shell is reinforced as per BS:4485, the wind load factor of 1.4 reduces to 1.0 if the wind speed increases to 46 m/sec. i. e. by 18%. Statistically it means that for a return period of wind of 50 years, the risk level increases from 0.63 to 0.97, or alternatively for a risk level of 0.63, the return period of wind reduces from 50 years to 14 years. This indicates that a proper assessment of wind speed is very much essential for the design.

TABLE - IX

Wind speed (m/sec)	Wind load Factor
39	1.400
40	1.331
42	1.207
44	1.100
46	1.006

3.3 Wind Induced Vibration

For large size cooling towers, the possibility of wind induced vibrations need to be investigated. The natural frequency is inversely proportional to the size, and it drops more rapidly due to increased shell thickness which is essential to provide the required factor of safety against buckling. For towers over 160 m height. the lowest natural frequency is generally below 1 Hz, and in such cases the design should take account of dynamical amplication factor for wind load based on aero-elastic model testing. To overcome this problem, it is found that by providing horizontal stiffening rings around shell, 4 or 5 in numbers along the tower height, the factor of safety against buckling could be provided without reducing the natural frequency. The ring stiffeners are located in the region of large buckling deformations of the unstiffened shell. The size of the rings is usually 5-6 times the shell thickness as the depth, and about 0.5m as the breadth. The shell around stiffening rings is designed for additional circumferential and meridional moments due to wind and temperature loading. Such towers with stiffening rings have already been built in Germany and the USA. Figure 5 shows the natural frequencies of a 165.5 m high tower of ISAR II nuclear power plant (Ref. [10]) in Germany, for both unstiffened shell and shell stiffened with 3 rings. It is seen that there is a marked improvement in the value of natural frequency with the ring stiffened shell. Figure 6 shows the mode shapes in buckling and vibration for the same tower.

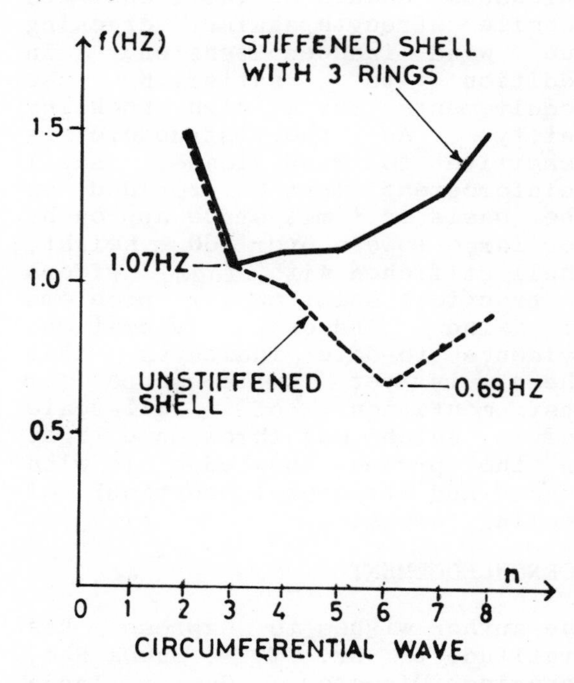

Figure 5: Natural Frequencies [10]

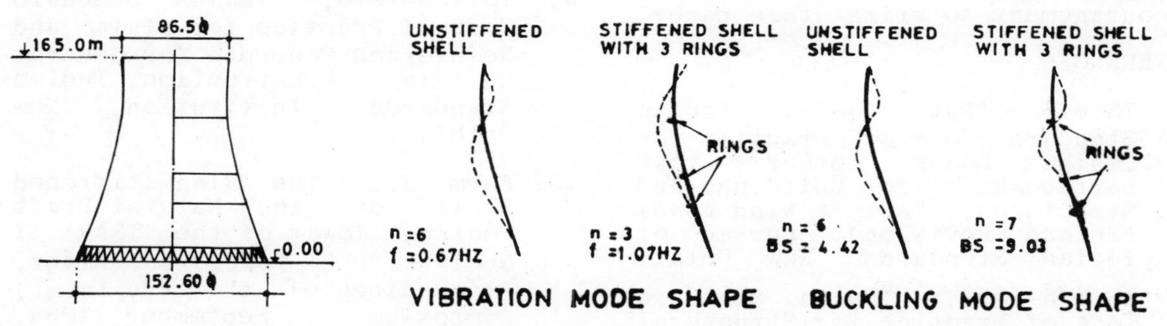

Figure 6: Vibration and Buckling Mode Shapes [10]

4. CONCLUSION

Cooling towers are undoubtedly one of the large civil engineering structures where wind forms the major applied loading in design. For analysing these structures, proper assessment of wind pressures and a clear understanding of the structural behaviour under asymmetric wind load are very much essential. The towers with increased roughness by providing meridional ribs, offer an economical solution, particularly in the high wind zones. The shell thickness should be based on its tensile strength against cracking due wind induced tension, in addition to satisfying the requirements for a high buckling safety. As the structure is sensitive to wind loads, shell reinforcement must be provided on the basis of limit-state approach. For large towers over 160 m height, shell stiffened with rings, offers a practical solution for problems of wind induced vibration. Evidence to-date indicates that there is yet ample scope for instrumentation of full-scale towers which may throw more light on the present knowledge of wind loads and structural behaviour of cooling towers.

ACKNOWLEDGEMENT

The author wishes to express his gratitude to Dr. T. N. Subba Rao, Managing Director, Gammon India Limited, Bombay, for his encouragement to write this paper.

REFERENCES

1. IS:875 (Part 3)-1987 Indian Standard Code of Practice for Design Loads (other than Earthquake) for Buildings and Structures, Part 3 Wind Loads (Second Revision). Bureau of Indian Standards, New Delhi.
2. IS:875-1964, Indian Standard Code of Practice for Structural Safety of Buildings Loading Standards (Revised), Indian Standards Institution, New Delhi.
3. Niemann H. J., Reliability of Current Design Methods for Wind Induced Stresses, Proceedings of the 2nd International Symposium, September 1984, Ruhr University, Bochum.
4. IASS Recommendations, Working Group Nr.3, Recommendations for the Design of Hyperbolic or Other Similarly Shaped Cooling Towers, Brussels, 1977.
5. Bautechnik bei Kuhlturmen, Teil 2 : Bautechnische Richtlinien (BTR) - VGB - Kraftwerkstechnik GmbH, Essen, 1979.
6. ACI-ASCE Committee 334, Reinforced Concrete Cooling Tower Shells - Practice and Commentary, Report ACI 334-2R-84, ACI Journal, November-December 1984.
7. IS:11504-1985, Indian Standard Criteria for Structural Design of Reinforced Concrete Natural Draught Cooling Towers, Indian Standards Institution, New Delhi.
8. BS:4485:Part 4:1975 Specification for Water Cooling Towers, Structural Design of Cooling Towers, British Standards Institution, London.
9. IS:456-1978, Indian Standard Code of Practice for Plain and Reinforced Concrete for General Building Construction, Indian Standards Institution, New Delhi.
10. Form J., The Ring-stiffened Shell of the Natural Draft Cooling Tower of the ISAR II Nuclear Power Plant, Proceedings of the 2nd Intnl. Symposium, September 1984, Ruhr University, Bochum.

Wind Loads on Structures, 1990

STUDY OF WIND LOADING ON ANTENNA DISHES

Prem Krishna *P.N. Godbole* *B.S. Reddy* *P.D. Porey*

Department of Civil Engineering
University of Roorkee
Roorkee, INDIA

SYNOPSIS

The paper summarises the detailed wind tunnel investigations carried out on rigid paraboloidal antenna dish models under a simulated boundary layer wind flow. The main parameters studied were the geometric ratio (f/D ratio) and orientations of antenna dish with respect to wind. Detailed pressure coefficient data has been generated which has further been utilised to obtain force and moment coefficients. A brief description on study of a latticed antenna dishes is also included.

1. INTRODUCTION

During the last few decades, there has been a rapid growth in the field of microwave communications, particularly so with the advent of satellite communication systems in the early sixties. These developments have resulted in the increased construction of large earth station antenna systems all over the world, including in India. The specific purposes for which these antenna systems are constructed, apart from microwave communication, are for radio astronomical observations and radar applications. An antenna dish is a vital element in these antenna systems, the main function of which being to radiate and receive the radio waves which propagate in space.

The size of a dish, represented by its diameter, may vary from as small as about 3 meters to a few hundreds of meters depending on the different communicational parameters, such as the operational wavelength, and the specific purpose for which the system is used. Generally, small dishes of 3-5 meter diameter are required for local television transmission; satellite communications require 10-60 meter diameter dishes and astronomical observations require very large dishes, or arrays of smaller dishes.

As in most cases of structural engineering design, the two requirements for antenna systems are the achievement of adequate strength and rigidity;

the latter being important in this case so as to maintain the geometrical accuracy of the reflector surface, as well as its spatial position. To attain the above objective a good assessment of loading is needed backed up by a procedure of accurate analysis and design of members. The different loads involved herein are : gravity, wind, seismic and thermal. While the evaluation of thermal and gravity loading is straight forward and a deterministic one, it is the determination of wind or seismic loading which presents a special problem. Antenna dishes being light in weight, seismic loading is unlikely to be important. In fact it can be said without much debate that wind loading, which is the subject of this paper, is the most important for antenna structures.

The evaluation of the wind loading requires mainly the information on the design wind speed and wind pressure data, preferably in a non-dimensional form of pressure coefficients. This has to be done either by adopting the design values based on Codal provisions as available, or, from measurements made in a wind tunnel or in the field. However, the Codes make only a cursory coverage of the subject. Further, a review of the literature on wind loading aspects of antenna structures reveals that there is only scanty information available. A limited amount of wind pressure data in the non-dimensional form of pressure and force coefficients has been presented by Cohen [1], Fox [2] and Sachs [3]. The pressure coefficient data is available only for a very few orientations of the dish with respect to the direction of the wind flow. These orientations may not include those which could be the most critical ones for evaluating the maximum wind forces. Likewise, the available aerodynamic force-moment data relates to only some specific dishes and as such may not be fully acceptable as a general basis for design computations. The two courses open, to obtain further information on wind loads are, (i) study of antenna dishes in the wind tunnel, and (ii) measurements in the field on prototypes. Field studies are not only more difficult to control but are expensive and time consuming as well. As such wind tunnel testing is perhaps the most expedient course open.

In India the only earlier work carried out on the subject as per the knowledge of the authors has been at the Indian Institute of Science, Bangalore. The present paper reports a study carried out to reduce the gaps in the existing available research information. The details of the experimental work related to the determination of wind pressure data, including the evaluation of force or moment coefficients for paraboloidal antenna dishes is described herein. These dishes are the most extensively used in the three principal fields of their applications, i.e. radio astronomy, radar and microwave communication systems.

2. EXPERIMENTAL PROGRAMME

The major part of the paper describes the study on solid dishes while a brief description (only sec. 5) is given for another study on a latticed dish. The solid models did not relate to any specific prototype, and were designed considering the typical geometrical proportions. The size of the models were chosen keeping in view the blockage problems and the Reynolds number issue.

Wind tunnel investigations were performed in an open-circuit type boundary layer wind tunnel at the University of Roorkee. The tunnel has a test cross-section of 2.00m x 1.85m at the upstream end, varying uniformly to 2.10m x 2.15m at the downstream end over a test section length of 15 m. The desired velocity profiles with the required turbulence characteristics were obtained in the tunnel with the help of floor roughening devices, a barrier wall and vortex generators.

The size of the solid dish model was fixed at 400 mm, mainly on the basis of an acceptable level of blockage in the wind tunnel, as mentioned above. The blockage, based on the frontal projected area of the dish surface with respect to the test cross-section of the wind tunnel amounted to about 2% and was thus within the acceptable limits. As such the blockage effects were neglected. The simulation of the back-up structure for the dish was not included in this study, because the back-up structure is expected to have only a small effect on the pressure coefficients, forces and moments [4,5]. Reynolds number values in practice will fall between 10^7 and 10^8 for such dishes. Values of this order could not possibly be obtained in the tunnel. However, attempts were made to minimise Reynold number effects by producing sharp edged antenna dish models. Tests were done at a Reynolds number of 4.4×10^5.

3. PRELIMINARY TESTS

It was considered necessary to undertake certain preliminary tests to help in understanding the significance of the many test parameters involved, to formulate the test programme and streamline the procedure. As such, a few preliminary wind tunnel tests were conducted on a rigid paraboloidal antenna dish model with 400 mm dia.,62.5mm depth and 3 mm thickness. The parameters studied during these tests were elevation angle (α), azimuth angle (β), positioning of the dish above the tunnel floor (g), length of metallic tubes at the pressure taps on the dish, sampling time of pressure record and roughness of the dish surface. 48 pressure tappings were provided on the dish surface to record the pressure measurements (see Fig. 1). Metallic tubes were inserted into the pressure tappings and affixed to the same. These were then connected through flexible tubing to the measuring instrument. The mounting system to which the model dish was attached, was duly shielded and permitted the rotation of the dish so that it could be placed in the desired orientation.

The trial model was tested in a simulated boundary layer flow developed in the wind tunnel using 3 vortex generators and a barrier wall placed near the entrance to the test section. A pressure transducer M.K.S. Baratron was used to measure the differential of pressure at each of the tappings and a suitably chosen reference point. Side-to-side symmetry was assumed to reduce the number of measurements needed. Pressures measured were reduced to non-dimensional coefficients, as described below.

$$C_{p1} = \frac{p_1 - p_o}{\frac{1}{2} \rho V^2} \quad (1)$$

$$C_{p2} = \frac{p_2 - p_o}{\frac{1}{2} \rho V^2} \quad (2)$$

and

$$C_p = C_{p1} - C_{p2} \quad (3)$$

where

p_1 = mean surface pressure on concave face of the dish,

p_2 = mean surface pressure on convex face of the dish,

p_o = free stream reference pressure,

C_{p1} = mean concave surface pressure coefficient,

C_{p2} = mean convex surface pressure coefficient, and

C_p = pressure difference coefficient.

The reference dynamic pressure used to reduce the pressure data into the above non-dimensional form was calculated with respect to the wind velocity measured at 22.5 cm. height above the tunnel floor.

The important conclusions derived from this study were :

i) The pressure coefficients are non-uniform over the surface of the dish and, as expected, are strongly dependent upon the orientation angles α and β of the dish.

ii) The variations in the parameters g, wind speed, and length of metallic tubes have negligible effect on the magnitude of pressure coefficients.

iii) The degree of surface roughness along the dish surface has significant effect on the magnitude of pressure coefficients, which in the present case decreased as much as 30% as the degree of surface roughness was increased.

iv) A sampling time of 10 seconds is sufficient for averaging the pressure, beyond which the average value becomes asymptotic to the averaging time.

4. DETAILED WIND TUNNEL INVESTIGATIONS

Based on the above preliminary tests, a detailed test programme was undertaken for tests in the wind tunnel. The parameters which were varied for this study are the orientation angles and the geometric ratio, f/D, where f is the focal depth and D is the diameter of the dish.

The orientation angles α and β were varied at regular interval of 15° in the range of 0 to 90° and 0 to 180° respectively. The 79 test orientations thus obtained ostensibly covered the range of angles expected in practice. The geometric ratio f/D, was varied to give a fair representation to its usual range of 0.25-0.50. These ratios are 0.25, 0.33, 0.40 and 0.50. All the models were made from 3 mm thick aluminium sheet and their profiles are shown in Fig. 2. It was ensured that the dish surface is as smooth as possibe and that all the dishes have a similar surface. The models were provided with pressure tappings at 48 locations on the dish surface and were attached to the mounting system as earlier for the preliminary tests.

Models were tested in a simulated boundary layer wind flow typical for a comparatively open terrain, where antenna systems will generally be expected to be installed. Such a terrain could be described as one with well scattered obstruction having heights generally from 1.5 to 10 m, in consonance with the terrain category-2 of the recently revised Indian Standard Code related to wind loads [6]. The above wind flow was developed in the tunnel by trial and error and with the help of 5 vortex generators, a barrier wall and 25 wooden blocks of 100 mm size which were arranged in a format as shown in Fig.3. The corresponding velocity profile and the turbulance intensities are shown in Fig. 4. A photograph of typical dish is shown in Fig. 5. The above wind flow conditions were set and measured with the help of Hotwire anemometer, pitot-static tube and M.K.S. baratron.

Out of the four models fabricated for testing purposes, detailed wind tunnel tests were carried out on only one antenna dish model with f/D ratio 0.40. The wind tunnel tests on other models were carried out only for certain critical directions of the dishes which were arrived at on the basis of the results of the wind tunnel tests on the first model, and the analysis of the data thus obtained.

Mean pressures* were recorded on both faces of the dish, i.e. concave and convex faces, separately. Typical results in the form of mean pressure-difference coefficients (C_p) for the six critical directions for the model dish with f/D ratio 0.4 are presented in Fig. 6.

The wind pressure data was also reduced to the form of overall forces

---

* Fluctuating pressures were measured too for a few cases for purposes of carrying out a dynamic analysis of the back-up structure, not reported in this paper.

and moments, with respect to the wind axes and body axes systems. Typical results are presented in a non-dimensional form of force and moment coefficients in Figs. 7 and 8, which describe the variation of drag and yawing moment with respect to the position of the dish.

The general pattern of results compares well with those given earlier by Fox [2] and Sachs [3,5], but a direct quantitative comparison is not feasible since their exact test conditions are not known.

5. MEASUREMENTS ON A LATTICED DISH MODEL

In continuation of the detailed study described in Secs. 2 to 4 on solid dishes, wind tunnel measurements were made on a 1/50 scale rigid model of a 45 m diameter antenna dish of latticed steel work (see photograph in Fig.9) for a Giant Meter Wave Radio Telescope Project. The model was fabricated in Bangalore and tested in the Wind Tunnel at University of Roorkee. Purpose of the exercise was to determine total force coefficients in the along wind (F_x) and the across-wind (F_y) directions for different orientations of the dish in a boundary layer profile corresponding to an open terrain. Measurements were made with a mean velocity of 16 m/s in the tunnel at a height of 1 m above the floor (power law coefficient = 0.1). The azimuth angle (α) was varied from 0° to 180° and the elevation angle (β) from 10° to 110°. The maximum force F_x occured for $\alpha = 10^o$ and $\beta = 180^o$ and that for F_y occured at $\alpha = 20^o$ and $\beta = 45^o$. The maximum value of F_y was about 40% of the maximum value of F_x.

6. CONCLUSIONS

1. The study made on solid dishes is comprehensive and has made it possible to determine pressure, force and moment coefficients related to a wide range of variable parameters. Only sample results are given in this paper.
2. It is seen that the maximum values of the significant parameters occur at one of the following critical orientations of the dish as given in Table 1.
3. The correlation between results reported earlier with that of the present study is good both qualitatively as well as quantitatively.
4. It appears possible that if a comprehensive programme of measurements is undertaken to cover different variables parameters including the approach terrain conditions, the study could be expected to yield dependable design data for application in a majority of cases of antenna structure design, obviating the need for tests in many cases.

TABLE 1 - CRITICAL ORIENTATIONS

Sl. No.	Critical Direction	Related Parameters
1.	$\alpha = 0^o$, $\beta = 0^o$ (Horizon- position)	Peak mean concave pressure coefficient(C_{p1})
2.	$\alpha = 0^o$, $\beta = 90^o$	Peak yawing moment coefficient (C_{ym})
3.	$\alpha = 30^o$, $\beta = 0^o$	Maximum drag force coefficient (C_D)
4.	$\alpha = 30^o$, $\beta = 30^o$	Peak mean pressure-difference coefficient (C_p)
5.	$\alpha = 90^o$, $\beta = 0^o$ (Zenith position)	Peak pitching moment coefficient (C_{pm})

7. ACKNOWLEDGEMENT

The paper is based primarily on the study made by B.S. Reddy for his doctoral work carried out under the supervision of Prem Krishna and P.N. Godbole. The help received in the experimental work from R.P. Gupta and other staff at the wind tunnel is gratefully acknowledged.

8. REFERENCES

1. Cohen, E. and Suh, S.S., "Calculation of Wind Forces and Pressures on Antennas", Annals of New York Academy of Sciences, Vol. 116, 1964.

2. Fox, N.L., "Load Distribution on the Surface of Paraboloidal Reflector Antennas", Internal Memorendum CP-4, Jet Propulsion Laboratory, California Institute of Technology, U.S.A.

3. Sachs, P., "Wind Forces in Engineering", Pergamon Press, 1978.

4. Blaylock, R.B. et.al., "Wind Tunnel Testing of Antenna Dish Models", Annals of New York Academy of Science, Vol. 116, 1964.

5. Richards, C.J., "Mechanical Engineering in Radar and Communications", Von Nostrand Reinhold Company, London, 1969.

6. IS:875 (Part 3) - 1987 "Indian Standard Code of Practice for Design Loads (other than Earthquake) for Buildings and Structures-Part-3 Wind Loads", Bureau of Indian Standards, New Delhi, India.

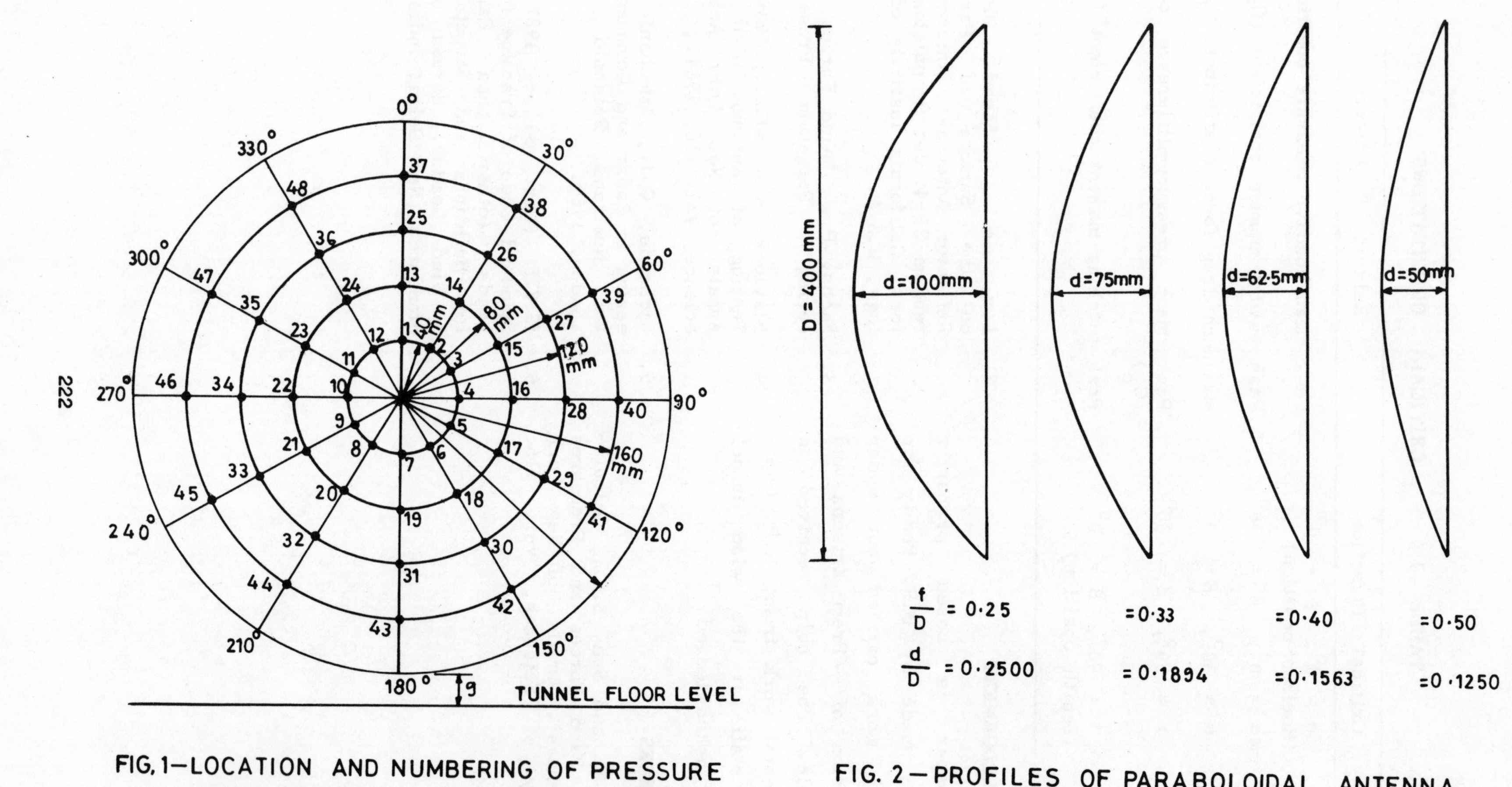

FIG. 1—LOCATION AND NUMBERING OF PRESSURE TAPPINGS ON PARABOLOIDAL ANTENNA DISH MODELS

FIG. 2—PROFILES OF PARABOLOIDAL ANTENNA DISH MODELS

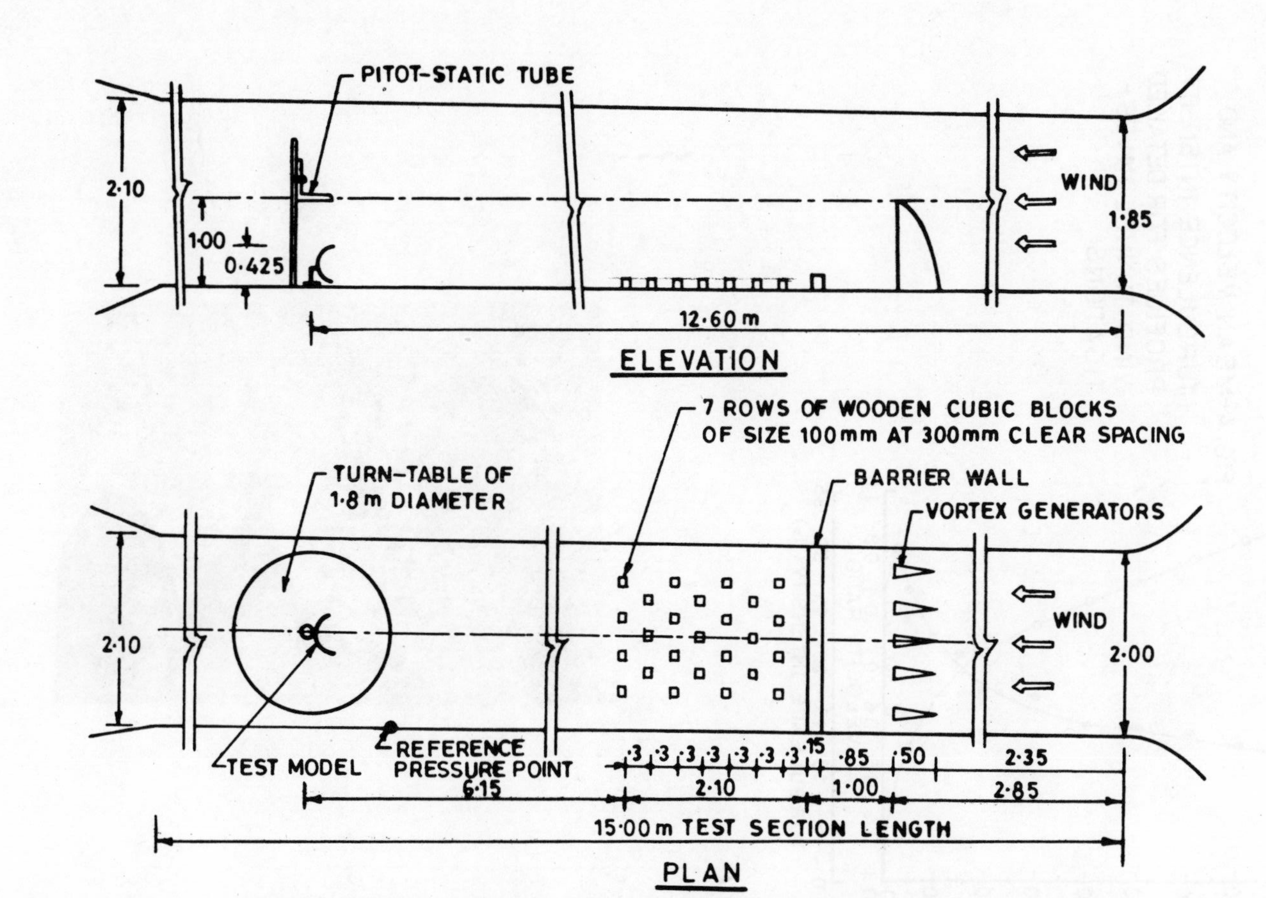

FIG. 3—ARRANGEMENT IN THE WIND TUNNEL TEST SECTION

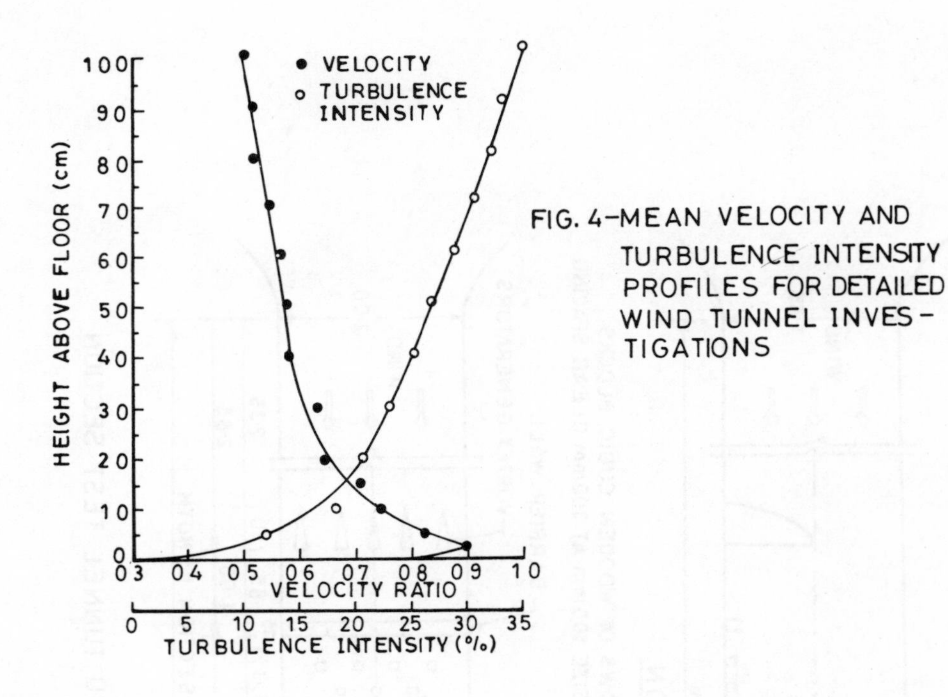

FIG. 4-MEAN VELOCITY AND TURBULENCE INTENSITY PROFILES FOR DETAILED WIND TUNNEL INVESTIGATIONS

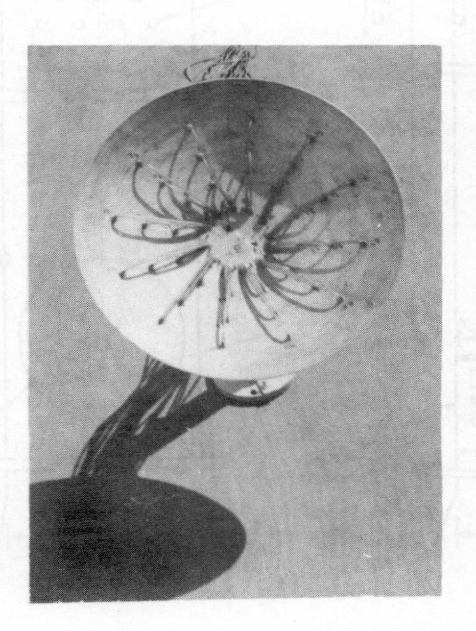

FIG. 5 - SOLID DISH MODEL WITH PRESSURE TUBINGS

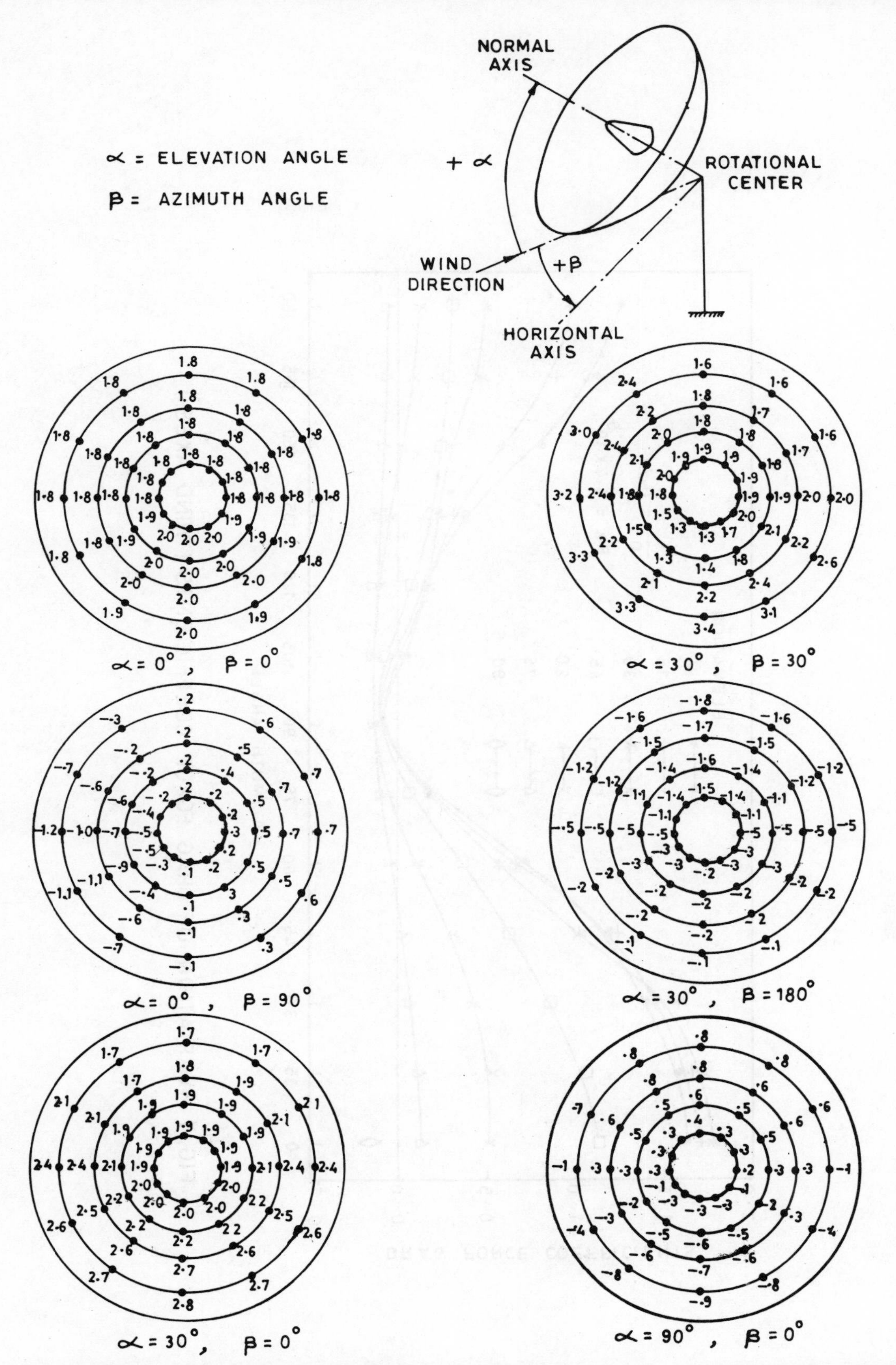

FIG. 6 — MEAN PRESSURE DIFFERENCE COEFFICIENTS FOR PARABOLOIDAL ANTENNA DISH (f/D = 0·40)

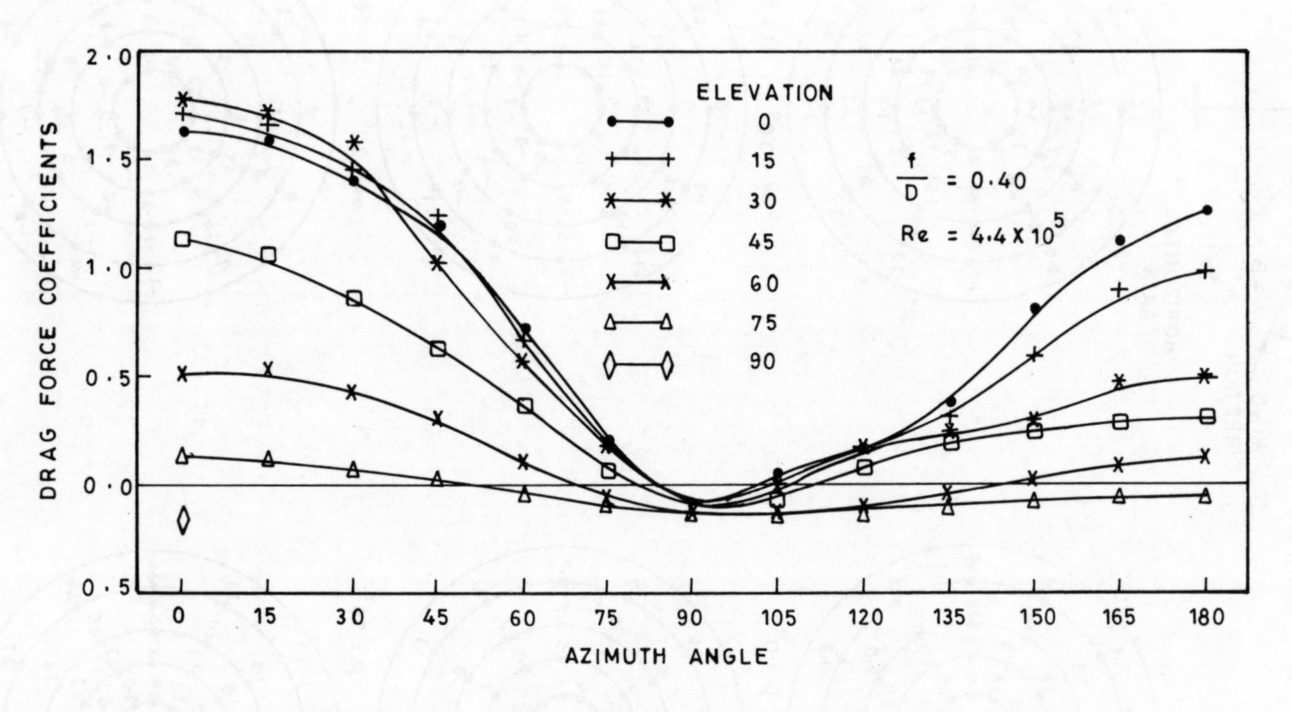

FIG. 7 —VARIATION OF DRAG FORCE COEFFICIENTS (WIND AXES)

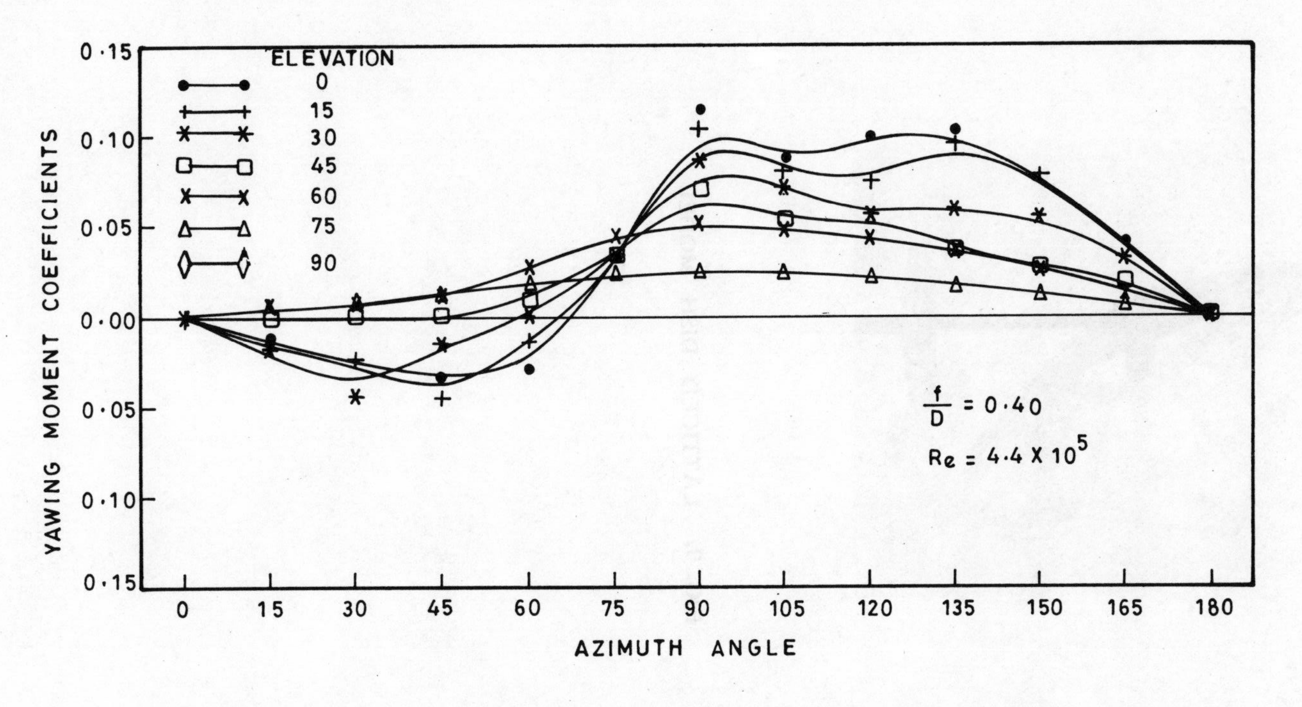

FIG. 8 — VARIATION OF YAWING MOMENT COEFFICIENTS (WIND AXES)

FIG. 9. LATTICED DISH MODEL

Wind Loads on Structures, 1990

AEROELASTIC MODEL TESTS ON CABLE-STAYED SLURRY PIPE BRIDGE OVER THE RIVER HOOGHLY

D. Yadav   A. K. Gupta   S.K. Jain   P. Dayaratnam

Indian Institute of Technology
Kanpur, INDIA

SYNOPSIS

Wind Tunnel investigation on a cable-stayed bridge have been presented. Tests were conducted with a full bridge model in different stages of erection. The maximum strains in the deck and pylon were monitored for varying wind velocities and directions of the flow. Their effects on the mean and oscillatory amplitude of the structural response were studied. The bridge with open deck construction was found to have stable aeroelastic characteristic.

INTRODUCTION

Cable-stayed bridges are slender and flexible structures and are susceptible to induced vibrations in natural winds. At times these oscillations can produce more movement than the design loads for the bridge. The aeroelastic effects include vortex induced oscillation, flutter, buffeting, torsional divergence and galloping instability in the presence of self excited forces [1,2]. Besides the deck, the bridge pylon and cables are also affected by the aeroelastic phenomenon.

The interactive forces of the natural wind with the bridge structure are complex. The wind has spatial and temporal randomness. Even minor differences in sectional configurations have been known to strongly affect the behaviour of the overall structure. A systematic treatment of the bridge aerodynamics is difficult and empirical methods are resorted to for analysis [3]. Under these circumstances model wind tunnel tests are pertinent in development of new designs, as well as, for performance evaluation of existing structures.

The experimental investigations should predict the behaviour of the complete bridge along with the intermediate construction stages as these may represent critical conditions due to incomplete support system.

This paper reports the aeroelastic investigation for a proposed 360 m span cable-stayed bridge. Tests were conducted with three construction stages - pylon alone, half bridge and full bridge. The following sections present the problem background, model development, test description, results, discussion and the conclusions drawn.

BACKGROUND

The proposed cable-stayed bridge is to support four ash slurry pipes from Bandel Thermal Power Station across the river Hooghly to the ash disposal location. It has an overall span of 360 m. The bridge is supported by thirty four cables from two pylons of 46.4 m height. Some features have been specially adopted in the design to improve its aerodynamic stability. The deck has been left uncovered as the bridge is not open to traffic. Disconnected walkways are provided on the sides to be used during inspection and repairs. This has been adopted as the uncovered deck generates less severe flutter forces compared to a covered deck. The parapets provided at the deck level are perforated to break up the wind flow continuity. Truss deck stingers are used in lieu of plate-girders to deter the vortex excitation. Stiffening truss chords and bracings are provided in the deck and the pylon frames to improve torsional stiffness. The truss mountings are projecting out board of the main frame to decrease the moment instability. The overall arrangement of the bridge is shown in Figure 1 with some structural details presented in Figure 2.

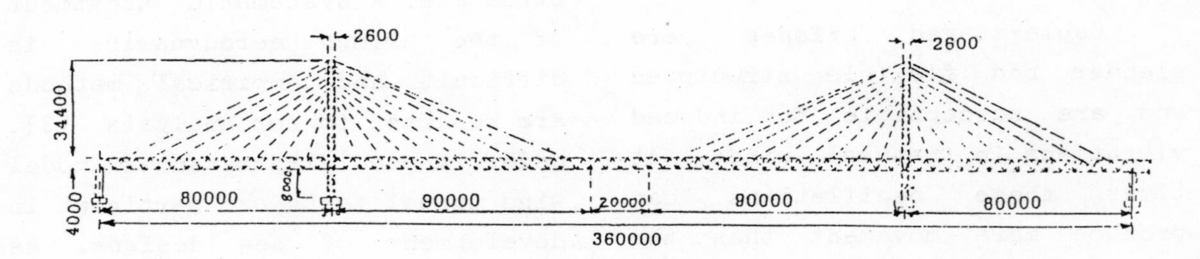

Fig. 1. Prototype Cable-Stayed Bridge

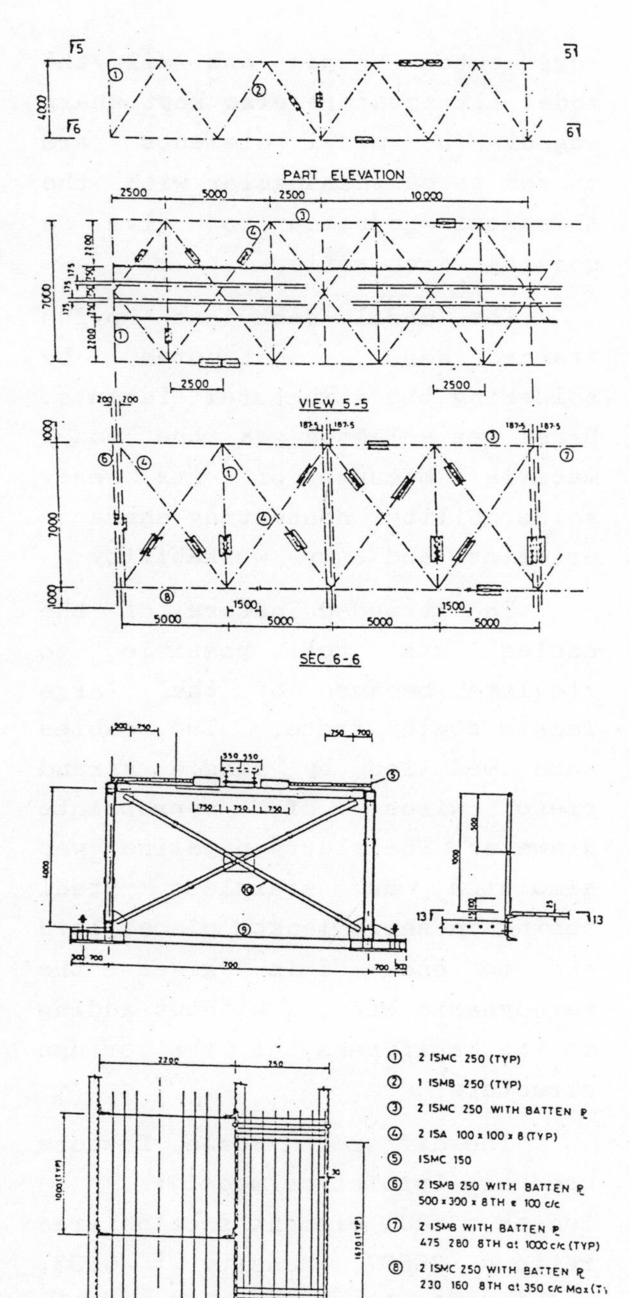

Fig. 2. Bridge Structural Detailes

Three types of model tests are common for bridge aeroelastic studies - full bridge model, sectional model and taut strip model [2]. Out of these the full bridge model has been adopted for this study. This allows the simulation of geometrical, mass distribution and mode shape similarities between the model and the prototype.

The bridge is made up of truss frame work with standard structural members. As the dependence of wind induced forces for sharp edged members on the turbulence level is very small, in practice, the wind tunnel tests are conducted in smooth flow available in the aeronautical wind tunnel.

MODEL DEVELOPMENT

The wind tunnel used has a test section of 900mm X 600mm size. This constrains the length scale factor to be 400 to enable the full bridge model to be accommodated inside the test section. Use of this large scale factor with thickness of the structural elements gives sizes too small for practical fabrication. Further, the requirement of good workability of the model material for a proper fabrication forces the model and prototype materials to be different. These result in the need to have unequal scaling factors in the thickness dimension in comparison to the length and width dimensions.

The requirements of complete model similarity to the prototype needs the scaling factors for all evolved forces to be the same for the model study. Comparison of the different force scale factors give a set of operative modelling rules [4]. However, because of the large length scale factor and conflicting nature of some of the requirements, all the rules cannot be satisfied simultaneously. As a compromise the rules controlling the main aspects of the study are satisfied first, while the remaining are attempted to be followed as far as possible. For the present study sectional solidity ratio, stiffness and nature of the outer shape of the structural element are considered important for simulation. Though complete sectional details cannot be duplicated by the model, these three properties are simulated. The solidity ratio is maintained by having uniform scales for the two dimensions facing the flow. The stiffness is modelled by confirming the area ratio of the structural element sections in the model to that of the prototype. The bridge elements have sharp edged outer shapes and all the model elements are also kept sharp edged. The model elements are chosen to be rectangular with the thickness adjusted to give a constant area ratio.

The model deck and pylon frames were constructed by soldering the structural elements. Brass was selected as the model material because of its easy solderability, nonrusting characteristics and easy workability.

The stranded nature of the cables was not possible to simulate because of the large length scale factor. The cables were modelled by single strand piano wires of appropriate diameter. The slurry pipeline was simulated by stainless steel tubing in short length pieces laid end to end. This gives the aerodynamic effects without adding to the stiffness of the bridge structure.

The different scale factors used are (prototype/model): length = 400, element section area ratio = 26667, density = 0.93, modulus of elasticity = 2; cable density = 1, cable elasticity = 1; wind velocity for static loading = 3.46, wind velocity for dynamic loading = 8, frequency = 0.02, amplitude = 400, strain = 1, stress = 2.

The values of the solidity ratio for the prototype, and adopted for the model, are :

bridge in elevation = 30%

bridge deck = 20%

pylon in elevation = 50%

Figure 3 shows some elements used in the model fabrication.

TESTING TECHNIQUE

The wind tunnel testing of the bridge model was conducted for three construction stages - pylon alone (Fig.4), half bridge deck without pipelines while the deck ends remained unsupported (Fig.5) and full bridge with pipelines and all supports in position (Fig.6). The strains on the structure were measured by using variable resistance strains gauges. The responses monitored are pylon

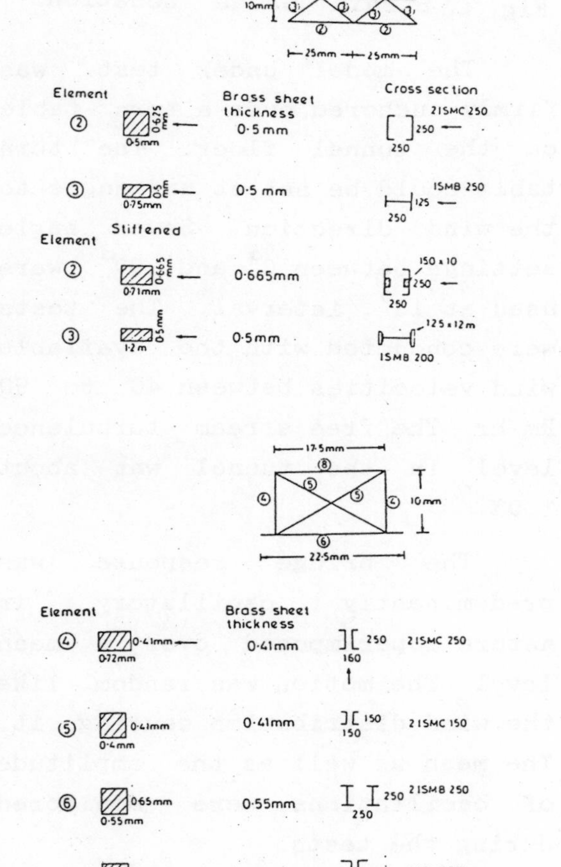

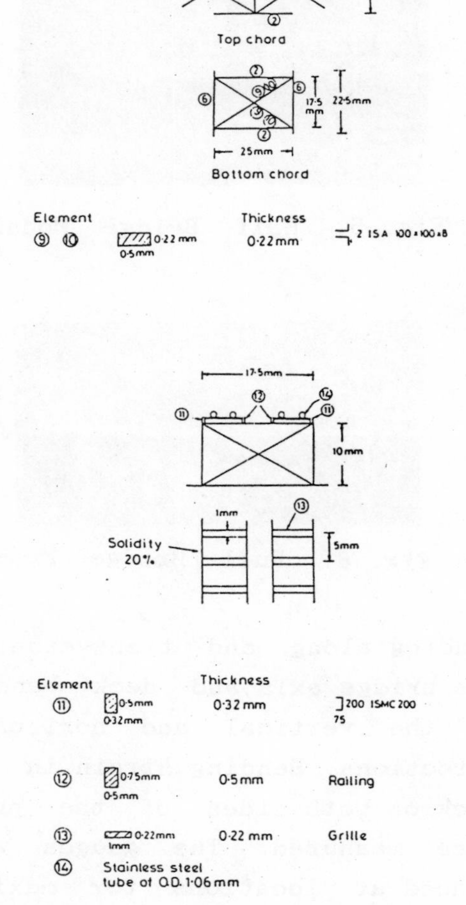

Fig. 3. Elements of Bridge Model

Fig. 4. Pylon Model

Fig. 5. Half Bridge Model

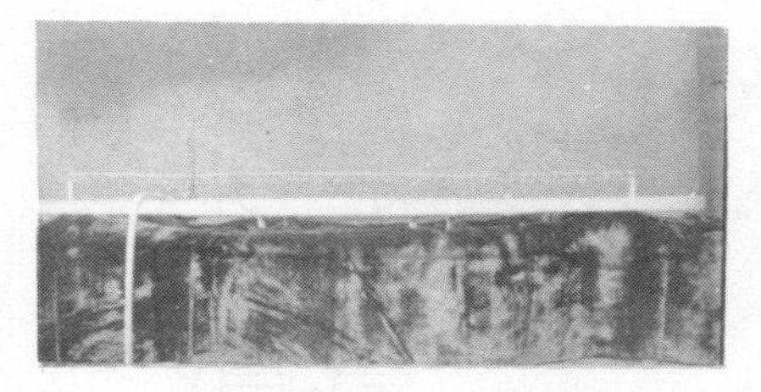

Fig. 6. Full Bridge Model

bending along and transverse to the bridge axis and deck bending in the vertical and horizontal directions. Bending strain in the deck on both sides of the pylon were measured. The gauges were placed at locations for maximum sensitivity. Their positions are shown in Figure 7.

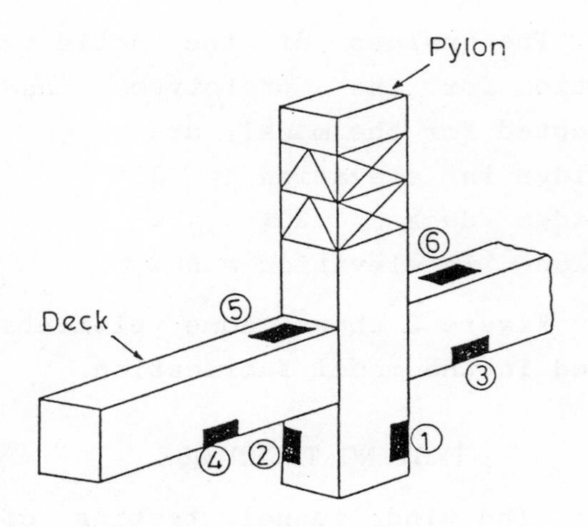

Fig. 7. Strain Gauge Locations

The model under test was firmly anchored over a turn table on the tunnel floor. The turn table could be set at any angle to the wind direction. Seven angle settings between 0° and 90° were used at 15° interval. The tests were conducted with the available wind velocities between 40 to 90 km/hr. The free stream turbulence level in the tunnel was about 1.0%.

The bridge response was predominantly oscillatory in nature superimposed over a mean level. The motion was random like the wind distribution causing it. The mean as well as the amplitude of oscillations were monitored during the tests.

A cathetometer with an accuracy of 0.01 mm was used to observe the deflection in the bridge deck.

RESULTS AND DISCUSSION

The mean strain and dynamic strain amplitudes for the tests with the pylon alone and the half bridge deck were small for the different flow directions. No untoward aeroelastic interaction was observed and the outcome of these tests are not being presented.

The results for the full bridge model are presented in figures 8 and 9 for the pylon and deck strains [5]. The deflection observed by the cathetometer were very small throughout the test.

Figure 8 shows the mean strain response with the six gauges for the seven flow directions used. Both the pylon and deck mean responses have, in general, a tendency to increase with the wind velocity. The pylon in bending transverse to the deck for flow angle setting of 30°, has a resonance peak at 250 km/hr prototype wind velocity. The tendency for resonance for the pylon response for other settings, if present, are very weak. The mean vertical response of the deck (gauges 5 and 6) have a weak resonance indication for most flow directions in the higher velocity range. In all cases the strain magnitudes are low.

Figure 9 shows the deck and pylon dynamic strain amplitudes for the airflow directions used. The magnitude of response for the end section of the deck (between the pylon and the river bank) was too low in most cases and has not been presented. The pylon experienced multiple resonances of varying magnitudes at different velocities for the different angular settings. However, no pattern is decernable. The bridge deck dynamic strain amplitude does not show any prominent resonance in the response. These have a steady increase in magnitude with increasing wind velocity.

The maximum predicted mean prototype stress at the wind speed of 204 km/hr specified for that location are:

Pylon –

- transverse to deck axis - 232.5 MPa for deck at 30° to wind
- along deck axis - 45.6 MPa for deck at 45° to wind

Deck –

- Horizontal - 71.6 MPa for deck at 75° to wind
- Vertical - 48.4 MPa for deck at 90° to wind

The maximum predicted prototype dynamic stress amplitude are:

Pylon –

- transverse to deck axis - 90.2 MPa for deck at 75° to wind
- along deck axis - 50.8 MPa for deck at 60° to wind

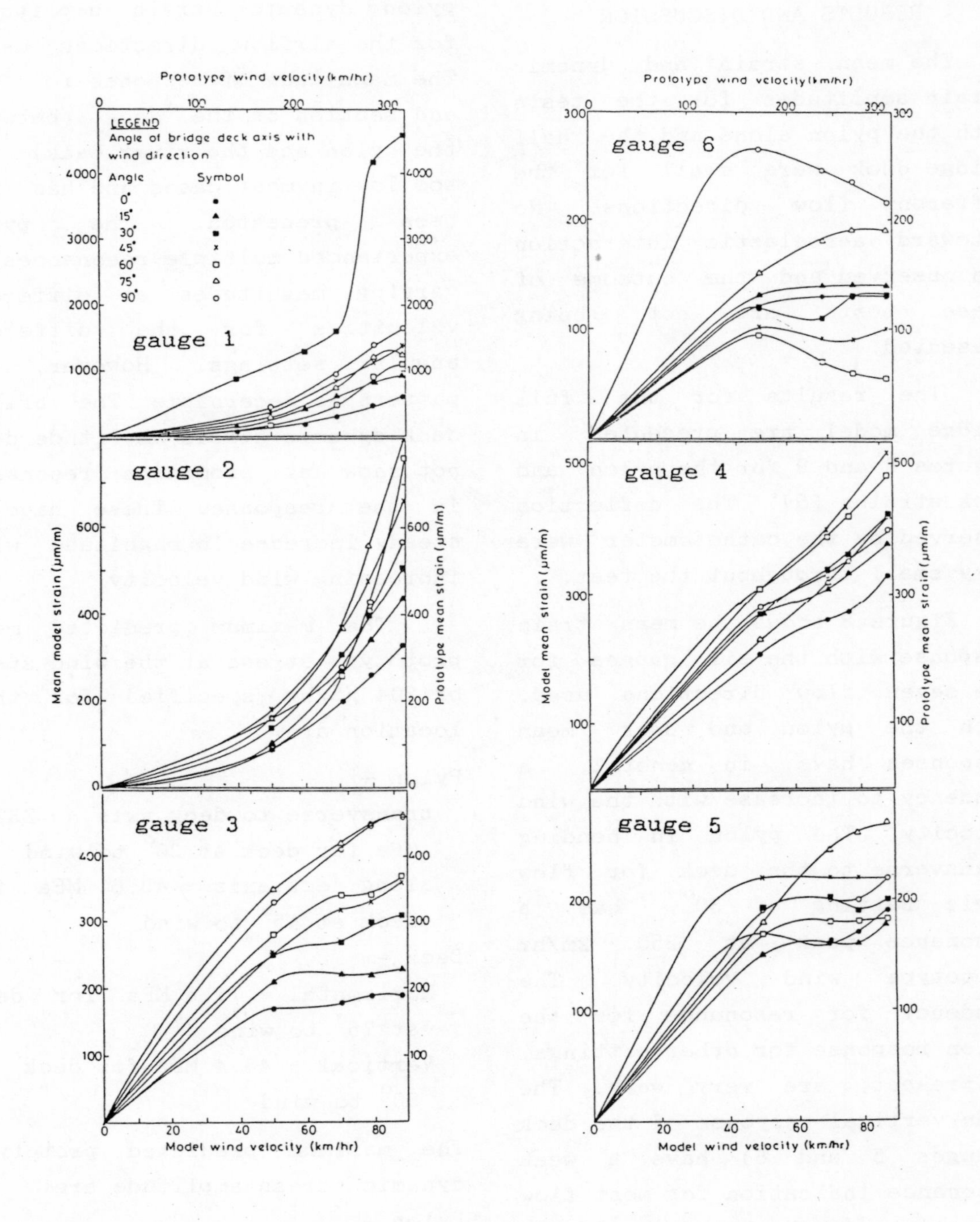

Fig. 8. Bridge Mean Strain Response to Wind

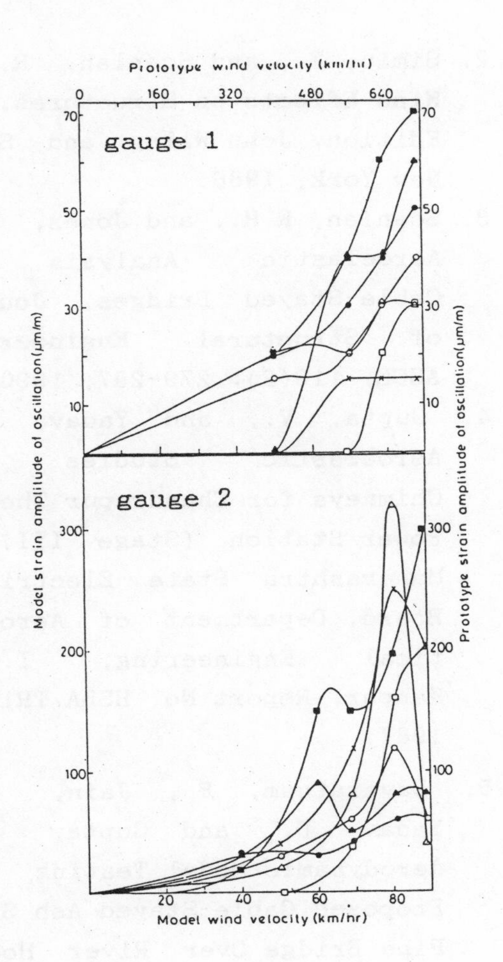

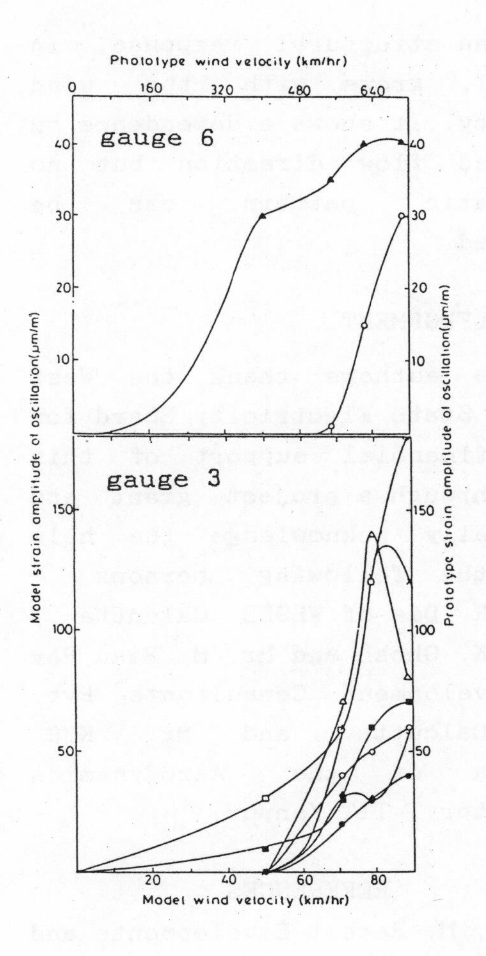

Fig. 9. Bridge Strain Response Amplitude to Wind

Deck –

Horizontal - 19.9 MPa for deck at 30° to wind

Vertical - 11.6 MPa for deck at 15° to wind

CONCLUDING REMARKS

Model tests for mean and dynamic stress response of the cable-stayed bridge was conducted for different stages of erection. Size limitation of the test section restricted the length scale factor to a relatively large value. The simulation problem for the model was tackled by using sectional area ratio factors for the deck and pylon truss members.

The bridge with the open deck construction has structurally stable aeroelastic characteristics in the range of the specified design wind velocity for its partial and fully erected stages.

The mean structural response, in general, grows with the wind velocity. It shows a dependence on the wind flow direction but no systematic pattern can be observed.

ACKNOWLEDGEMENT

The authors thank the West Bengal State Electricity Board for their financial support of this work through a project grant and gratefully acknowledge the help from the following persons - Mr. A.R. Das of WBSEB Calcutta ; Mr. D.K. Ghosh and Mr. M. Basu Ray of Development Consultants Pvt. Ltd. Calcutta; and Mr. K.S. Muddapa of the Aerodynamics Laboratory, IIT Kanpur.

REFERENCES

1. Ito, M. Recent Developments and Trends: Dynamic Response - Bridge, Transmission Lines and Roof Membranes, **Proceedings, 5th International Conference in Wind Engineering,** Vol. 2: 803-806, 1980.
2. Simiu, E., and Scanlan, R. H. **Wind Effects on Structures,** 2nd Edition, John Wiley and Sons, New York, 1986.
3. Scanlan, R.H., and Jones, N.P. Aeroelastic Analysis of Cable-Stayed Bridges, **Journal of Structural Engineering, ASCE**, 116(2): 279-297, 1990.
4. Gupta, V., and Yadav, D., Aeroelastic Studies for Chimneys for Chandrapur Thermal Power Station (Stage III) of Maharashtra State Electricity Board, Department of Aeronautical Engineering, I.I.T. Kanpur, Report No. HSLA TR1-87, 1987.
5. Dayaratnam, P., Jain, S.K., Yadav, D., and Gupta, A.K., Aerodynamic Model Testing of a Proposed Cable-Stayed Ash Slurry Pipe Bridge Over River Hooghly Near Bandel Thermal Power Station, Technical Report, Department of Aerospace Engineering, I.I.T. Kanpur, 1989.

SALIENT FEATURES OF THE INDIAN WIND LOADING CODE AND ITS BACKGROUND

M.C. Tandon

Managing Director

Tandon Consultant Private Ltd.
17, Link Road
Jangpura Extension
New Delhi 110 014, INDIA

SYNOPSIS

The recently published Indian Wind Loading Code is an attempt to catch up with the progress made by other International Codes. The previous code which was published a quarter of a century ago presented serious limitations in the context of present day applications. The paper identifies some of the more important shortfalls of the previous code before highlighting the main features of its latest version.

INTRODUCTION

Success in satisfactory codification of wind loading on structures has remained elusive so far. In most cases codification has followed, not preceded structural failures or distress.

The present day structures have an ever increasing tendency towards forms which are wind-sensitive because of their slenderness, flexibility, size and lightness. Added to this isthe introduction of a broader range of materials and the subjection of the material to a higher range of stresses. These factors have demanded a more realistic, if not, precise definition of wind loading.

Updating of some Internation Codes of practice, notably the British (Ref 3), Australian (Ref 4), Canadian (Ref 5), American (Ref 6) and French (Ref 7) has been affected fairly frequently over the last two decades and the present verions incorporate most of the advances made in terms of understanding wind characteristics and its effect on structures. These have been highlighted in a previous paper (Ref 8).

The scenario in India has been sluggish in relation to codification, with the shortfalls of the last version IS:875-1964 (Ref 1) surfacing ever so frequently in recent times. With the publication of the recent update IS:875 (Part 3) - 1987 (Ref 2), an attempt has been made to catch up with other International Codes and provide to the Indian structural engineer adequate guidelines for arriving at wind loading for design purposes.

This paper identifies some of the more important shortfalls ofthepREVIOUS code (Ref 1) which formed the basis of the modifications incorporated in the recent version (Ref 2). Thereafter, the paper also highlights some of the main features of the new code.

2.0 SHORTFALLS OF THE 1964 CODE

The code, which was published a quarter of a century ago has provided reasonably good guidance to structural engineers for most of the simple structures designed in this period. In most cases, its provisions have been safe, if not conservative.

A look at some of the recent International Codes (Refs 3,4,5,6,7) however, clearly indicates the tremendous advances made in wind engineering in the lastfew years. Not only is there an improved knowledge of the characteristics of wind, the recent trends, innovations and requirements of structural design, demand a much more accurate definition of wind forces. Viewed from this background, the shortfalls of IS:875-1964 (Ref 1) must be understood in today's context. In the revision of the code an attempt was made to overcome the obvious shortfalls and update the information to present-day knowledge of wind engineering.

Some of the more important shortfalls that had surfaced over the years and provided the main impetus to the revision of the 1964 code are briefly summarised below:

a. The wind maps of the 1964 Code indicated wind pressures with and without consideration of short duration squalls. The squall, which represents the averaging time of possibly 5 minutes or so by itself has no real significance unless it is correlated to winds of shorter averaging time. Of interest to structural designers are short duration gusts with averaging time of say 3 to 15 secs.

b. No indication was available about the return period in the 1964 Code. The present international trend by and large is towards designing normal structures for a return period of 50 years. For wind-sensitive structures and for structures with post-disaster functions, the return period is often increased to 100 years. As a matter of fact, for structures of exceptional importancethe return period could logically be even higher to reduce risk levels still further.

c. The wind map zoning in the 1964 Code represented the state of knowledge of the wind climate and measurements taken some 30 years ago. Several studies have since been published based on increasingly available data which have been summarised in Ref 9. It was abundantly clear that the wind zoning both over-estimated and under-estimated the occurence of extreme wind speeds in different parts of the country. The revision of the wind map, it was felt, was long overdue.

d. The wind instruments used by meteorologists measure directly the wind velocity with an averaging time of about 3 secs and not the pressure exerted by it. The wind pressure is evaluated by the equation $p = KV^2$, where K is a constant. The value of K ostensibly used in the 1964 Code was found to be higher by 25% as compared to other International Codes.

e. The variation of wind velocity with height and character of surrounding terrain has important implications in structural design. The 1964 Code specified constant pressures upto a height of 30 m and thereafter a variation corresponding to the equation:

$$P_z = P_{30} * (Z/30) B$$

where,

P_{30} = basic wind pressure which is constant upto height of 30m

Z = height in metres above ground of point under consideration

B = power index, with a value of 0.2

The generally accepted norm in International Codes is to take a constant velocity upto 10 m height and thereafter a variation given by the following equation to obtain the velocity profile :

$$V_z = V_{10} * (Z/10) A$$

where,

V_{10} = velocity at 10 m above ground

Z = height in metres of point above ground of point under consideration

A = power index whose value varies from 0.16 to 0.35 depending on the terrain; the increasingly larger indices corresponding to increasing roughness of terrain

Considering that the power index of IS:875-1964 was applicable to pressures, the corresponding index for velocity would only be 0.1, which is a very low value indeed. While structures of low height may not be affected significantly by the low value of the power index, the 1964 Code would yield pressures which are unable for tall structures in the upper region where they are most serious. Not only did the code not indicate variations with terrain roughness, it was silent about wind pressures above 150 m above ground. With tall structures already reaching heights of 300 m in the country and still taller structures on the drawing board, the shortfalls of IS:875-1964 became particularly evident for tall structures.

f. For tall, long span and slender structures a "dynamic analysis" of the structure is essential. Wind gusts cause fluctuation forces on the structure which induces oscillations. The severity of the wind-induced loading depends on the frequency of vibration and the damping of the structure. The oscillations are induced in both "along-wind" direction as well as "across-wind" direction. The "along-wind" response of the structure is accounted for by a magnification factor (often called the "gust factor") applied to the static or direct forces. The "across-wind" response requires a separate dynamic analysis. The 1964 code did not give any guidance on these aspects of structural design of tall structures, where the need for careful evaluation of wind effects is of paramount importance.

g. The 1964 Code did not indicate any correlation between the structure size and wind loading. The magnitude of the load imparted on a structure by a gust depends, not to an insignificant extent, on the size of the structure. Neglecting the dynamic response of the structure due to wind induced oscillations in the present agreement, the larger the structure size, the smaller the static wind load. This conclusion can also be reached directly by a physical understanding of the phenomenon. A spatial correlation between the gust and the structure determines the magnitude of the wind load. The larger the structure, the longer should be the gust duration for it to envelop the entire structure. However, we know that the longer the averaging time of the gust, the lower the wind speed (and consequently the wind load).

h. Structures can be of many shapes, particularly when we combine the whole gamut of possibilities in concrete and structural steel. The "direct or static" wind force on a structure is given by the equation F = S * A * P, where S is the Shape factor, A is the projected area in the plane perpendicular to the wind direction and P is the wind pressure. The 1964 Code gave some values of the shape factor for the basic shapes (circular, octagon and square), but this information needed to be extended to other shapes with various wind directions. Also, some of the values needed to be reviewed as they differ from those in some International Codes. One example was the circular shape for which the shape factor can increase from 0.5 to 1.2 as the height/width ratio and roughness increases. The value suggested in the 1964 Code was 0.7 over the whole range. Again, for steel structures of different plan shapes, shielding configurations and solidity ratios, no guidance was available, and it became essential to either consult other International Codes or take advice from aerodynamic experts for many of the tall structures of this type in the design stage.

i. Many industrial buildings in India are being designed with metallic and asbestos sheet claddings for roof and walls. Apart from outmoded fastening and weather proofing systems used in conjunction with such claddings,the inadequate codal provisions have also been responsible for the unsatisfactory performance of the buildings. It was felt that the codal stipulations should be updated in relation to pressure coefficients (internal and external) for various configurations of buildings, wall openings and wind directions. Additionally, a need was felt to define higher local external negative coefficients to cater to concentrations near edges of walls and roofs, where the failures are primarily initiated.

3.0 MAIN FEATURES OF THE NEW CODE

3.1 The Wind Map of India

A revised basic wind speed map of India has been presented in Fig 1 of the new Code. The country has been divided into 6 regions with basic wind speeds varying from 33 m/sec to 55 m/sec for a 50 year return period and a probability of exceedence level of 63%. The indicated wind speed characteristics are in line with other International Codes (for example see Ref 3, 4), viz, short duration gusts averaged over a time interval of about 3 seconds applicable to 10 m height in "open" terrain. The code recognises that there is not enough meteorological data to draw isopleths (lines joining places of equal velocity) on the wind map of India as seems to be the practice in other International Codes.

The revised map when compared to that of Ref 1 shows that in general, in the following areas of the country, the wind velocity was previously under-estimated :

- North : North-eastern part of Jammu and Kashmir
- West : Coastal areas of Kutch in Gujarat
- East : Tripura and Mizoram

The wind map indicates updated supplementary information (1877 to 1982) relating to both total number of cyclonic storms as well as those of severe category.

The revised wind map has come about as a result of the work of many papers published over the years on the subject of extreme wind speeds in India (Ref 9).

3.2 Design Wind Speed

The design wind speed is dependent on the geographical basic wind speed, return period, boundary layer roughness, height above ground, structure size and local topography. The format of the equation of design wind speed adopted in the new Code is similar to the one in Ref 3 and is as follows:

$Vz = Vb * K1 * K2 * K3$, where,-(1)

Vz = design wind speed at height z in m/sec

K1 = risk coefficient

K2 = terrain, height and structure size factor

K3 = local topography factor

3.2.1. Return periods

The coefficient K1 is identified in the Code for return periods of 5,25,50 and 100 years along with recommendations regarding "mean probable design life of structures" of various types (refer Table 1 of the Code). For general buildings and structures of permanent nature the return periods have been identified as 50 years as in most international Codes with the notable exception of Ref 5, which is based on a 30 year return period.

3.2.2 Terrain roughness and structure size

The coefficient K2 has been indicated at different heights in a convenient tabular form (refer Table 2 of the Code) which identifies the terrain category (1,2,3 or 4) and class (i.e. size) of structure (A, B or C). The background of the numerical values in the Table is available in Ref 10 and are based largely on Ref 11. The coefficient K2 assumes different values (refer Table 33 of the Code) when dealing with wind speeds averaged over one hour, which are required for evaluating dynamic effects.

3.2.3 Local topography

The coefficient K3 allows for undulations in the local terrain in the form of hills, valleys, cliffs, escarpments and caters to both upwind and downwind slopes. Details for evaluation of K3 are included in the Code in Appendix 3.

3.3 Velocity - Pressure Relationship

One of the important modifications in the code relates to the correction regarding velocity v/s pressure relationship so as to bring it in line with other International Codes (Ref 8). In the earlier version of the Code (Ref 1), the pressure was over-estimated by about 25% for a given wind velocity. The modified equation now reads as follows:

$$P_z = 0.6 V_z^2 \quad \text{--------- (2), where,}$$

Pz = design wind pressure at height z in N/sq mm

Vz = design wind speed at height z in m/sec

The wind effects in terms of static loading on the structure is determined from design pressures as indicated in the next section.

3.4 Wind Pressure and Forces on Buildings/Structures

The code stipulates requirements for calculations for wind loading from three different points of view :

- The building/structure taken as a whole
- Individual structural elements such as roofs and walls
- Individual cladding units such as sheeting and glazing including their fixtures

The wind loading is given in terms of pressure coefficients Cp and force coefficients Cf and can be determined by the following equations:

$$F = C_f * A_e * P_d \quad \text{------ (3)}$$

$$F = (C_{pe} - C_{pi}) * A * P_d \quad \text{- (4)}$$

where,

F = wind load

Cf = force coefficient

Ae = effective frontal area obstructing wind

Pd = design wind pressure

Cpe,Cpi = external and internal pressure coefficients

A = surface area of structural elements

The presentation in this part of the code follows closely that of Ref 3.

3.4.1. Force coefficients

Force coefficients are applicable to the building/ structure as a whole as well as to structural frameworks which are temporarily or permanently unclad. In both these situations equation (3) is to be employed.

For evaluating force co-efficients for the clad building/ structure as a whole, the code gives guidance for a variety of plan shapes and height to breadth ratios (refer Table 23 of the Code). It also indicates extra forces occuring due to "frictional drag" on the walls and roofs of clad buildings.

For evaluating force co-efficients for unclad buildings/ stru^tures full frameworks as well as individual members are comprehensively covered by the code. The frameworks included are those that are single (i.e. isolated), multiple or in the form of lattice towers. The effects of "shielding" in parallel multiple frames and different solidity ratios have been incorporated along with global force coefficients for lattice towers (refer Tables 28 to 32 of the Code). Force coefficients for individual members of various structural shapes and wires/cables have been separately indicated (Tables 26 and 27 of the Code).

3.4.2. Pressure coefficients

Pressure coefficients are applicable to structural elements like walls and roofs as well as to the design of cladding. The calculation process implies the algebraic addition of Cpe and Cpi to obtain the final wind loading by the use of equation (4). The Code indicates both these coefficients separately for a wide variety of situations generally encountered in practice.

Pressure coefficient Cpe depend on wind direction, structure configuration in plan and its height versus width, ratio and, characteristics of roof and its shape. Tables 4 to 22 of the Code are devoted exclusively to the determination of Cpe. For the specific design considerations relating to cladding, local coefficients have been separately shown delineating the areas at the edges of walls and roofs where high concentration of negative pressure is often found to exist.

Pressure coefficient Cpi are largely dependent on the percentage of openings in the walls and their location with reference to wind direction. The Code indicates Cpi as a range of values with a possible maximum (i.e. positive pressure) and a possible minimum(i.e. negative pressure) with the provisio that both the extreme values would be examined to evaluate critical stresses in the concerned member. Three cases have been specifically indicated for arriving at Cpi:

- openings upto 5% of wall area
- openings 5% to 20% of wall area
- openings larger than 20% of wall area (including buildings with one side open)

The last case is of particular interest in determining wind loading for structures such as aircraft hangars which have wide full height openable doors forming one side of the enclosure (refer Fig 3 of the Code).

3.5 Dynamic Effects

The Code incorporates a new section relating to dynamic effects, recognising the advent of a large number of tall, flexible and slender structures in the country's sky-line. Structures which require investigation under this section of the Code are defined as:

- height to minimum lateral dimension of the structure is more than 5.0.
- structures with a first mode natural frequency of less than 1.0 cycles/sec.

Guidelines are given in the Code for the determination of natural frequency of multistoreyed buildings and the designer is cautioned regarding certain structural responses such as cross-wind motions, interferences of upwind obstructions, galloping, flutter and ovalling. The Code encourages use of wind tunnel model testing, analytical tools and reference to specialist literature when wind induced oscillations of the structure reach significant proportions.

3.5.1. Along-wind effects

The gust factor approach has been suggested by the Code for the evaluation of along-wind structural response. The methodology is similar to Ref 4, which in turn is not very different from Ref 5. A study of Ref 11 reveals that the provisions of the Indian Code incorporate the latest codification efforts in the international field.

The calculations involved for the along-wind effects have been reduced to a simple equation given below:

$$Fz = Cf * Ac * Pz' * G \quad \text{--- (5)}$$

where,

F_z = along-wind static loading at height z, applied on effective frontal area A_c of strip under consideration

P_z' = design pressure at height z corresponding to hourly wind speed

G = gust factor

One of the important departures for along-wind dynamic effects as compared to static effects is to approach the problem of design pressure evaluation from the route of hourly wind instead of short duration gusts as was done in equations (1) and (2) earlier. Table 33 of the Code indicates the revised factors K_2 which have to be applied to the basic wind speed (short duration gusts) of Fig 1 for arriving at the hourly mean wind speed and consequently the corresponding design pressures, P_z'.

The gust factor G, which depends on natural frequency, size and damping of the structure and on the wind characteristics can be evaluated directly from parameters obtained from graphs given inthe Code (refer Figs 8 to 11 of the Code).

3.5.2. Across-wind and other effects

The Code highlights that the method of evaluating across-wind and other components of structural response to wind loading are fairly complex and have not as yet reached a stage where successful codification can be attempted. It further encourages the designer to consult specialist literature as well as other Indian Standards as appropriate. It would not be out of place to mention here that separate standards dealing with reinforced concrete chimneys and natural draft cooling towers, steel chimneys and over-head power lines and transmission line towers already exist and that these standards are under active review in the light of the provisions of the new wind code.

4.0 REFERENCES

1. IS:875-1964, "Code of Practice for Structural Safety of Buildings : Loading Standards". Indian Standards Institution March 1964.

2. IS:875 (Part 3) - 1987, "Code of Practice for Design Loads (other than earthquake) for Buildings and Structures - Part 3 : Wind Loads". Bureau of Indian Standards, Feb 1989.

3. BSCP 3, "British Standard Code of Practice, Basic Data for the Design of Buildings. Chapter V : Loading - Part 2: Wind Loads". British Standards Institution, 1972.

4. AS 1170, "Minimum Design Loads on Structures. Part 2 : Wind Forces". Standards Association of Australia, 1983.

5. "National Building Code of Canada. Part 4. Design". National Research Council, Ottawa, 1975.

6. ANSI A58-1-1982, "American National Standard, Minimum Design Loads for Buildings and other Structures". American National Standards Institute, 1982.

7. NV 65, "Regles Definissant Les Effets de la Neige et du Vent sur les Constructions" Editions Eyrolles, Paris, 1975.

8. Tandon, M.C. "Wind loads on Structures : Looking at and Beyond some International Codes of Practice". Bridge and Structural Engineer (IABSE-ING), June 1983.

9. Rao, G.N.V., "Studies on New Wind Zone Demarcation for India". Asia Pacific Symposium on Wind Engineering, Roorkee, Dec 1985.

10. Venkateswarlu B., Arunachalam S., Shanmugasundaram J, Annamalai G., "Variation of Wind Speed with Terrain Roughness and Height".

11. DR 87163, "Draft Australian Standard. Minimum Design Loads on Structures. Part 2: Wind Forces". August 1987.

STUDIES ON WIND SPECTRA AND CODAL PROVISIONS

V.R. Sharma — Research Scholar, Civil Engg. Deptt.
K. Seetharamulu — Professor, Civil Engg. Deptt.
K.K. Chaudhry — Professor, Applied Mech. Deptt.

Indian Institute of Technology
New Delhi 110 016, INDIA

SYNOPSIS

Bureau of Indian Standards in revised wind loading code IS: 875 (Part 3) - 1987 [1], has incorporated Gust Factor Method as an alternative for design of flexible structures. The gust spectra and the related terms used in the recommended procedure are not based on measurements pertaining to wind climate in Indian Sub-Continent. The mean wind spectra and gust spectra obtained from the wind measurements are reported. The application of mean wind spectra and gust spectra in Gust Factor Method for design of civil engineering structures is discussed.

INTRODUCTION

Wind velocity is random in nature and it imparts critical loading to high rise buildings and flexible structures in general. For convenience of analysis wind velocity is broken down into mean component and fluctuating component. While mean velocity component is assumed to result in static wind pressure and corresponding steady deflection, the fluctuating component gives rise to dynamic amplification.

The specific information required about wind at a building site for structural design purposes is the wind structure comprising of mean velocity profile and turbulence properties and statistics of wind climate which is accounted for by extreme wind speeds predicted from long term records. Representation of wind in frequency domain is accepted as a rational approach for incorporating random behaviour for design purposes.

Revised wind loading code of India IS: 875 (Part 3) - 1987, like its predecessor is based on peak winds. However, Gust Factor Method has been incorporated as an alternative method for flexible structures only. In this method basic pressures are computed based on mean winds which are further enhanced by Gust Factor to arrive at design pressures. Gust spectra and other related terms used in Gust Factor method incorporated in IS Codal provisions are not based on measurements pertaining to wind climate in Indian sub-continent. The work on measurements for mean winds and gusts is in progress at IIT Delhi and accordingly mean wind spectra together with gust spectra covering the entire range of frequencies have been obtained. Mean wind spectra is useful for arriving at design mean wind speed required for computing basic pressure. Gust measurements and related spectra are helpful in arriving at Gust Factor.

MEAN WIND SPECTRA

Experimental Observations And Results

Three cup generation anemometer "PRICOL" Model WMG 100 fitted at the top of the mast on the roof of a three storey building at IIT Delhi, at a height of 19.2m above ground level, is used for velocity measurements. The response time of anemometer being 15 sec, output can be taken as 15 sec mean wind speed. The wind speed was recorded at different hours.

The data so collected was used for obtaining mean wind spectra. Two approaches (a) Graphical Method, and (b) Discrete Fourier Transform (DFT), as illustrated by Sharma et al [[2] were used to convert wind data from time domain to frequency domain. The results obtained from the two methods have been found to be in agreement. The power spectrum as obtained from the two methods for four different sets of observations (Set 1: April 1987 to June 1987, Set 2: July 1987 to September 1987, Set 3: October 1987 to December 1987, and Set 4: January 1988 to March 1988) covering a full year are shown in Fig.1 through Fig.4. The spectral peaks are noticed corresponding to the time periods of four days, one day, twelve hours and eight hours. The highest peak obtained is for a time period of one day. The spectral peak corresponding to twelve hour time period is the second highest. The other spectral peaks are of lesser magnitude. The comparison of spectra obtained for four sets indicates that spectral ordinates for sets 1 and 4 are much higher

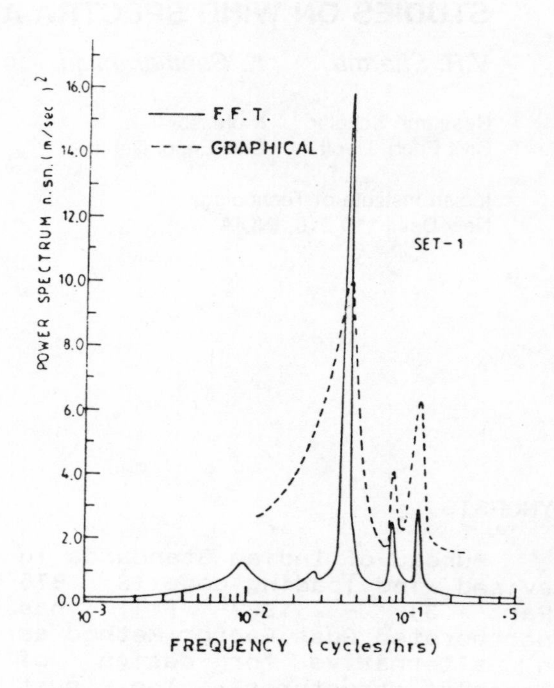

FIG.1. MEAN WIND SPECTRA

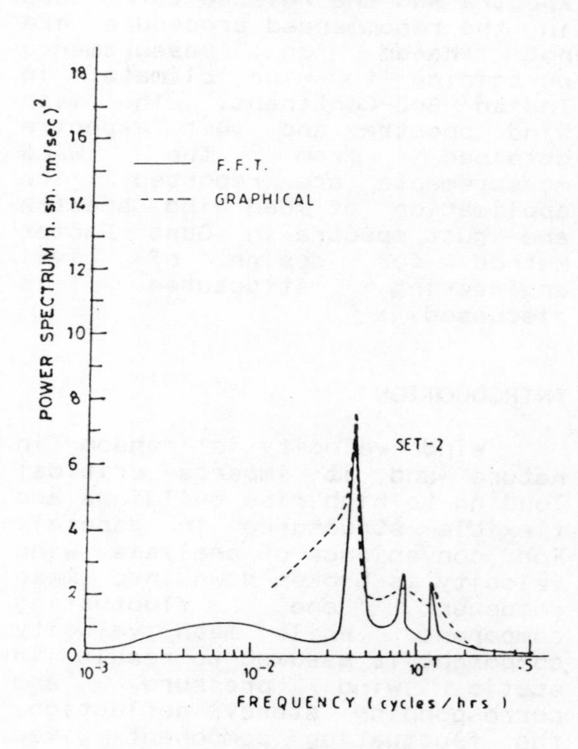

FIG.2. MEAN WIND SPECTRA

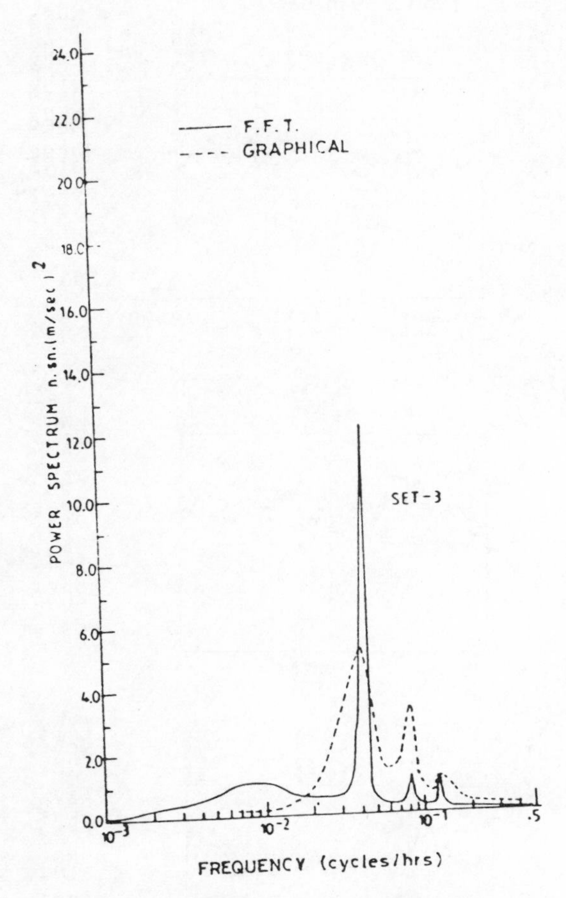

FIG.3. MEAN WIND SPECTRA

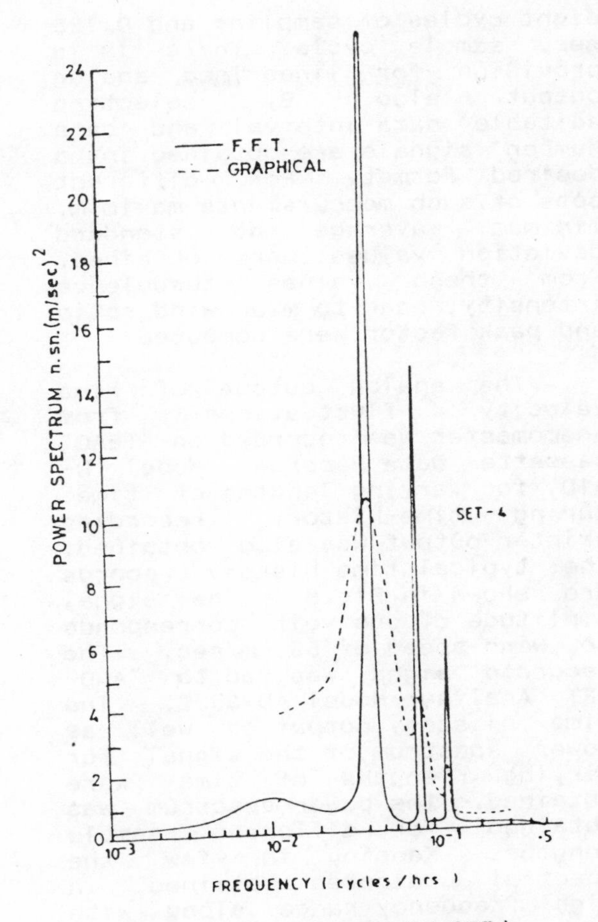

FIG.4. MEAN WIND SPECTRA

than those obtained for sets 2 and 3 respectively.

__Application of Mean Wind Spectra__

The presentation of mean wind velocity in the form of spectra provides important information in respect of energy at different frequencies at which the spectral values are the highest. This form of presentation involving parent population is more direct than the existing mathematical models for extreme mean winds. The spectra obtained for low frequencies, therefore, results in a more rational static analysis than the one based on extreme mean winds. The design mean wind speed so obtained can be used for computing basic wind pressure required in Gust Factor Method of analysis.

GUST SPECTRA

__Experimental Observations and Results__

Gust measurements were carried out with KANOMAX Anemomaster Model 6611. It is a constant temperature type thermal anemometer with large size liquid crystal display and printer facilities. It has a response time of 0.2 sec. and the moving average value is displayed after eight cycles of sampling and 0.125 sec. sample cycle. There is a provision for linearised analog output also. By selecting suitable data interval and data number signals are obtained in a desired format. For different sets of such measurements maximum, minimum, average and standard deviation values were obtained. From these values turbulence intensity, peak to mean wind ratio and peak factor were computed.

The analog output of wind velocity fluctuations from anemomaster was recorded on "TEAC" cassette Data Recorder Model R-61D, for varying lengths of time. During time-history recording printer output was also obtained. The typical time history records are shown in Fig.5. The signal amplitude of one volt corresponds to wind speed of 59 m/sec. The recorded signal was fed to "AND" FFT Analyser Model AD 3522. The time hjistory output as well as power spectrum of the signal for varying lengths of time were obtained. The power spectrum was obtained for different sample lengths. Keeping in view the spectral ordinates obtained in high frequency range along with frequency range of civil engineering structures, 10 hz range (20 sec sample length) was selected. For this range instantaneious spectra as well as averaged spectra for varying duration of time on various days at different hours were obtained.

Two typical records of 33 minutes and 18 minutes duration and spectra obtained for these are reported here. Fig.6 corresponds to averaged spectra for continous 33 minutes whereas Fig.7 and Fig. 8 are averaged spectra corresponding to 1st 16.5 minutes duration and 2nd 16.5 minutes duration of the record respectively. The time history reported is with DC component filtered out and is for the last

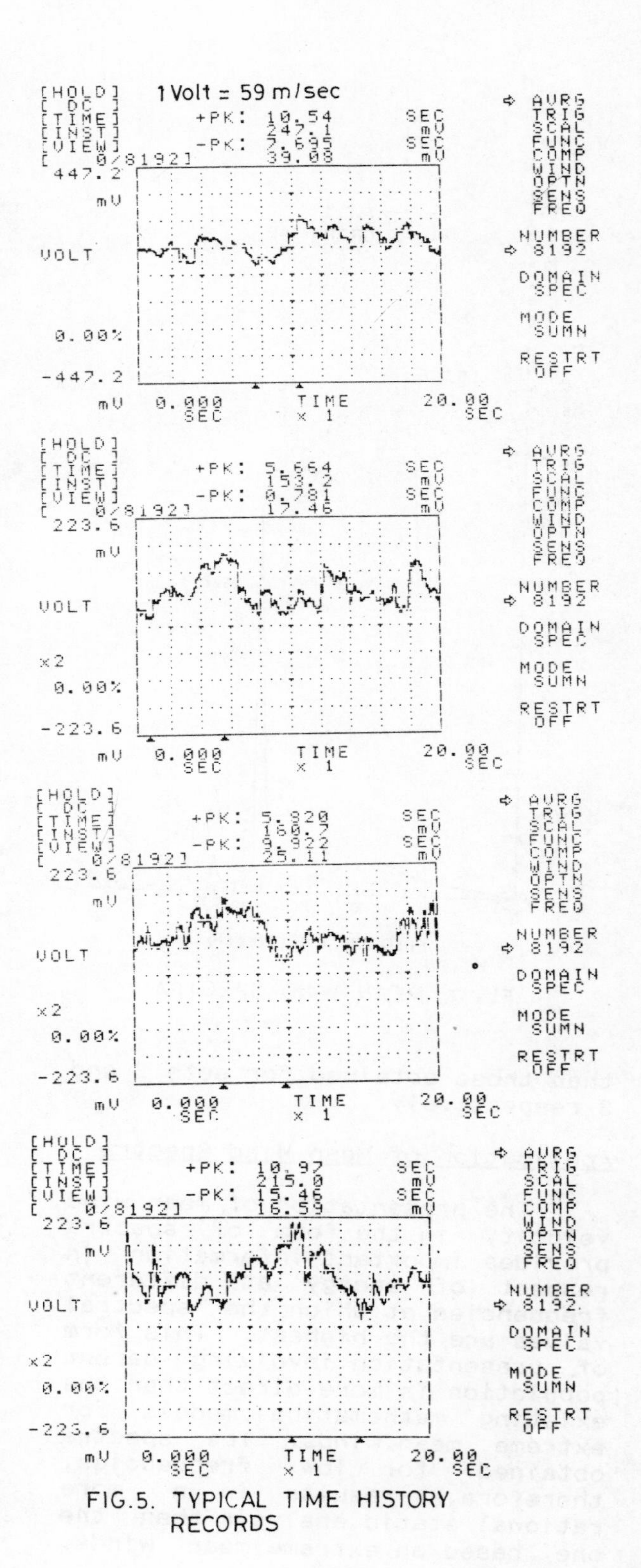

FIG.5. TYPICAL TIME HISTORY RECORDS

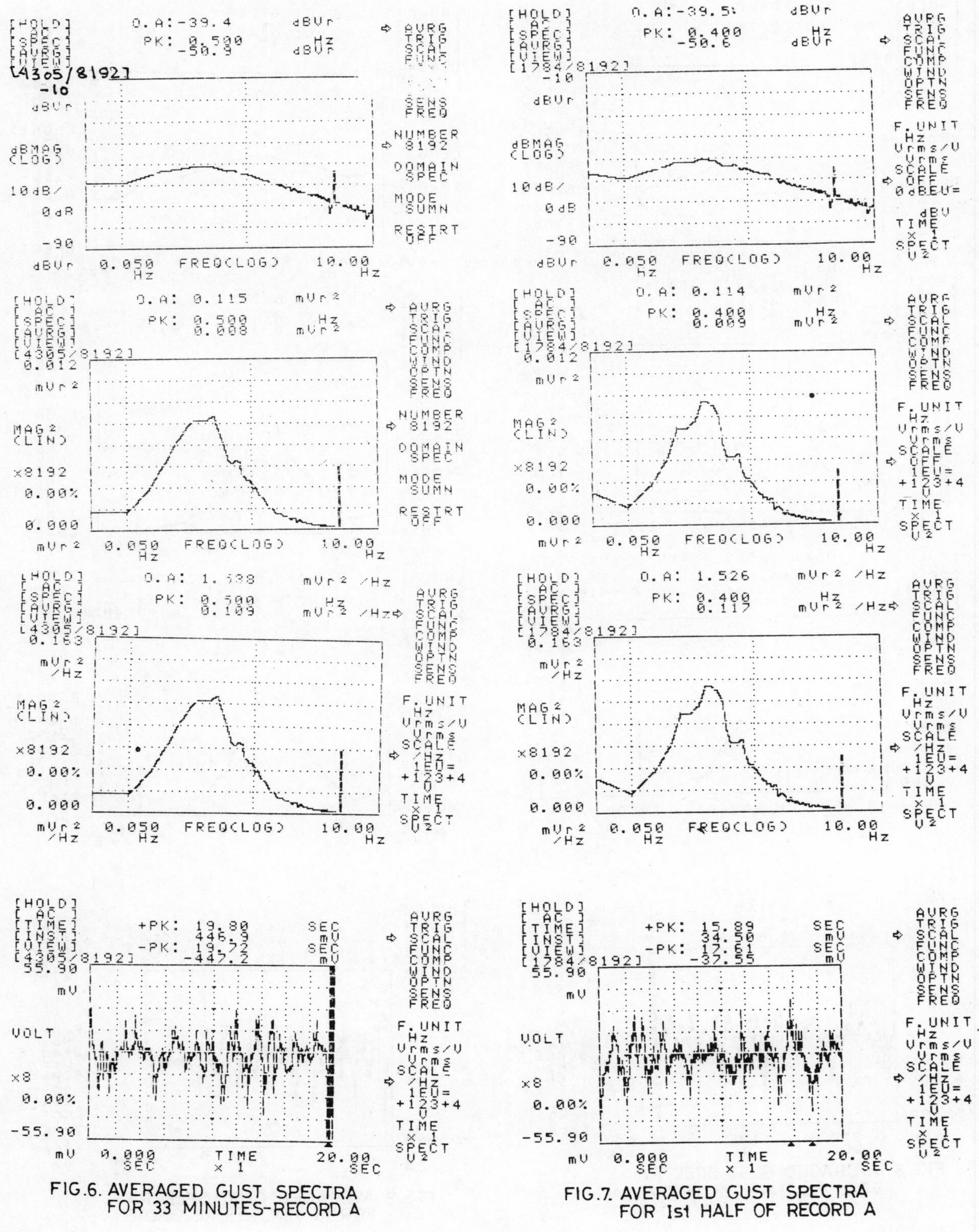

FIG.6. AVERAGED GUST SPECTRA FOR 33 MINUTES-RECORD A

FIG.7. AVERAGED GUST SPECTRA FOR 1st HALF OF RECORD A

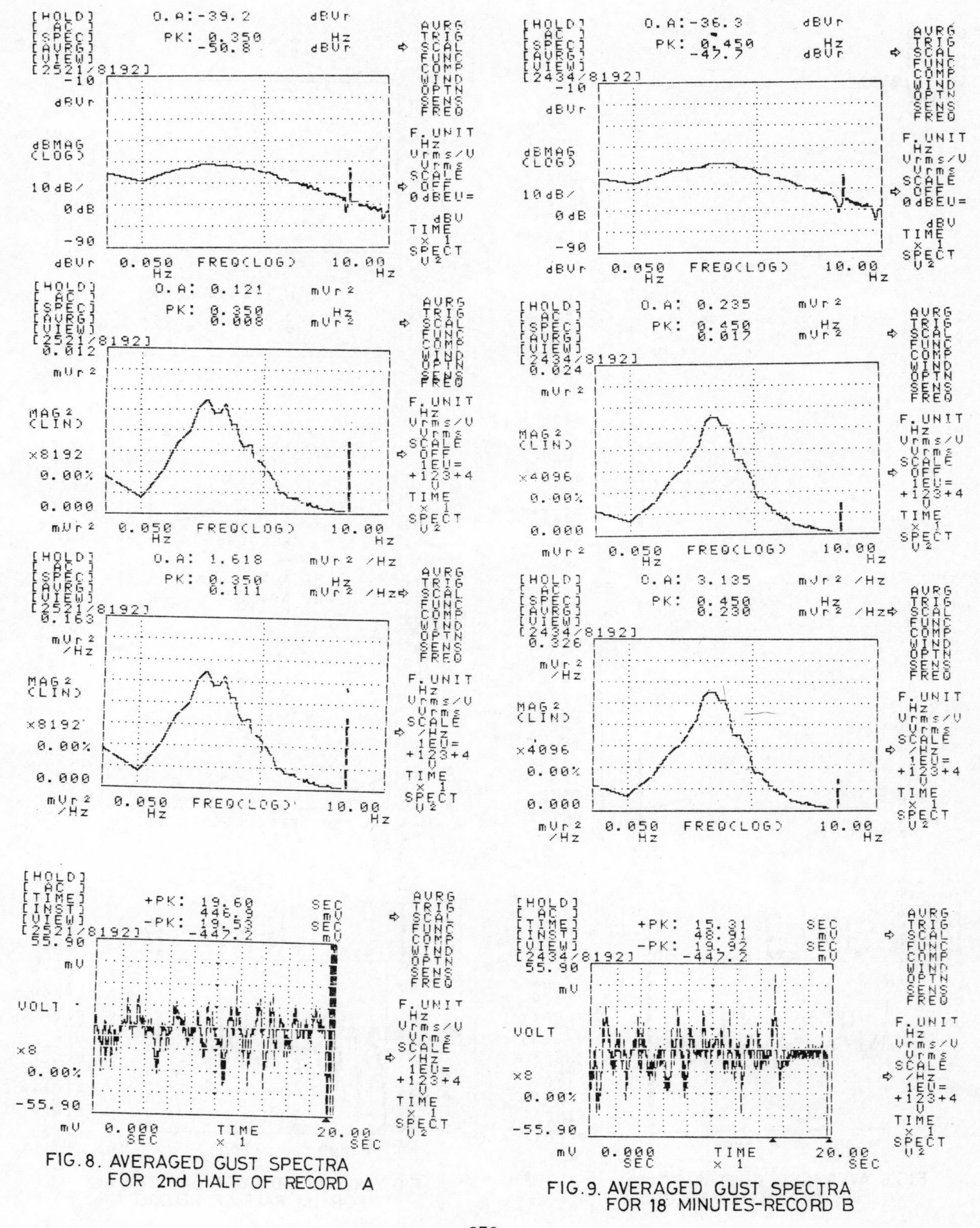

FIG.8. AVERAGED GUST SPECTRA FOR 2nd HALF OF RECORD A

FIG.9. AVERAGED GUST SPECTRA FOR 18 MINUTES-RECORD B

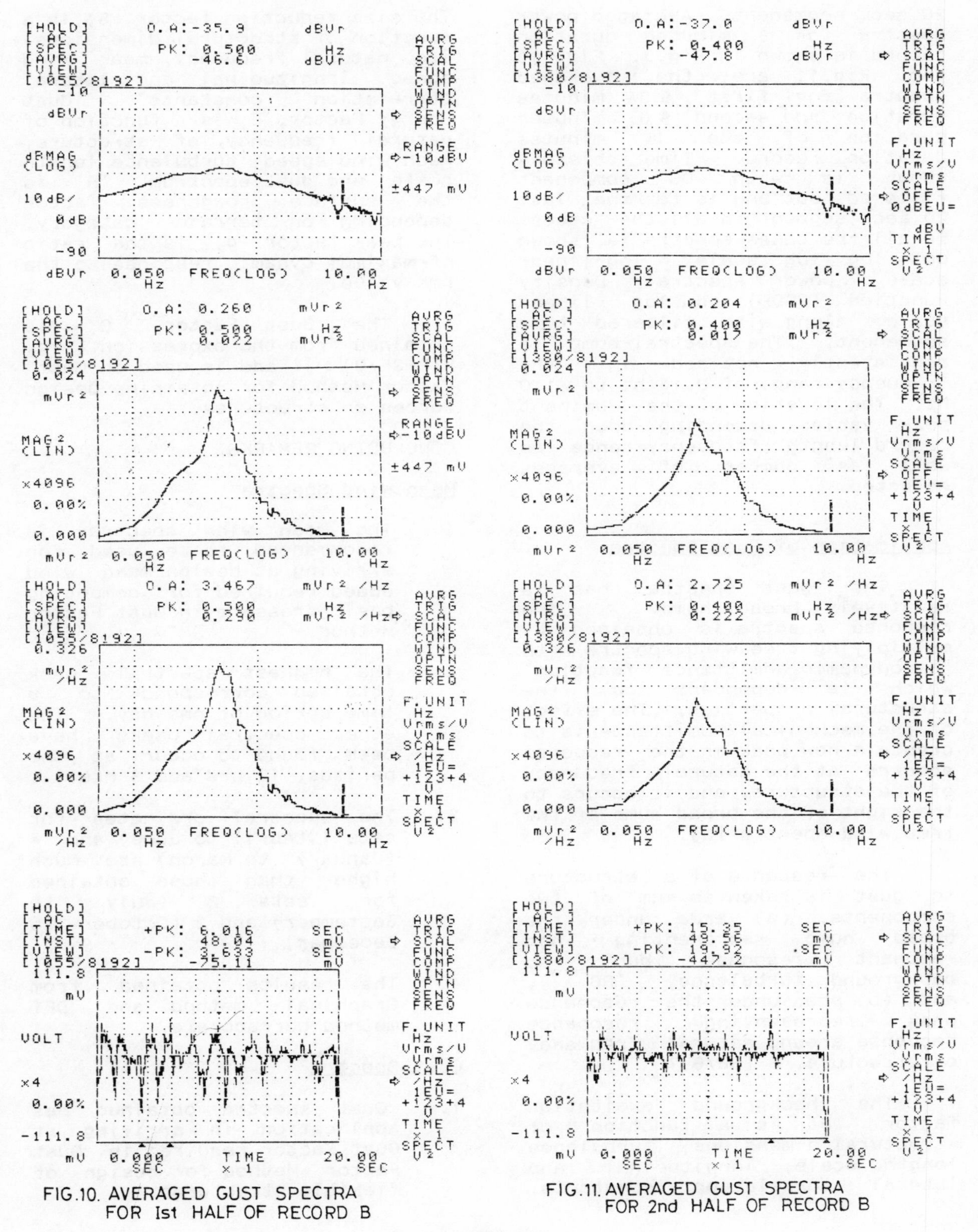

FIG.10. AVERAGED GUST SPECTRA FOR Ist HALF OF RECORD B

FIG.11. AVERAGED GUST SPECTRA FOR 2nd HALF OF RECORD B

20 sec. segmenent. Averaged power spectra for 18 minutes duration record is shown in Fig.9. Fig. 10 and Fig.11 are the averaged spectra for first 9.0 minutes duration and second 9.0 minutes duration of the 18 minutes duration record. Time history shown is with DC component filtered out and is for the last 20 sec. segment in all the cases. In all the cases spectra is shown on log log scale, log-linear scale, power spectral Density Function (PSDF) on log linear scale, along with filtered time segmenent. The spectral hump is consistently noticed in the frequency range of 0.16 hz to 1.0 hz. The location of the dominant peak varies depending upon the record length, frequency range as well as number of averages selected.

Application of Gust Spectra

The gust spectra has a relatively broad hump. The response spectra is obtained by multiplying free wind spectra with "aerodynamic admittance function" which is dependent on the structural properties. The effect of the mechanical admittance is to create a new peak in the response spectra at the natural frequency of the structure, and it occurs to the right of the broad hump of the free wind spectra [3].

The response of a structure to gust is taken as sum of two components: (a) area under the broad hump representing non resonant response due to background turbulence, $(RB^{1/2})$, and (b) area under the resonance peak representing resonance response around natural frequency of structure, $R\ (SE/\beta)^{1/2}$

The background excitation factor 'B' is a function of structural dimensions, turbulence length scale, longitudinal and lateral correlation constants. The size reduction factor, S, is a function of structural dimensions, its natural frequency, mean wind speed, longitudinal and lateral correlation constants. Gust Energy Factor, E, is a function of natural frequency of structure, mean wind speed, turbulence length scale and gust spectrum. 'R' is the surface roughness factor depending on terrain category. The peak factor, g_p, is the ratio of maximum dynamic response to the rms value.

The Gust Factor, G, is obtained from the expression $[1+g_p R(B+SE/\beta)^{1/2}]$, and is used in Gust Factor Method for obtaining design forces on structures.

CONCLUDING REMARKS

Mean Wind Spectra

1. The mean wind spectra is recommended to be used for arriving at design mean wind speed required for computing basic pressure in Gust Factor Method.

2. The highest spectral peak obtained corresponds to a time period of one day. The ether spectral peaks have been found to occur at the periods, 12 hrs and 8 hrs.

3. The spectral ordinates for sets 1 (April to June) and 4 (January to March) are much higher than those obtained for sets 2 (July to Septemper) and 3 (October to December).

4. The results obtained from Graphical method and DFT method corraborate.

Gust Spectra

1. Gust spectra obtained has application in arriving at Gust Factor required in Gust Factor Method for design of flexible structures.

2. The spectral hump has been found to be in the frequency range of 0.16 hz. to 1.0 hz.

3. The location of the dominant peak varies depending on the record length, frequency range as well as number of averages selected. For the spectra reported, Fig. 6 through Fig. 11, the peaks occur at the frequencies, 0.5 hz, 0.4 hz, 0.35 hz, 0.45 hz, 0.5 hz and 0.4 hz respectively.

REFERENCES

1. IS: 875 (Part 3) - 1987. "Indian Standard Code of Practice for Design Loads (other than earthquake) for Buildings and Structures, Part 3, Wind Loads" Bureau of Indian Standards, New Delhi, 1985.

2. Sharma, V.R., K. Seetharamulu and K.K. Chaudhry. Spectral characteristics of Mean Winds in Delhi. Proceedings of symposium on PMMND held at New Delhi, Aug. 2-4, 1989, pp. 4-55 to 4-60.

3. Commentary B on Part 4 of the National Building Code of Canada, 1980, Part 4, Wind Loads, Canada. National Research Council of Canada, Ottawa, Canada, 1980.

POST-DISASTER CYCLONE DAMAGE SURVEYS AND THEIR IMPACT ON DESIGN OF STRUCTURES

J. Shanmugasundaram *S. Arunachalam* *T.V.S.R. Appa Rao*

Structural Engineering Research Centre
Madras, INDIA

The coastal regions of India are battered frequently by severe cyclones. The annual average cost of damage caused by cyclones in India is estimated to be around Rs.200 crores. Conducting post-disaster field surveys and documentations of cyclonic wind data, typical failures of structures, etc., provide invaluable information that will help in developing suitable analytical models for determining structural response. Salient details of the post-disaster survey conducted after a severe cyclone which hit Kavali, Andhra Pradesh, on 9th November 1989 have been presented. From the failure analysis of a R.C. pole, the lower bound of the cyclonic wind speed was estimated to be equal to 57.5 m/s (207 kmph) and this compared well with the wind speed of 52 m/s (187 kmph) estimated by the India Meteorology Department. Some of the design implications based on observations on damage suffered by buildings and structures have also been discussed.

INTRODUCTION

Many nations suffer from loss of life and large scale destruction of property (crops, dwellings, and other structures) due to natural hazards such as tropical cyclones and earthquakes. The natural calamities considerably retard the national, regional or global development. Tropical cyclones are considered to be the worst devastating natural hazard when viewed in terms of their occurrence, intensity and areas affected. Tropical cyclones originate between 5 and 30 degree latitudes on either side of the Equator and affect islands and coastal regions. The cyclones are characterised by a revolving wind system with a mean wind speed of 33 m/s or more, covering an area of more than 800 km across. The Indian Ocean is one of the six major cyclone-prone regions of the world. In India, cyclones frequently occur in April and May and between October and December. It has been noticed from the past, that about 80% of the total cyclones in India occur along the east coast. Such cyclones occur frequently, almost every year, with a high wind speed of about 250 kmph ($\sim$ 290 kg/m^2), causing considerable havoc to different types of buildings and industrial structures, power transmission and communication systems etc. The annual average cost of damage caused by cyclones in India is estimated to be around Rs.200 crores, which at times, has exceeded even Rs.400 crores.

However, at present, the state-of-knowledge about various characteristics of cyclones and their disastrous effects on buildings and structures, is very much limited. Since the problems in this area are highly complex in nature, owing to many uncertainties involved, studies based on theoretical models alone will not be adequate to understand the realistic nature of wind and to predict the behaviour of structures in cyclonic storms. Surveys on structural damage caused by cyclones (post-disaster field surveys) will provide invaluable information that will help in developing/validating analytical methods for determining structural response. Documentations of cyclonic wind data, typical failures of structures etc., enable in identifying various factors influencing the damage to structures by cyclonic wind forces.

In this paper, a brief review of the findings from post-disaster damage surveys conducted by SERC, Madras, and various organisations in India and abroad is presented. Typical failures of buildings and other structures caused by the severe cyclone which hit Kavali, Andhra Pradesh, on 9th November 1989 were surveyed and analysed by the authors, and a case study is reported in this paper. The impact of knowledge/experience gained in this area, on improvement of structural design of buildings and structures to resist cyclonic forces is discussed.

EARLIER STUDIES ON CYCLONE DAMAGE SURVEYS

Recognising the cost of structural damage caused by frequent cyclones/hurricanes and the need to improve current structural engineering practice towards mitigation of disastrous cyclone effects, during the past two decades, various organisations around the world have documented wind-induced damage on a comprehensive and systematic basis. Some of these organisations include James Cook University of Queensland, in Australia, the Building Research Establishment (BRE), in U.K., Iowa State University, Texas Tech University, National Bureau of Standards, National Research Council, in U.S.A., Structural Engineering Research Centre, Madras, and University of Roorkee, Roorkee in India. The Institute for Disaster Research, Texas Tech University has conducted on-site building damage documentation in 58 separate windstorm events during the past 20 years. Based on such documentation work, Mehta et al., [1] have presented rational methods for assessment of wind speeds from analysis of damaged structures and for understanding the performance of buildings subjected to extreme winds. The Building Research Establishment, U.K., has carried out post-disaster surveys in various developing countries, such as Mauritius, Philippines, Hongkong, Sri Lanka, Caribbean, to document the structural damage due to cyclones [2]. Based on this investigation, BRE has recommended specific construction details which emphasize the importance of strong connections between components to maintain structural integrity during wind storms [3]. The National Research Council (Chiu et al.,) [4], U.S.A. has surveyed structural damage due to hurricane Iwa, which hit Hawaii on November 23, 1982. It was reported that most of the wind damage to timber-framed buildings and wood-framed single and multiple family dwellings could be attributed towards inadequate fastening of the roof covering, poor anchorage of the roof systems to the walls, and weak connections of the walls to their foundations. The failure of two transmission line towers at Delhi during the 1978 tornado was studied to assess the likely wind speeds of tornado [5]. SERC, Madras, has been actively conducting field surveys on cyclone damage to various buildings and structures, with a view to mitigating cyclone disaster in the coastal regions of India. For example, scientists of this Centre carried out field surveys on the damage suffered by residential and industrial structures during the November 1977 severe cyclone, which hit twice within a span of two weeks in areas around Nagapatinam, in Tamil Nadu and around Machilipatnam, in Andhra Pradesh.[6] Similar field surveys on damage to buildings/dwellings and other industrial structures, were also conducted by this Centre, during subsequent cyclones in 1979, 1981 and in 1983, but the extent of such damage caused to structures were less severe. The severe damage caused to several buildings and the collapse of a 300 m^3 capacity elevated R.C.C. water tank during the 1984 severe cyclone at Sriharikota, Andhra Pradesh were also examined by scientists of this Centre.

DAMAGE OBSERVATIONS - A CASE STUDY

A post-disaster field survey on damage suffered by various buildings and structures during the severe cyclone which hit Kavali, a small town in Nellore district, in Andhra Pradesh, on 9th November 1989, was carried out by SERC, Madras. The above cyclone has caused widespread destruction to both life and property. An area of 8 km radius in and around Kavali, in Nellore district was severely battered while some regions in Prakasam district also suffered extensive damage. According to official sources, the cyclone caused 48 deaths and over 500,000 people in 545 villages were affected in both districts and over 100,000 people were rendered homeless. The gale wind speeds, when the cyclone crossed the land, were estimated to be between 220 kmph (61 m/s) and 240 kmph (66.7 m/s).

The maximum wind speed of the cyclone at the time of crossing the land as estimated by the India Meteorological Department was 187 kmph (52 m/s). From the terrain features of the Kavali town, it may be said that the central region of the town can be categorised as a low sub-urban area (category 3), while the surrounding villages represent an open terrain type (category 2), as per the terrain classifications of Indian Standards on wind loads [7].

It was observed from the survey that the cyclone caused much havoc to a large number of buildings and structures, ranging from well-engineered type of structures such as communication towers, large industrial sheds, water tanks to non-engineered type of structures such as common man's dwellings, small sheds, compound walls. In addition to complete destruction of houses with thatched roofs, almost all buildings with pitched roofs covered either with A.C. sheets or G.I. sheets sustained heavy damage due to blowing off of the claddings. It was observed that the failure of roof claddings was primarily due to pulsating forces of uplifting suction, low dead weight of the cladding materials ($\sim$ 185 N/m^2) and due to poor strength of the connections in effectively holding down the claddings to the rafters/ purlins. Particularly, it may be mentioned that in many of the low-rise industrial structures surveyed, the A.C. claddings were tied to the rafters/purlins by means of J-bolts (instead of U-bolts) and in some cases they were tied with simple nailing, without proper washers, etc., even near the eaves, which is vulnerable to uplifting forces. Pitched roofs with Mangalore tiles or country tiles, even with mortar strips sustained serious damage to claddings, because of lightweight of the tiles and of the inadequate spacing between the mortar strips (Fig.1). However, it was observed that a continuous coat of mortar over the tiles had protected the cladding to perform satisfactorily against upward suction forces as shown in Fig.2. Failures of brickwalls of low-rise industrial sheds were observed, in many cases, and this is attributed to their inadequate strength to resist lateral wind forces. This led to subsequent falling down of the roof trusses supported on brick walls. Total collapse of the structure was observed when the trusses were not adequately braced (Fig.3). Failures of well-engineered structures caused by the cyclone include complete collapse of a 12.6 m high overhead service reservoir (40,000 ltrs. capacity) which was under construction, failure of a 22 m span tubular truss system of the Andhra Pradesh State Road Transport Corporation bus depot (Fig.4) and collapse of a 91 m high microwave tower (Fig.5).

DESIGN IMPLICATIONS

From the observations on damage suffered by buildings and structures caused by the cyclone at Kavali, the following design implications may be noted.

i) Building frames and masonry walls are required to be designed to have adequate lateral strength to withstand cyclone forces, in addition to having strength to resist vertically acting dead and live loads.

ii) Residential buildings with one to two storey height, and with R.C.C. roofs, performed well during the cyclone. This is attributed to the fact that the heavy dead load of the R.C.C. roof (2400 N/m^2) adequately resists the uplifting suction forces and the buildings possess adequate strength to resist lateral forces.

iii) In pitched roof buildings asbestos sheets are more popular than G.I. sheets for roof cladding. J-bolts are conventionally used to tie down the roof cladding to the purlins. Cyclonic forces are highly fluctuating in nature and the loads transmitted to the J-bolts cause flattening of the bolts which weakens the strength of the connection, in effectively holding down the cladding sheets to the purlins. By providing U-bolts in place of J-bolts at closer intervals, failure of J-bolts due to stretching can be avoided and damage due to blowing-off of cladding sheets can be considerably reduced.

iv) The eaves regions in low-pitched roof buildings are critical and a large proportion of damage are initiated in those regions. By providing continuous M.S.flat ties in the region, the initial and cumulative damage to the roof cladding can be considerably minimised, if not totally eliminated, as shown in Fig.6. Similarly, by providing a continuous coat of mortar over tiles, the resistance of cladding to suction forces can be increased.

v) Cyclone damage to many of the industrial buildings, is aggravated by the presence of large openings. Failures of rolling shutters led to unintended opening which would contribute towards increase in internal pressure. Provision of steel plates across the width of the rolling shutters can improve the resistance of the rolling shutters against wind forces.

vi) In many of the large-span industrial structures, the total destruction has been triggered by failure of supporting brick walls. By providing a continuous R.C.C. beam over the walls, the lateral resistance of the brick walls can be highly improved. Alternatively, the roof truss systems can be directly supported on R.C.C. columns, with infilled brick panels.

vii) Progressive collapse of roof truss systems can be arrested by providing suitable bracing of the roofing system.

viii) Timing of certain stages of construction of special structures may be suitably chosen to avoid extra risk of being subjected to extreme winds during these periods. Special/additional bracing of scaffoldings and structural systems during construction may be provided.

ix) Observations on failures of typical structures/components during post-disaster surveys provide information for estimating the maximum cyclonic wind speeds that would have been faced by the structures. Wind speeds estimated from failure data would be helpful in corroborating with any measured values, if available, and in further improvement of knowledge in this area.

x) The maximum wind speeds that were estimated during the severe cyclone vary between 55 m/s to 70 m/s, and these values are higher than those recommended in the Indian Standards on wind load criteria [7]. This implies that systematic collection of more cyclonic wind speed data and estimation of characteristic wind speeds based on the statistical analyses are essential.

ASSESSMENT OF WIND SPEED FROM DAMAGE

Instruments presently being employed to record wind speeds are not capable of recording maximum cyclonic wind speeds due to their limitations. By analysing the type and extent of damage suffered by typical structures, it is possible to reasonably assess the maximum wind speed during the cyclone, that could have caused the failure. The accuracy of the above type of analysis in back-calculating the maximum wind speed from the damage depends on various factors such as the type of the structure, construction material, gust sensitivity of the structure, influence of surrounding structures, if any, and values of terrain conditions assumed in the analysis. It is preferable to select a suitable structure (for the failure analysis) for which reliable data is available for computing the strength of the structure. Further, it is desirable to consider a severely damaged structure with a simple mode of failure so that it is amenable for analysis with least number of assumptions. In the present case, a typical failure of a R.C. communication pole in bending is considered (Fig.6).

The R.C. pole was 6.06 m high above the ground level. The cross section dimensions of the pole were 10 cm x 20 cm. The reinforcement provided was 3 numbers of 12 mm dia. HSD bars on each side. The bending was observed to have occurred at a height of 1.76 m above the ground level. The ultimate bending moment carrying capacity of the R.C. pole at the failure section was computed as equal to 2378 kg.m. The grade of concrete was assumed as M15. However, variation in this parameter was found to have weak influence on the moment carrying capacity. The pole was idealised as a cantilever structure. The failure was assumed to have been caused by the along-wind forces of the cyclone. The dynamic effects of gusts were considered using the gust factor method. Trial and error procedure was employed to arrive at the bending moment due to wind forces which was nearly equal to ultimate moment carrying capacity of the section. Based on this, a mean hourly wind speed of $\overline{V}_{10}$ = 39 m/s yielded a bending moment due to external wind forces equal to 2453 kgm and the difference between these two critical moments was only about 3%. It was, hence concluded that a minimum wind speed of $\overline{V}_{10}$ = 39 m/s should have occurred to cause the failure of the pole. The above wind speed can be expressed as equivalent to a 3 sec gust speed of 57.5 m/s (207 kmph), using Lam and Lam equation. This compares well with the value of 52 m/s (187 kmph) estimated by the India Meteorology Department (IMD) based on satellite/radar photographs.

CONCLUSIONS

Severe cyclones occur in India almost every year causing extensive damage to various buildings and structures, besides killing people. At present, our state-of-knowledge on characteristics of severe cyclones, their interaction effects on buildings and structures etc., is far from satisfactory. The need to conduct further detailed surveys on damage to structures due to cyclonic storms and to document typical failure analyses of buildings and structures, is emphasised in this paper. Details of the post-disaster survey, carried out by SERC, Madras, on damage to different structures caused by the severe cyclone which hit Kavali in Andhra Pradesh are briefly described. Based on the above survey observations, design implications towards mitigation of structural damage due to cyclones are presented. The failure of a R.C. pole in bending caused by this cyclone was analysed taking into account the available moment carrying capacity of the failure section and the effects of along-wind forces due to the cyclone. The lower bound of the cyclonic wind speed that could have caused the failure was found to be equal to 57.5 m/s (207 kmph) and this compares well with the cyclonic wind speed of 52 m/s (187 kmph) estimated by IMD, based on satellite/radar photographs.

ACKNOWLEDGEMENTS

The authors are thankful to Shri N.V.Raman, Director, Structural Engineering

Research Centre, Madras, for the kind permission to publish this paper. The assistance rendered by the staff of SERC, Madras is gratefully acknowledged.

REFERENCES

1. Mehta, K.C., Minor, J.E., and Reinhold, T.A. Wind Speed-Damage Correlation in Hurricane Fredric. Journal of Structural Engineering, ASCE, 109(1), 1981.

2. Alan Mayo, "Cyclone-Resistant Houses for Developing Countries", Building Research Establishment Report, Gartson, U.K., 1988.

3. Cook, N.J. The Designer's Guide to Wind Loading of Building Structures, Part 1, Butterworths Publishers, London, U.K., 1985.

4. Chiu, A.N.L., Escalante, L.E., Mitchell, J.K., Perry, D.C., Schroeder, T.A. and Walton, T., "Hurricane Iwa, Hawaii, November 23, 1982, "National Research Council, National Academy Press, Washington, D.C., 1983.

5. Krishna, P., and P.K. Pande. Failure of the AIR Transmission Towers in the Delhi Tornado of March 1978. Journal of the Institution of Engineers (India), 60(3), 1979.

6. Venkateswarlu, B., and Muralidharan, K., "A Report on Structural Damage Due to Cyclone in Andhra Pradesh," SERC, Madras, 1978.

7. IS:875 (Part 3) - 1987 Indian Standard Code of Practice for Design Loads for Buildings and Structures, Wind Loads, Bureau of Indian Standards, New Delhi, 1989.

FIG.1 SEVERE DAMAGE TO TILE CLADDING

FIG.2 PROTECTION OF TILED ROOFING BY PROVIDING CONTINUOUS COAT OF MORTAR

FIG.3 FAILURE OF THE TRUSS SYSTEM DUE TO LACK OF BRACING

FIG.4 COLLAPSE OF THE ROOFING SYSTEM OF A 22 m SPAN BUS DEPOT

FIG.5 COMPLETE COLLAPSE OF A 91 m HIGH MICROWAVE TOWER

FIG.6 FAILURE OF A R.C. POLE IN BENDING

DAMAGES OF VARIOUS STRUCTURES DUE TO CYCLONIC WINDS — CASE STUDIES OF 1989 AND 1990 CYCLONES IN ANDHRA PRADESH, INDIA

Narendra Verma — Deputy Director and Head
V.K. Gupta — Scientist
R.K. Garg — Scientist

Rural Building & Environment Division
Central Building Research Institute
Roorkee, INDIA

SYNOPSIS

In India, at an average, one cyclone takes place almost every year in the Bay of Bengal and strikes either Andhra Pradesh, Tamil Nadu, Orissa or West Bengal. They cause large scale destruction, damage to properties and buildings, deaths, injuries and socio-economic upheaval in the area. There has been two severe cyclones i.e., in Nov. 1989 (Nellore District) and May 1990 (Krishna District) that struck the coastal areas of Andhra Pradesh (A.P.). Central Building Research Institute (CBRI), Roorkee undertook studies on failures of different buildings and structures as a result of high wind speed, stormsurge and heavy downpour. The paper describes the failure mechanism of few typical structures like houses, industrial and poultry sheds, godowns, school building, cinema hall, communication system etc. The problem areas with regard to the effect of wind and performance of buildings are also indicated.

INTRODUCTION

The natural events like Earthquakes, Fire, Floods, Tsunamis, Cyclones, Stormsurge etc. occur in different parts of the world. One or the other happens every year in different areas. Cyclones are the most destructive kind of storms which frequently strike coastal belt of India with varying degree of fury. Its frequency in the Bay of Bengal is about four times higher than that in Arabian sea. The months of May, June, October and November are the months of probable cyclone in Bay of Bengal. Some of them may be severe cyclones and many a time associated with stormsurge or heavy downpour causing flooding in many parts. Cyclones are the vortices in the atmosphere having a core (called

*Deputy Director & Head of Rural Buildings & Environment Division
**Scientists, Rural Buildings & Environment Division
CENTRAL BUILDING RESEARCH INSTITUTE, ROORKEE, INDIA

'Eye') of extremely low pressure and light winds surrounded by strong winds of nearly circular isobars. The maximum wind speed of severe cyclone in India, lies in the range of 200 to 250 km/hr. Area of their influence also vary according to their diameter and severity. They cause wide-spread damage to property, life and communication system resulting in devastation.

The Andhra Pradesh, India, was recently struck by the two severe cyclones. One of the cyclones struck in Nellore district (Kavali Division) in Nov. 1989 and the other in large coastal belt in May 1990. The Central Building Research Institute (CBRI), Roorkee undertook studies of these areas after few days of the cyclone to know about the failure mechanism of various buildings, structures, communication systems etc. The paper describes the various failures and their probable causes.

NOV. 1989 CYCLONE IN KAVALI DIVISION, NELLORE DISTT. (A.P.)

The cyclone warnings were received well in advance by the District authorities about the probable dates of cyclone striking the Nellore district and the preparations were initiated accordingly. The entire district was divided into different zones and a senior officer was appointed as Zonal Officer to prepare, organise and administer the programme of creating awareness among the people, informing them about the safer places and pressing the required number of transport facilities (lorries) into service for evacuating the people on receiving the final warning from the Indian Meteorology Department (IMD). The final warning was received on 8th Nov.'89 and 100 lorries were used to evacuate 6000 people to 10 Relief Camps and thousands moved to the nearby cyclone shelters.

The cyclone storms with gales speed of about 250 km/hr. crossed the Kavali coast (Nellore distt.) at about 11.30 PM on 8th Nov. and at 1.30 AM on 9th Nov. '89.(Fig.1). It caused havoc in 490 villages covering an area of 2831 sq.km. The most significant part of this cyclone was that its damaging effects were observed upto a distance of 100 - 120 km from the coast in few Taluks of Nellore and Ongole districts (Fig.1). The villagers, during discussions, informed that the effect was more severe when the second part of the cyclone i.e., after 1.30 AM on 9th Nov. struck the area from opposite direction.

The cyclone left behind a trail of devastation affecting human lives, livestock, private properties, public properties and large number of houses. Almost all the thatched huts within the affected areas were fully or partly damaged. Tiled houses and GI sheet roofs were severely damaged. The industrial buildings, sheds with brick and steel columns were also damaged. It was noted that even the old and very strong buildings within the town were razed to ground while in the same area, buildings constructed with almost the same material, were left unaffected.

Thousands of old and big trees were uprooted. Many of them were sheared from a height of about 1.2 m by torsional effect of cyclonic wind. Even in the big gardens all the Cashewnut trees were flattened in the same manner. Entire electrical network including high tension lines towers, poles, transformers were uprooted/bent/broken. A Microwave Tower, designed to withstand a wind storm of 250 km/hr. was also razed to ground. Though the preparations made to evacuate the people as a result of advance warning, reduced the number of deaths to 42 only, the loss to livestock and poultry was very high. The cyclone storms apart from hitting the coastal town of Kavali caused serious damage in about 484 villages in Eleven Mandals of different divisions.

MAY 1990 CYCLONE OF ANDHRA PRADESH

A severe cyclonic storm was spotted in Bay of Bengal on 4th May 1990 and it developed fully on 6th May centered about 450 km south-east of Madras. It was expected to cross between Nellore (A.P.) and Nagapattinam (Tamil Nadu). But instead it hit the Krishna, Guntur, West Godavari, Vishakhapatnam areas of Andhra Pradesh on 9th May 1990. Movement of cyclone and its affects were as follows (Fig.2).

May 6 : Cyclone storm lay centered at 8.30 A.M. about 450 km south-east of Madras and was likely to intensify further and move in a north-western direction.

May 7 : Cyclone moved towards coastal area of A.P. and was expected to reach Krishna and Prakasham Districts in next 48 hours.

May 8 : The cyclone, considered to be the severest to hit the south coast since 1977, was likely to cross A.P. coast between Nellore and Masulipatnam late to-night.

May 9 : The severe cyclonic storm in Bay of Bengal hit the Andhra coast between Masulipatnam and Krishnapatnam at 6.30 P.M. It lashed the coast with a wind speed of 250 km/hour flatening hundreds of thatched houses and uprooting trees and electric installations along a long stretch of coast.

May 10 : Cyclonic storm crossed Andhra coast last evening and left devastation in coastal Andhra Pradesg. Ninety villages in Deviseema area of Krishna District had remained cut off since 9th May evening. Vast areas of Bandar Koduru, Raipally, Nizam Patnam, Deviseema remained submerged under sea water.

CYCLONIC DATA

Cyclonic storm crossed the Andhra coast between Machillipatnam and Krishnapatnam at 6.30 P.M. on 9th May, 1990. The tidal wave was about 7m high and swashed 50 km radius under a thick sheet of water. Wind speed was 250 km/hour or more at the time of crossing and damage warning signal of 10 (the maximum) was posted.

Devastating Effects

The devastation caused by the cyclone of May, 1990 was unparallel in the history of A.P. but number of deaths were reduced due to timely action as a result of advance warning and better preparedness. Approximately 26 lakhs people were affected. In all 430 people died. Guntur district which received maximum impact in terms of areas accounted for 198 deaths while Krishna reported 90 casualities. The deaths in other districts were reported as Godavari (30), Vishakhapatnam (40), Prakasham (37), West Godavari (13), Nellore (3), Vijayanagaram (7), Mehboobnagar (1), Khammam (2). In all, 3059 villages in 113 Mandals were affected in different districts and were hit by the cyclone. Damage to crop and property has been estimated at over Rs. 1000 crores.

It was reported by the Indian Meteorology Department that the devastating effects were much more than the cyclone that inundated Devi taluk of Krishna district in November 1977 which killed 10600 people. The tidal wave was also 7 metre high against 6 metre in 1977 and the rain continued for many days against only for two hours in the earlier case. The wind speed in the core of cyclone was more than 250 km/hr. The second eye of the storm developed enroute a phenomenon that persisted for unusually long period with its furocity. The cyclone harvested the entire mango crop of 16,000 hectares over night in the upland areas of Krishna, West Godavari, Guntur districts. It has not left even leaves. Citrus, Banana, Betal and vegetables also suffered extensive damage in Krishna and Guntur districts. Poultry farms suffered irreparable damage. The cyclone killed lakhs of chicks and damaged the sheds of the prosperious poultry industry of the Challapalli. Sheep and buffalos died in large number. All 800 hectares under praun and fish culture suffered heavy damage. Each hectare of pond now needs re-investment of Rs.25,000/-

DAMAGES IN VARIOUS BUILDINGS

There has been large scale failures and damages in different buildings like houses, industrial and poultry sheds, storage godowns, communication network and plantation. They experienced a combination of uplift and lateral wind forces. Few of the typical failures, described below, were identified to make detailed analysis to arrive at some conclusion about the probable causes of their failure.

Houses

More than two lac houses including huts were damaged. Most of them were built with mud walls and thatched roofs with low height. The roofs were pyramidal in shape. The roofs were blown off, walls collapsed and houses flattened (Fig.3) The huts having properly anchored supprting structures with columns, roof tied with straw ropes in different directions (Fig.4) and without walls, could survive.

A number of houses were built by the Government and voluntary agencies after 1977 cyclone. They had either burnt brick, concrete block, stone block or precast R.C.C. panels for walling and insitu or precast R.C.C. roofs. Almost all the roofs were found leaking and showed heavy corrosion of reinforcement. Some of the roofs were blown off (Fig.5) while few failed due to excessive corrosion (Fig.6). This was primarily due to inadequate thickness of roof slab, inadequate cover to reinforcement, absence of water proofing treatment on the top and cement plaster at the bottom (i.e., ceiling), use of saline water for construction and prevailing corrosive conditions in coastal areas.

In one of the old houses, roof of the room and verandah was constructed with wooden rafters and Mangalore pattern tiles. The tiles were covered with a thick layer of lime/cement concrete (Fig.7). The roof of the varandah was uplifted by wind pressure due to its inadequate fastening with columns. This left the verandah corner as freely vibrating component.

A Symmetrical Structure (Small Canteen)

The structure is symmetrical about both the axis having hipped roof (Fig.8). The roofing tiles are covered with lime concrete. The openings are small in size thereby allowing less air to penetrate. The structure remained intect with nominal damage while other buildings in the vicinity were completely damaged.

School Building

The school building constructed with stone walls and tile roof covered with lime concrete, had H-shaped plan (Fig.9). The wind blew perpendicular to the length of central part. The air was trapped between two wings. The wind pressure built up and blew off the roof leaving behind only the wooden rafters and walls (Fog.10).

Cinema Hall

The cinema hall having tubular trusses and asbestos sheet roofing collapsed in spite of being surrounded by buildings on three sides (Fig.11). On its one side there was a tall and long building while on another rows of single storey houses with streets perpendicular to the hall. The diversion of wind by the housing rows forced it to pass through the street which in turn produced tunnel effect thereby enhanced the wind speed to strike the cinema hall with higher force. The trusses were fixed in columns with only 60 cm long anchor bolts. The upper part of the column was pulled aside by the wind forces acting on roof trusses and both of them crumbled down. The remaining walls collapsed later (Fig.12).

Industrial Sheds, Poultry Farms And Godowns

The structural systems adopted were either brick columns and wall, or R.C.C. column and wall with G.I. sheet roofing over steel trusses. The roofs of these buildings were blown off. The G.I. sheets were torn apart due to inadequate number of J-bolts at vulnerable point and absence of wind ties. The trusses were pushed aside due to their improper fixing with columns. The walls with unreinforced brick columns also collapsed (Fig.13) while those with reinforced columns

(Fig.14) or R.C.C. columns could partly survive (Fig.15).

Steel Frame Structures

The buildings with steel frame structure were made with steel columns and tubular trusses without bracings. The absence of bracing was the main cause of failure of these buildings (Fig.16) while all the members except roofing sheets were intact and rigidly fixed together.

Communication System

Power supply, road, rail and telecommunication systems were badly disrupted. Almost entire network of 11 KV and 33 KV lines was snapped. Telephone and electricity poles were uprooted or twisted under the impact of storm. All the electrical poles including those supporting transformers collapsed by bending and twisting. This failure was common to steel I-section (Fig.17), prestressed concrete and GI pipes (Fig.18) used for the purpose. The probable cause of this failure may be non-uniformity in sectional dimensions about two axis of I-section and prestressed concrete poles and the inherrent weakness of thin skinned circular G.I. pipes. The other interesting phenomena noted is that all the poles were bent at a height varying from 1.2 to 1.5 metre. The reason of this typical failure need more detailed study. Road communication was disrupted and hundreds of lorries were stranded between Vijayawada and Vishakhapatnam for a week or more. The condition of interior roads was very bad.

Trees

The trees with thick trunk were broken at the height about 1.3 metre (Fig.19) by shearing as a result of twisting. Even in the big gardens, all the trees were twisted and broken from the same height. Trees in relatively sandy soils were uprooted.

Microwave Tower

The microwave tower was expected to withstand the wind speed of 250 km/hr. The steel section at joint near the base sheared off (Fig.20) and the tower collapsed (Fig.21). It is possible that the detailing of revets and plate at this joint was not proper and has resulted in the failure. On the contrary, it is also possible that the wind velocity might have been much more than the design speed.

Cyclone Shelters

A large number of cyclone shelters were constructed in coastal areas of Andhra Pradesh after 1977 cyclone to provide accommodation to masses at the time of cyclone. These shelters were made by different voluntary organisations such as Tata, AWARE, Red Cross Society, European Economic Community etc. and Govt. of Andhra Pradesh. Different shapes e.g., rectangular, circular, octagonal were adopted. All of them are framed structures mde of R.C.C. beams and columns using M15 concrete. They served their purpose during the cyclone. However, large scale corrosion was observed in almost all R.C.C. columns, beams and slabs of these shelters. Many of these shelters are now in bad condition and need major repairs or else will have to be declared unsafe after few years. In some cases almost entire reinforcement of columns and beams have corroded and structure is standing on filler walls only (Fig.22). This was due to inadequate cover to reinforcement.

CONCLUSIONS

Certain problem areas with regard to wind effects on the buildings and their performance may be identified as follows.

1. Single storey houses lack strength and have little stiffness and are susceptible to collapse under the lateral and uplift forces.
2. Thatch roofs, the common type of traditional structure, are the first casualities during cyclone. However, the few properly tied with straw ropes and fixed with columns could survive. This practice needs to be propagated in the area.
3. A weak link in components or cladding fails under wind loads and leads to progressive failure. Detailing of different joints and connections according to the potentialities of material used for construction, importance of the structure and expected wind loads may bring down the number of failures.
4. Steel structures without wind bracings and G.I. or A.C. sheet roofing without adequate number of J-bolts and wind tie are susceptible to collapse.
5. Joint between column and roof supporting structures like trusses, ballis, rafters, beams etc. is most vulnerable point and need special care in its design, detailing and construction.
6. Symmetrical structures with pyramidal or hipped roof and less openings have better chances of survival.
7. The planning pattern and street arrangement is an important parameter to reduce tunnel effect in the built up areas.
8. Corrosion of reinforcement due to saline and humid conditions is a predominant phenomena that effects the life and strength of structure.
9. Torsional failure of electric poles from a particular height irrespective of the material used is a typical phenomena. Similarly, the shearing of the trees from almost the same height as that of poles indicates that the knowledge about the behaviour of wind forces on such structures is not fully known. It is also possible that the wind effect is more severe above this height.

RECOMMENDATIONS

The study revealed that there are many questions that cannot be answered and many failures that cannot be explained on the basis of presently available knowledge. The twisting and bending of steel poles from a particular height, failure of trees by shearing from a typical height, the survival of few structures in the vicinity of the severly damaged or completely collapsed buildings are few of the examples. It is clear that the damages to buildings are caused by a number factors such as wind velocity in excess of design wind speed, design deficiencies or defective construction, lack of understanding of wind induced forces and effect of surrounding buildings and structures. These factors are important in obtaining the correlation between wind speed and damages. It is essential that wind speed be ascertained through analysis of wind speed data and analysis of damaged strctures.

The model studies should also be conducted in wind tunnel by simulating almost identical conditions to ascertain and confirm the correlationship between the field failures and laboratory experiments. There is a need to instrument few typical buildings in cyclone prone areas to record the measurement of various forces acting during the cyclone. The effects of shape, size, planning pattern (spacing of buildings) and forms on wind flow in and around structures, when carried out, will result in the identification of their suitability.

These studies may be undertaken at the places where suitable capabilities are already available or a central facility may be created to take up such studies.

ACKNOWLEDGEMENT

The paper is published with the kind permission of the Director, Central Building Research Institute, Roorkee and the studies reported in the paper were conducted under the regular research programme of the Institute. The authors are grateful to all Andhra Pradesh State Authorities, Engineers and Public who helped, explained and guided the team to the remote and difficult locations. The authors are also grateful to Dr. B.G. Rao, a senior Scientist of the Institute for his help in the study. The assistance rendered by the staff of the Rural Buildings and Environment Division, CBRI, Roorkee is acknowledged with thanks.

REFERENCES

1. National Newspapers from May 1, 1990 to May 15, 1990.
2. Report of the Cyclone Review Committee, May 1984, Vol. II, DST, New Delhi.

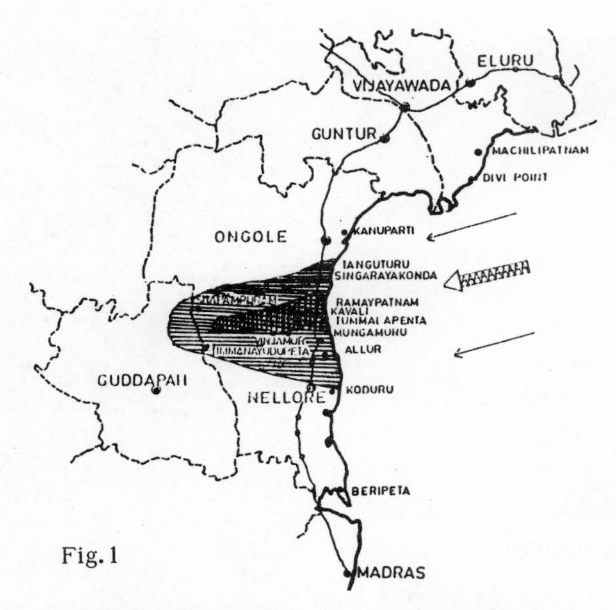

Fig. 1

AREAS AFFECTED BY NOV.'89 CYCLONE

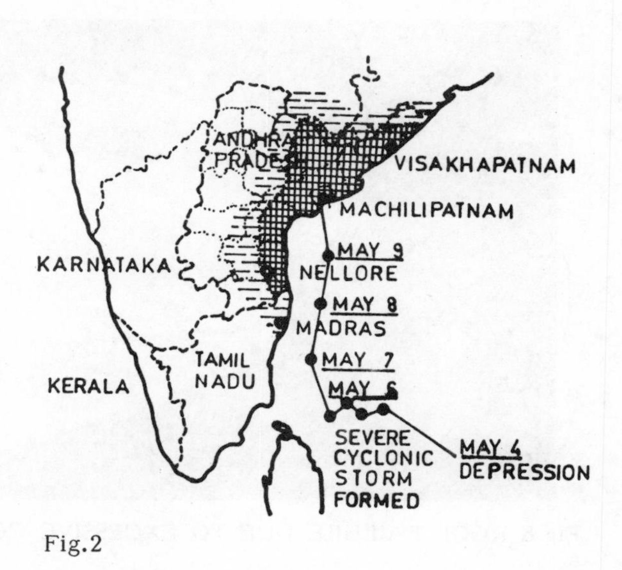

Fig. 2

AREAS AFFECTED BY MAY '90 CYCLONE

Fig.3 STRUCTURAL COLLAPSE OF A THATCHED ROOF

Fig.4 THATCH ROOF WITH STRAW ROPES

Fig.5 COLLAPSE OF PERMANENT R.C.C. HOUSES

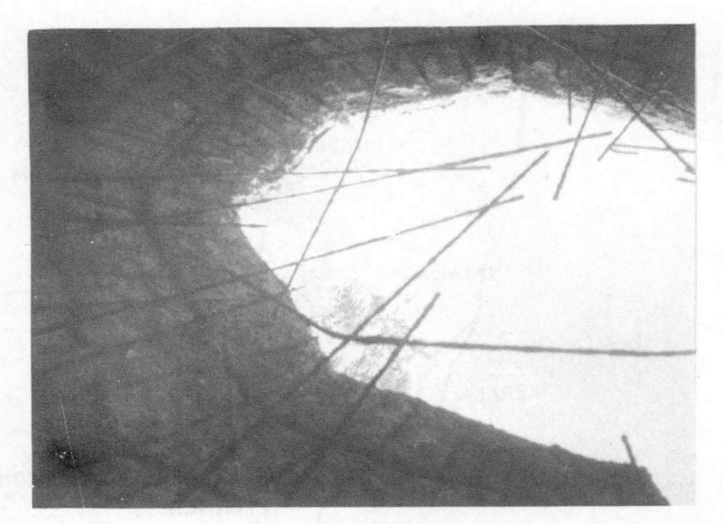

Fig.6 ROOF FAILURE DUE TO EXCESSIVE CORROSION

Fig.7 VARANDAH FAILURE

Fig.8 A SYMMETRICAL STRUCTURE

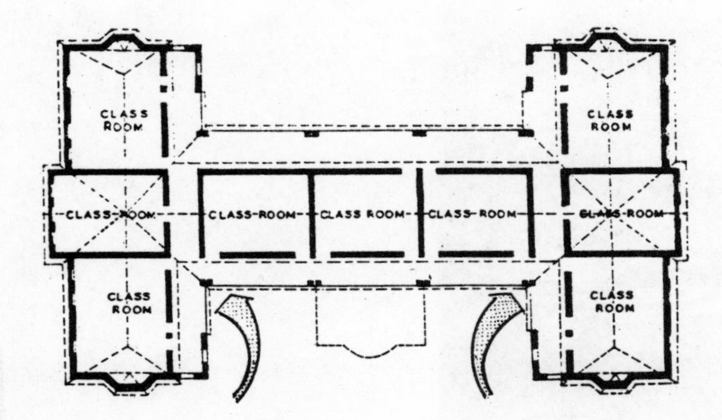

Fig.9 WIND FORCES ON SCHOOL BUILDING

Fig.10 ROOF BLOWN OFF (SCHOOL BUILDING)

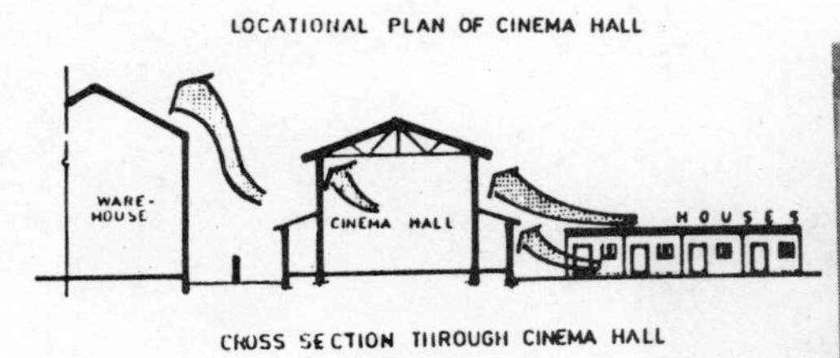

Fig.11 WIND FORCES ACTING ON CINEMA HALL

Fig.12 DAMAGED CINEMA HALL

Fig.13 POULTRY SHED

Fig.14 FCI GODOWN

Fig.15 STRUTURE WITH R.C.C. COLUMNS AND WALL

Fig.16 STEEL STRUTURE WITHOUT BRACINGS

Fig.17 TWISTING OF ELECTRIC POLE

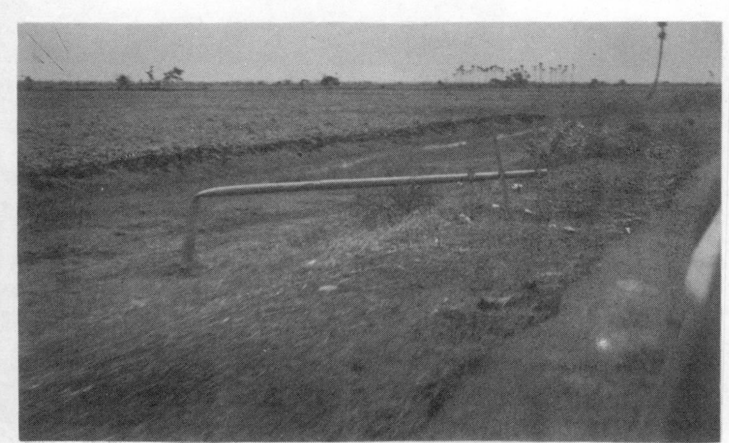

Fig.18 FAILURE OF A TELEPHONE POLE

Fig.19 TWISTING AND SHEARING OF TREES

Fig.20 SHEARING OF STEEL SECTION-MICROWAVE TOWER

Fig.21 MICROWAVE TOWER FAILURE

Fig.22 CORROSION IN R.C.C COLUMN

STATIC AND DYNAMIC WIND LOAD CONSIDERATIONS FOR A 45 M ANTENNA WITH WIRE MESH REFLECTOR

B. Janardhan *M.K.S. Yogi* *G.Swarup*

Tata Consulting Engineers
Bangalore 560 001, INDIA

Tata Institute of Fundamental Research
Pune 411 007, INDIA

SYNOPSIS

Design of large antennas for wind is rather complex. A cost-effective design becomes possible if a satisfactory antenna configuration and right structural shapes are selected. Proper values for aerodynamic parameters, drag coefficients and lift coefficients must be chosen to obtain reasonably good estimates of wind loads to ensure that a safe design with respect to dynamic and static wind effects is obtained. Corroboration of asssumed parameters and wind force coefficients with satisfactory experimentation, wherever possible, would enhance the confidence of the designer and the user of the antenna.

This paper discusses the above aspects and highlights the approaches used during the various stages of the 45 m antenna design.

INTRODUCTION

GMRT (Giant Metrewave Radio Telescope) is an array of 30 antennas arranged in a Y - pattern as shown in Fig. 1 . Fourteen antennas are located in a central square (1 Km x 1 Km) at the Y-arm junction. Six antennas are located on one arm of the Y & five antennas are located on each of the other two arms. Each arm is about 15 Km long. This configuration was chosen by TIFR (Tata Institute of Fundamental Research) considering radio astronomy research objectives.

The GMRT, presently under construction at Khodad village about 80 km from Pune, will be the largest steerable radio telescope

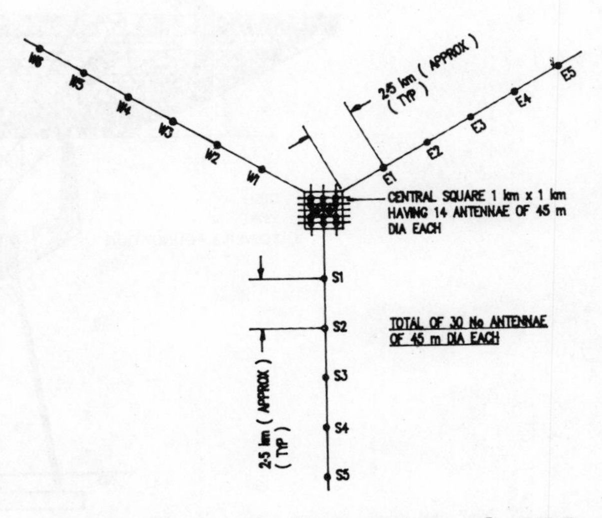

Fig 1 : Configuration of GMRT

of this type in the world. Each antenna is a paraboloidal dish of 45 m diameter on an EL-AZ mount having the required sky coverage. A stainless steel wire mesh, of three different wire spacings, is used for the reflector surface. Antenna can receive signals of five discrete frequencies in the 38 MHz to 1420 MHz frequency range.

MAJOR STRUCTURAL SYSTEMS OF THE ANTENNA

Fig. 2 shows the major structural systems of the 45 m antenna. The dish is supported on a yoke structure of steel plate construction. The conical tower, of RC construction, houses electronic equipment and supports the yoke structure on an azimuth bearing. The circular raft foundation is designed for soil conditions prevalent at the antenna location.

STRUCTURAL SYSTEMS OF THE DISH

Fig. 3 shows the structural systems of the dish. Lightning arrestor (LA) rods are supported on legs, each of TFTM (triangular frame of tubular member) construction. Each LA leg is supported near the top of a quadripod leg. The quadripod structure is supported at four points near the rim of the dish. The rotating feed cage structure, on which the the antenna receivers are mounted, rotates about the top beam of the quadripod Axis of rotation of the feed cage is parallel to the elevation axis of the dish. Quadripod legs are also of TFTM construction.

Back structure of the dish consists of sixteen PRF's (parabolic radial frames) of TFTM construction, joined at their outer ends by a circular rim of ΓFTM construction. Eight PRF's are terminated at their

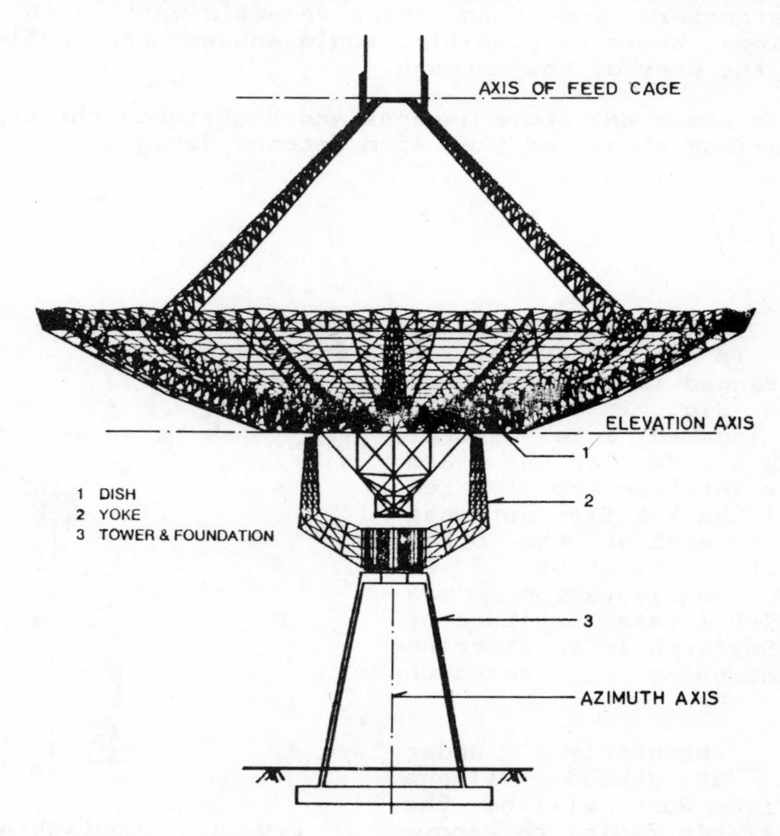

Fig 2 : Major structural systems of the 45 m antenna

inner ends on a circular hub, also of TFTM construction. The other eight PRF's are continued through the hub to terminate at a circular ring near the centre of the dish. Tension rods connect the bottom members of adjacent PRF's. Hub is supported at four points on the cradle structure.

The cradle structure is supported at two bearings on the elevation axis of the dish. A bull bear, integrally connected to the cradle structure, is moved about the elevation axis by the pinion of the elevation drive system, to point the dish in any desired direction. Elevation & azimuth drive systems are supported on the yoke structure.

Wire mesh of the reflector surface is stitched to the top wire ropes of the wire rope truss system. A typical outer wire rope truss sector is bounded by two adjacent PRF's, the rim and the hub. Two radial wire ropes, connected at their outer ends to the rim and inner ends to the hub, span across the top wire ropes of the circumferential wire rope trusses (CWRT). Top wire ropes supporting the wire mesh are connected through turnbuckles to adjustment studs on PRF's, rim and hub. Link wire ropes of the CWRT system are connected through turnbuckles to bottom nodes of the PRF's.

Inner rope truss sectors are similar to the outer rope truss sectors. Radial wire ropes are connected at their outer ends to the hub and inner ends to a circular ring near the centre of the dish.

CHOICE OF STRUCTURAL SYSTEMS

Stainless steel wire mesh reflector surface and its supporting wire rope truss system have a low self weight and attract a low wind load.

Two configurations of CWRT, shown in Fig. 4, were examined. Configuration in Fig. 4 (a) was chosen to minimise the number of wire rope elements and to have a rope truss system with least possibility of dynamic instability. Wire rope elements are pretensioned to ensure that they do not slacken in winds of speeds up to 24 m/sec. Some wire ropes slacken in winds of speeds between 24 to 37 m /sec.

Members of the dish are major contributors to self weight and

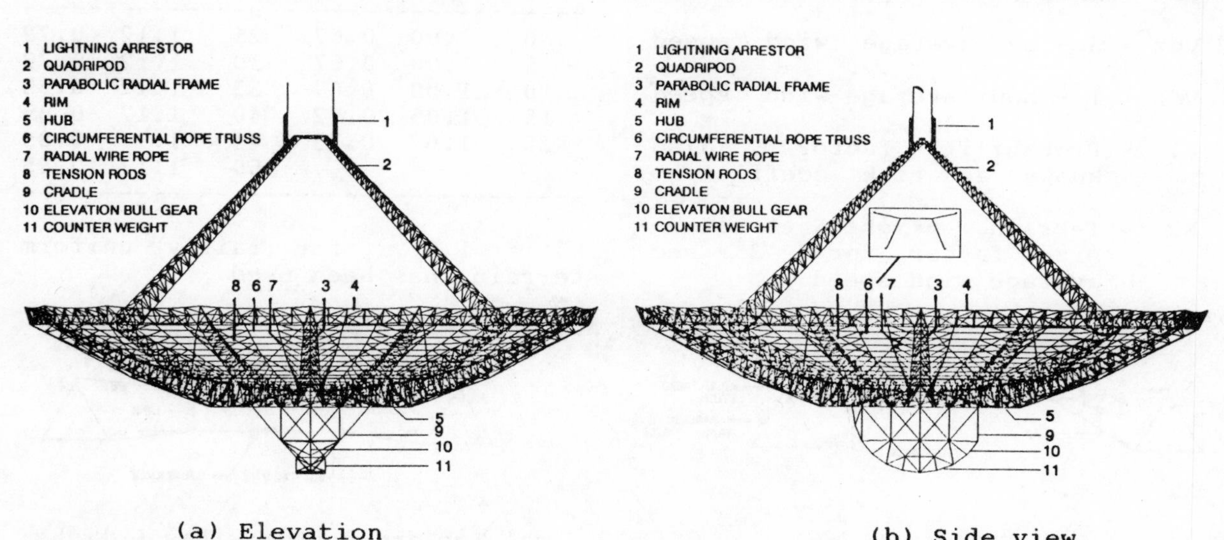

(a) Elevation (b) Side view

Fig 3 : Structural systems for the 45 m dish

wind load on the structure. Steel tubes, having low drag coefficient compared to other structural shapes, were selected to ensure that the dish is subjected to the least wind load. Self weight, mass, and mass moments of inertia for a tubular dish are also low. Cost savings for mechanical systems (elevation & azimuth bearings, drive systems, counter-weight etc.) were higher compared to the cost increase for a tubular structure, resulting in an antenna design of lower total cost.

Tension rods, connecting bottom chords of adjacent PRF's, were provided to improve torsional rigidity of the dish & to transmit forces from rim to hub.

WIND SPEEDS

Design wind speeds

Design wind speeds are given below:

$Vdz = K1 * K2 * K3 * Vb$ ----- 1

$\overline{V}dz = K1 * \overline{K2} * K3 * Vb$ ----- 2

where

Vb = Basic wind speed for site

Vdz = 3 - sec average wind speed

$\overline{V}dz$ = 1 - hour average wind speed

K1 = Probability factor , also known as risk coefficient

K2 = Terrain, height & structure size factor for 3 - sec average wind speed

$\overline{K2}$ = Terrain, height & structure size factor for 1 - hour average wind speed

K3 = Topography factor

Vb = 39 m / sec for Pune region [Ref. 1] , has been used.

K1 = 1.06 , for 100 years of design life, was used for feasibility study. K1 = 0.92, for 25 years design life for structures with low degree of hazard to life, was suggested for final design.

TIFR studied wind data for Pune and specified a 33 year design life for GMRT. K1 = 0.95, was therefore used for design.

Terrain, height & structure size factors, for terrain category-2 and class-A structure size (size less than 20 m) were used. These are given in Table - 1 .

TABLE - 1 : Terrain , height & size factors

Elev above GL (m)	$\overline{K2}$	K2	Elev above GL (m)	$\overline{K2}$	K2
0	1.00	0.67	25	1.12	0.79
5	1.00	0.67	30	1.12	0.79
10	1.00	0.67	35	1.17	0.85
15	1.05	0.72	40	1.17	0.85
20	1.07	0.75	45	1.17	0.85
			50	1.17	0.85

K3 = 1.0 , for fairly uniform terrain, has been used.

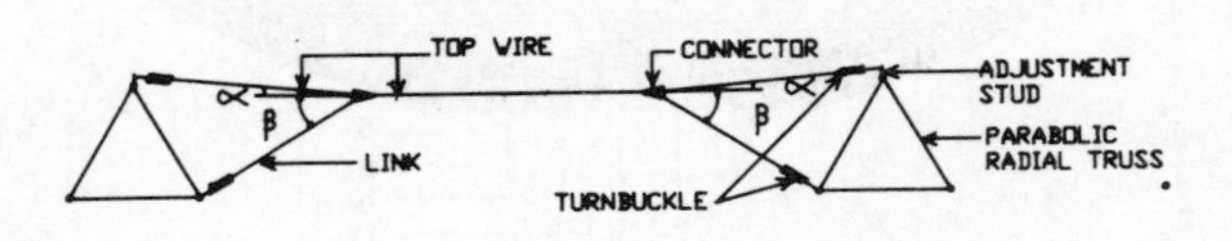

a) Top wire & links

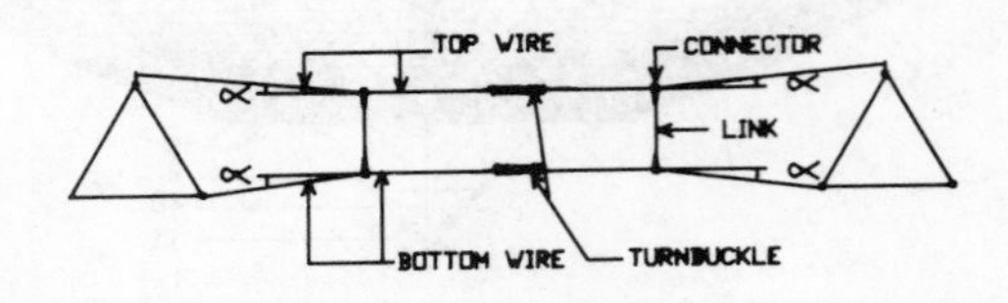

b) Top and bottom wires & links

Fig 4 : Configurations of circumfrential wire rope trusses

WIND PRESSURE

Wind pressure (N / sq m) is obtained from the following expression :

$$pz = .57 * Vz * Vz \qquad (3)$$

Factor of .57 is obtained for an air density of 1.106 Kg / cu. m . This is the air density assumed for GMRT site, having an elevation of 650 m above mean sea level.

WIND LOADS

Wind loads for following structural elements are considered :

- Tubular members & wire ropes
- RC tower
- Yoke & miscellaneous items like
  - Mesh support angles
  - Adjustment stud system
  - Rotating feed cage, feed support frames & feeds
- Wire meshes

Drag Coefficients - Cd

The Cd values used for design are generally according to Ref. 1. Experimental work was done to obtain Cd values for wire meshes.

Tubular members & wire ropes

Cd = 1.2 was used since a majority of members have $Vdz * D < 6$, & wire ropes are of fine stranded type with $Vdz * D < .6$. Vdz is design wind speed in m / sec & D is tube or wire diameter in m .

RC tower

Cd = 0.7 has been used for estimating wind loads on RC tower.

Yoke & miscellaneous items

Cd = 2.0 has been used for estimating forces on these elements.

Wire meshes

Spacings for 0.55 mm diameter wire for the three meshes are given in Table - 2 .

TABLE - 2 : Reflector wire meshes

Spacing	Dish region
20 mm * 20 mm	Outer 1 / 3 area
15 mm * 15 mm	Middle 1 / 3 area
10 mm * 10 mm	Inner 1 / 3 area

Curves 1, 2 & 3 of Fig. 5 show the data available for wire meshes of generally higher solidity ratios. Curve-4 was suggested for Cd values to be used on the following basis:

- Meshes have low solidity ratios; shielding between wires for wind parallel or perpendicular to mesh is therefore negligible.
- $Cd > 1.2$ for wind perpendicular to mesh.
- $Cd > 0.6$ for wind parallel to mesh; one could expect full blockage for vertical wires & negligible blockage for horizontal wires.

Experimental work was done at National Aeronautical Laboratory (NAL), Bangalore, to determine Cd values to be used for design. Two models, one for wind perpendicular to mesh and the other for wind parallel to mesh, were studied. Results of studies [Ref. 5 & 6] are given in Fig. 6. Curve-4 was derived using NAL test results & curve-3 of Fig.5. Cd values of curve-4 in Fig. 6 were used for design.

Lift Coefficients - Cl

Curves 1 & 2 of Fig. 7 show variation of lift coefficient with wind attack angle. Curve-2 was used for design of the GMRT antennas.

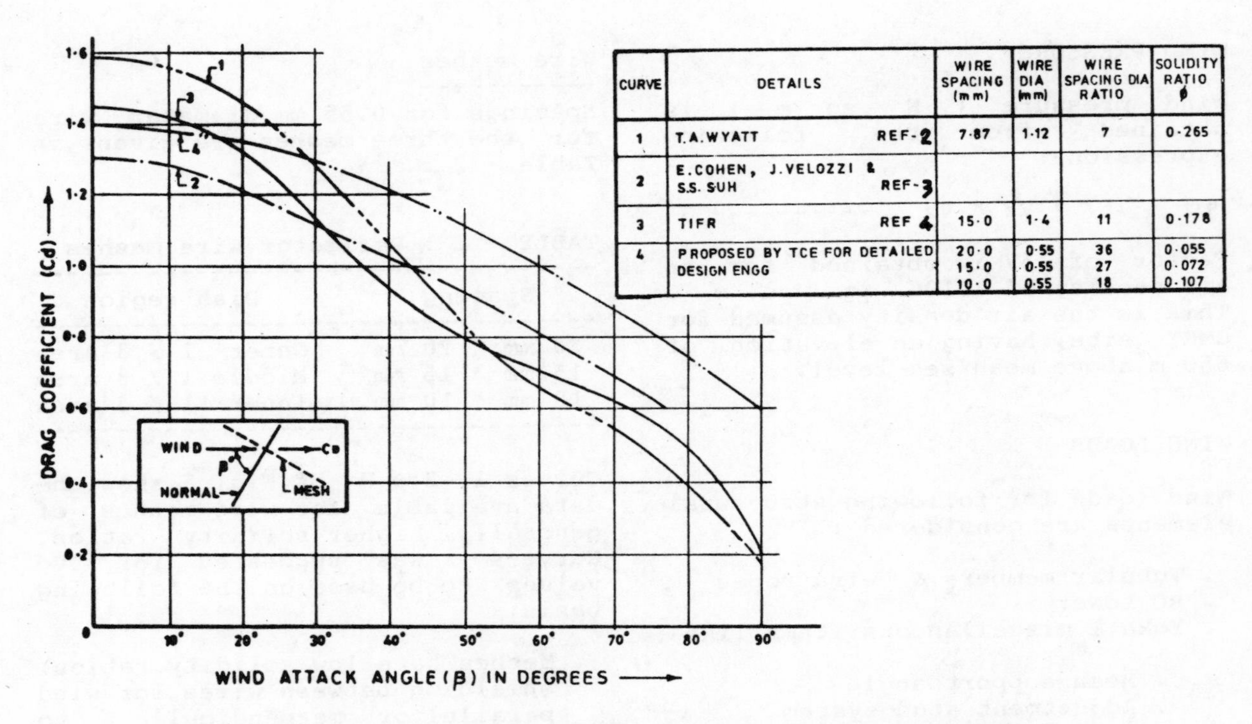

CURVE	DETAILS	WIRE SPACING (mm)	WIRE DIA (mm)	WIRE SPACING DIA RATIO	SOLIDITY RATIO φ
1	T.A.WYATT REF-2	7·87	1·12	7	0·265
2	E.COHEN, J.VELOZZI & S.S. SUH REF-3				
3	TIFR REF 4	15·0	1·4	11	0·178
4	PROPOSED BY TCE FOR DETAILED DESIGN ENGG	20·0 15·0 10·0	0·55 0·55 0·55	36 27 18	0·055 0·072 0·107

Fig 5 : Drag coefficient curves - available data

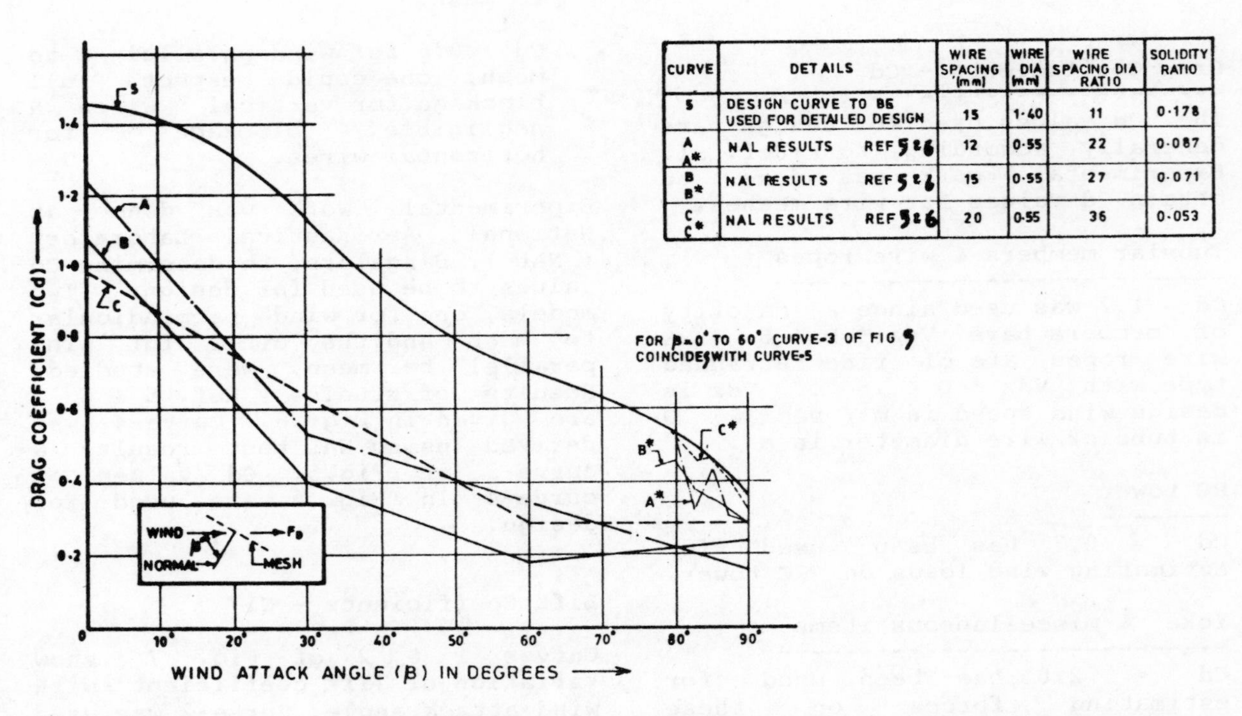

CURVE	DETAILS	WIRE SPACING (mm)	WIRE DIA (mm)	WIRE SPACING DIA RATIO	SOLIDITY RATIO
5	DESIGN CURVE TO BE USED FOR DETAILED DESIGN	15	1·40	11	0·178
A A*	NAL RESULTS REF 5 & 6	12	0·55	22	0·087
B B*	NAL RESULTS REF 5 & 6	15	0·55	27	0·071
C C*	NAL RESULTS REF 5 & 6	20	0·55	36	0·053

Fig 6 : Drag coefficient curves - NAL test results & design curve

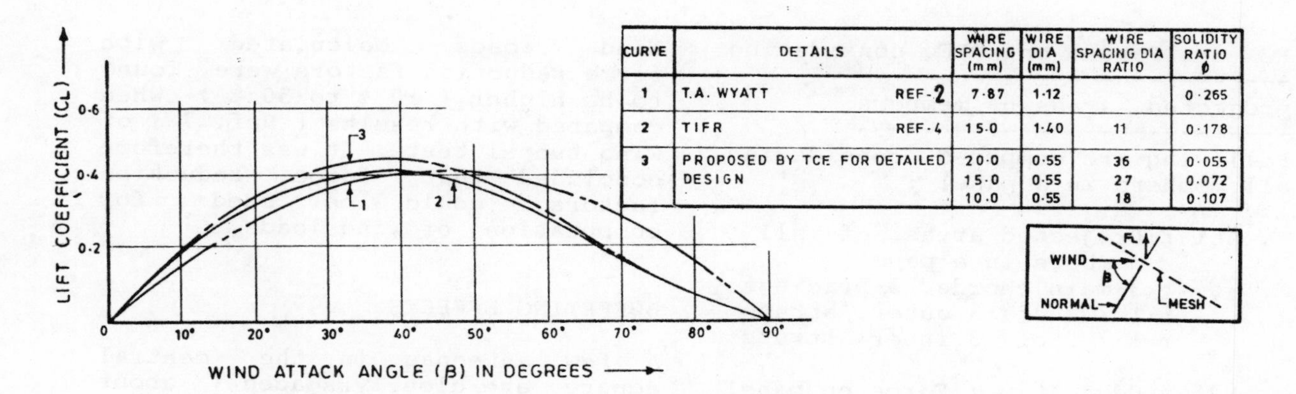

CURVE	DETAILS	WIRE SPACING (mm)	WIRE DIA (mm)	WIRE SPACING DIA RATIO	SOLIDITY RATIO φ
1	T.A. WYATT REF-2	7.87	1.12		0.265
2	TIFR REF-4	15.0	1.40	11	0.178
3	PROPOSED BY TCE FOR DETAILED DESIGN	20.0 15.0 10.0	0.55 0.55 0.55	36 27 18	0.055 0.072 0.107

Fig 7 Lift coefficient curves

SHIELDING EFFECTS

Shielding effects reduce the wind load that can act on a structure. Two types of shielding could be visualised :

- member / member shielding
- frame / frame shielding

Ref. 1 gives force coefficients for vertical equilateral triangular towers composed of round members. This data is condensed in Table-3.

TABLE - 3 : Force coefficients

Solidity ratio (front face)	Force coefficient (Subcritical flow)
0.05	1.8
0.10	1.7
0.20	1.6
0.30	1.5
0.40	1.5
0.50	1.4

Frame / frame shielding effects for the above towers are not available.

Shielding effects (member / member & frame / frame) for dish cannot be considered precisely because PRF's and other triangular frames :

- do not have an equilateral shape
- are not vertical or parallel to each other and have different orientations for different antenna positions

Force Reduction Factors

Wind load on the dish is computed in a convenient manner by considering projected areas of members on a plane perpendicular to the wind direction. Member / member and frame / frame shielding effects are considered approximately through force reduction factors.

Force reduction factor for the front condition is estimated considering a PRF which, incidentally, has the lowest solidity ratio. Force reduction factors for other conditions are estimated using judgement.

F1 - Force on PRF as per Ref. 1

Following are computed for the bottom face of each panel, i , of a PRF :

- Abi : Area blocked by members
  : 2-bottom chords, 1-bracing
  : 1 / 2 of outer strut
  : 1 / 2 of inner strut
- Aei : Overall area of a panel
- Abi / Aei : Solidity ratio
- Cf1i : Force coefficient from Table - 3
- F1i = Cf1i * Abi :Force on panel

F1 is obtained as a sum of forces , F1i, for the 12 panels of the PRF.

F2 - Force on PRF considering projected areas of members

Following are computed considering all members in a panel :

. Ati : Projected areas of all members in a panel
: 3-main chords, 3-bracings
: 1 / 2 of 3 outer struts
: 1 / 2 of 3 inner struts

. F2i = Cd * Ati : Force on panel

F2 is obtained as a sum of forces, F2i , for the 12 panels of the PRF.

Force reduction factor - front wind

Ratio F1 / F2 gives the force reduction factor. F1 / F2 = 0.75 for front wind.

Force reduction factor - side wind and other conditions

Frame / frame shielding would be maximum for the side wind condition. Force reduction factor for side wind condition is assumed to be :

0.9 * 0.75 = 0.68

Force reduction factors for other conditions are assumed to be given by the following expression :

0.68 + 0.09 * Cos A ---------- 4

where, A is the angle between dish axis and the horizontal plane. A is equal to 0 degrees for horizon position of the dish & 90 degrees for zenith (sky) position of the dish.

Comparison With Results of Wind Tunnel Studies

Wind tunnel studies were conducted at University of Roorkee (UOR) to determine wind forces on dish for various dish positions. Model (scale 1 : 50) was fabricated by TIFR .

Wind loads calculated with force reduction factors were found to be higher (10 % to 50 %) when compared with results [Ref. 7] of wind tunnel tests. It was therefore concluded that force reduction factors could be used for computation of wind loads.

BUFFETING EFFECTS

A few antennas in the central square are closely spaced (about 70 m apart) from radio astronomy considerations . It was felt that wind tunnel studies for these antennas would help to assess severity of buffeting effects, if any. These closely spaced antennas will be built after wind tunnel studies are completed. Design would be modified for higher wind loads, if required.

Wind tunnel studies are in progress at UOR to study these effects. Models for these studies were fabricated by TIFR .

PROTOTYPE DISH

Back structure, along with rope truss system and wire mesh reflector surface, was fabricated, assembled and hoisted on to a frame of 6 m height at the Radio Astronomy Centre of TIFR in Udhagamandalam. This prototype structure was built for the purpose of :

- Identifying and sorting out problems associated with fabrication, assembly & erection of the tubular steel welded structure as well as the rope truss cum wire mesh reflector system.
- Observing dynamic behaviour of rope truss cum reflector system under winds of high speeds during the monsoon period.
- Attempting to measure wind loads acting on the dish and correlate them with calculated wind loads.

It was found that the rope truss cum reflector system was stable under the high monsoon winds. NAL has provided instrumentation for measuring wind loads on the prototype dish. Measurements are to be completed .

DYNAMIC ALONG-WIND RESPONSE

Oscillations may be set up in the same direction as the wind by the gusty random nature of the wind. These oscillations are superimposed on the steady response given by the mean hourly wind. The combined effect accounts for the peak dynamic loading on the structure.

Ref. 1 gives the method for computing along-wind response for structures. This method, known as the GF (gust factor) or the GEF (gust effectivness factor) method, has been used to determine whether along-wind response should be considered for the GMRT antennas.

The 1-hour average wind velocity was used to compute the maximum possible value of the gust factor for the antenna structure. Two values of natural frequency, 1.5 Hz (obtained for a preliminary design of the antenna) & 1.0 Hz (lowest antenna structure frequency limit from servo system design considerations), were used to estimate the maximum possible value for the gust factor. An overall structural damping ratio of .01 [Ref. 1] was assumed for these calculations.

The maximum gust factor was found to be sufficiently low to conclude that the dynamic along-wind effects need not be considered.

DYNAMIC CROSS-WIND RESPONSE

Dynamic stress is induced in a round tubular member due to cross-wind oscillatory response (vortex shedding phenomenon) when wind flows over it. The three types of possible oscillatory responses are described below :

Narrow-band Oscillatory Response

This response has the following features :

- Large peak amplitudes, greater than 2.8 % of the tube diameter
- Vortex shedding force on member is amplified
- Response is harmonic in nature. Vortex shedding frequency tends to 'lock-on' to the closest member frequency (nj) over a small range of wind velocities, centred around member critical velocity for nj.
- Occurs when Reynolds number is less than about 300,000

Broad-band Oscillatory Response

This response has the following features :

- Smaller peak amplitudes, less than 2.8 % of the tube diameter
- Random nature of the frequency of induced fluctuating force on member; significant amount of energy is distributed over a large range of wind velocities, centred around member critical velocity Vcr for nj.
- Occurs when Reynolds number is greater than about 300,000

Response Switching

This is related to the gust effect and possibility of oscillations building up to their values within the gust duration. Response switching from narrow-band to broad-band oscillations and vice versa could occur when RMS amplitude of response is around 2 % of the tube diameter.

Computation of Cross-wind Response

ESDU-85038 [Ref. 14] has been

used for determining peak cross-wind deflection response for tubular members of the dish. Response switching effects have been ignored.

Aerodynamic data and boundary conditions used for members influence response considerably. Data and boundary conditions used are explained below.

Aerodynamic data

Some damping ratio values cited in the available literature are given in Table - 4 .

TABLE - 4 : Damping values

Source	Damping ratio	Remarks
Ref. 8	.0032	Steel tubes for telescope structure
Ref. 9	.0050	Steel tubes
Ref.10	<.0050	All-welded structures
Ref.11	.0032	Bolted steel lattice tower
Ref.12	.0030	Steel lattice towers
Ref.13	.0020	Unlined steel stacks
Ref.14	.0030	Steel tubes

Suggested values of damping ratio vary considerably. Underestimation of its effects could lead to an unsafe member / structure design. Tests were conducted by NAL on members of a prototype dish built by TIFR at Udhagamandalam. A damping ratio of .0042 was suggested by NAL for computing cross-wind response of members of the dish, after examining the test results.

Surface roughness of .000005 m, recommended for painted steel tubes [Ref. 13], has been used.

Boundary conditions

Members of the dish structure are welded at their ends. End conditions for members depend on the stiffness of other members meeting at the ends. Flexibility at the end of a member cannot be determined by a simple calculation procedure. It was therefore decided to calculate peak responses for clamped-clamped and clamped-hinged boundary conditions and ensure that the member is safe for both boundary conditions.

Ref. 14 gives calculation information for clamped-clamped & hinged-hinged boundary conditions. Peak response for clamped-hinged condition was obtained by interpolation.

Dynamic Bending Stresses

Dynamic bending stresses are computed using peak deflection response and mode shape characteristics for the corresponding boundary condition.

Fatigue Considerations

Fatigue effect is considered through reduced permissible bending stresses. Reduced permissible stress is obtained by multiplying normal permissible bending stress with a reduction factor. Reduction factors, for 2,000,000 cycles of stress reversals and for various ratios of maximum to minimum stress (for mild steel material conforming to IS:226), are taken from Ref. 15.

Criteria for Design

Member section chosen is checked for cross-wind dynamic bending stress along with associated static stress, considering appropriate reduced and normal permissible stresses. Stresses for a member are examined for the following wind speeds , as found appropriate :

. Vcr : Critical wind velocity at member natural frequency

. Vdz : Site design wind velocity for member

Off-peak dynamic response at Vdz is

determined from the peak dynamic response if Vcr > Vdz . This off-peak dynamic bending stress is considered with the associated static stress at Vdz to check adequacy of the member.

Member is checked for off-peak dynamic bending stress at Vdz, as explained above, when Vcr < Vdz . In addition to this, the member is also checked for the peak dynamic bending stress at Vcr along with the associated static stress at Vcr.

WIND INSTRUMENTATION FOR ANTENNAS

It would be useful to mount instruments on some of the antennas to be built for the GMRT project to obtain wind and structure behaviour data under actual wind conditions at site. A comprehensive testing program could be organised with interaction among scientists from TIFR, experts from educational institutions and experts from the community of practicing engineers.

ACKNOWLEDGEMENTS

The authors thank Tata Consulting Engineers and Tata Institute of Fundamental Research for permission to publish this paper. The authors also thank Indian Institute of Science, National Aeronautical Laboratory & University of Roorkee for their valuable association, inputs and wind tunnel testing work.

REFERENCES

1. Indian Standard Code of Practice for Design Loads (Other Than Earthquake) for Buildings and Structures, Part - 3 : Wind Loads, IS : 875 - 1987 .

2. T.A.Wyatt, Annals of The New York Academy of Sciences, 222-238, Vol 116, 1964 .

3. E.Cohen , J.Vellozzi & S.S.Suh , Annals of the New York Academy of Sciences , 161 - 209 , Vol 116, 1964 .

4. G. Swarup, Internal Communication TFR(BNG):162(11-1):TCE DES: :87, Dtd 1987-11-02, With Curves for Drag & Lift Coefficients for Wire Mesh Screens.

5. Wind Tunnel Testing of Screen Elements Used for a Radio Telescope Antenna (GMRT), Project Document FM 8723, National Aeronautical Laboratory, Bangalore, October 1987 .

6. Wind Loads on Sparse Meshes Used in Antennae of Large Telescopes , Project Document FM 8815, National Aeronautical Laboratory, Bangalore, Septembert 1988 .

7. Wind Tunnel Studies on 45 m Diameter Antenna Dish, University of Roorkee, February 1989 .

8. A 300 Foot High-precision Radio Telescope, Report No. 24, Appendix - 1, National Radio Astronomy Observatory, Green Bank, West virginia, USA, May 1969 .

9. Dansk Standard DS 410.2 , Dansk Ingeniorforenings Norm for Last Pa Bacrende Konstruktioner - 2 Vindlast, October 1977 .

10. Harris & Crede, Shock & Vibration Handbook , Mc Graw Hill Co., New York, 29-38, 2 nd Edition, 1976 .

11. Peter Sachs, Wind Forces in Engineering , Pergammon Press, Table 8.4, 2 nd Edition , 1978

12. The Response of Flexible Structures to Atmospheric Turbulence , Engineering Sciences Data Unit , 251-259 , Regent Street , London, W1R 7AD, ESDU - 76001, September 1976 .

13. Across-flow Response Due to Vortex Shedding : Isolated Circular Cylindrical Structures in Wind or Gas Flows, Engineering Sciences Data Unit , ESDU - 78006 , October 1978 .

14. Circular Cylindrical Structure : Dynamic Response to Vortex Shedding , Part - 1 : Calculation Procedures & Derivation , Engineering Sciences Data Unit , ESDU - 85038 , December 1985 .

15. Indian Standard Code of Practice for Steel Bridges , IS : 1915 - 1961.

TECHNICAL NOTES

BOUNDARY LAYER MODELLING FOR STUDYING WIND PRESSURE DISTRIBUTION ON LOW RISE BUILDINGS IN BUILT-UP AREAS

Ishwar Chand

Central Building Research Institute,
Roorkee, INDIA

Data on wind pressure distribution on low rise buildings in built-up areas have dual application, viz., (1) for estimation of wind load on building envelop and (2) for assessment of the magnitude of the aeromotive force causing natural ventilation in buildings. Hence accuracy and reliability of this data need due consideration in context with structural as well as functional design of buildings. Most of the available data on wind pressure distribution on buildings and structures have been generated through model studies in wind tunnels. Majority of these investigations were carried out in the wind streams having velocity gradients represented by power law with exponents having values 1/7, 1/4.5 and 1/3 for different types of terrain representing open country sites, sub-urban areas and heart of urban areas respectively. It has been demonstrated that the pressure distribution is significantly affected due to a change in the velocity gradient in the on blowing wind. Since the power law holds good at high elevations only, the existing data related to the tall and isolated structures can be confidently used in the design of high rise buildings. However, low rise buildings in built-up areas are exposed to wind fields whose characteristics are not yet fully known. Some of our measurements have also shown that the wind velocity gradient in this region does not obey the well known power law. The large variations in the planning and layout of buildings result in the wind flow patterns and velocity gradient therein which are too complex to be represented by any simple relationship. As such the application of the available data on wind pressure distribution will be erroneous in case of low rise buildings located in built-up areas. Hence to evolve realistic data, there is a need to develop experimental facilities with the capability of simulating the wind flow characteristics existing in the lower region of earth's boundary layer enveloping low rise buildings. In this context following aspects are worth considering:

(1) All the existing data on wind pressure distribution over buildings should be collected and feasibility of its application in the design of low rise buildings should be explored. It implies identification of situations in which the said data could be used without inviting any risk to the safety of the structure and to the well being of the occupants.

(2) Extensive studies on exploration of the characteristics of wind in the boundary layer region below the standard anemometric height of 10 metre should be carried out over various types of terrain. The findings should form a basis for evolution of flow simulation criterion for model studies in wind tunnels.

(3) Experimental techniques should be developed to simulate the flow with the aforesaid characteristics in the test sections of the existing wind tunnels. This amounts to adding to the capabilities of the existing facilities in use for studying problems on interaction of wind with buildings and structures.

(4) Some basic configurations (or modules) of plan forms should be evolved such that variety of plan forms may be deduced through various combinations of these configurations. Data on wind pressure distribution on each of these isolated units as also on their different combinations should be collected.

(5) Large number of case studies covering variety of plan forms should be carried out and bulk of data so collected should be made use of for validation test of laboratory data.

(6) Data from all possible sources should be pooled in and a data bank be created. It should be formatted in a way so that it may lead to some general design guidelines and may also help, through interpolotion and extrapolation of data, in prediction of wind pressure distribution for any desired plan form.

Wind Loads on Structures, 1990

SIMULATION OF THE SUB-URBAN NEUTRAL ATMOSPHERIC BOUNDARY LAYER IN A WIND TUNNEL

S. Arunachalam

Structural Engineering Research Centre
Madras, INDIA

Construction of tall buildings and other structures such as telecommunication towers, chimneys, suspension bridges, in recent years in India has generated a growing sensitivity of the design profession to the need for better understanding of the dynamic wind effects on such structures through wind tunnel testing on models. In order to investigate such studies in a wind tunnel, it is essential to simulate sufficiently deep boundary layers whose properties are representative of the atmospheric boundary layer (ABL) in nature. Although simulations of strong winds at laboratory level have been successfully accomplished in long test section wind tunnels by developing the required boundary layer over roughness elements, space and cost constraints led various researchers to develop methods to artificially thicken the boundary layer in short test section wind tunnels. Among such different simulation techniques studied, the methods suggested by Counihan (1) and by Standen (2) are widely referred to in the literature and are employed in many boundary layer wind tunnels. The system proposed by Standen consists of an array of triangular shaped spires, with or without splitter plates followed by floor roughness elements. The spire type of vortex generators are relatively simple from the fabrication point of view. This note briefly presents the details of the flow thickening devices, velocity measurements, evaluation and comparison of the results of an experimental study on simulation of the ABL corresponding to a sub-urban or wooded rough terrain type, to a model scale of 1:750, using Standen type of spires and roughness elements, carried out at the wind tunnel laboratory of the Ruhr Universitat, Bochum, West Germany, during the author's stay there as a DAAD Scholar.

The wind tunnel is of an open-circuit type and the size of the test section is 1.8 m (W) x 1.6 m (H) x 9.48 m (L). (Fig.1) The dimensions and spacing of the triangular shaped vortex generators with a splitter plate attached to them, and of the cubical roughness elements, used in this experiment, were selected according to the simple guidelines given by Irwin (3). The dimensions of the spires are as shown in Fig.1 and they were made of plywood material, 10 mm thick and were fastened to the tunnel floor with a 10 mm thick base plate and 12 mm dia. bolts at the four corners.

Based on the empirical equations given by Irwin, one finds that cubic roughness elements of 36 mm size can be selected with a spacing of 16.9 cm c/c (both spanwise and streamwise) to simulate boundary layer of a terrain type characterised by $\alpha = 0.28$ and $\delta = 75$ cm in the wind tunnel. However, in view of time constraint, the following three types of wooden block roughness elements, which were readily available were used in this investigation.

a) 36 mm x 36 mm x 36 mm and 36 mm x 36 mm x 18 mm blocks placed staggered 14.8 cm c/c in alternate rows.

b) 25 cm cubes placed at 7.5 cm both ways;and

c) 20 mm x 20 mm x 10 mm blocks placed at 9.0 cm c/c also both ways.

Hence it was expected that the influence of the roughness elements on α and on the turbulence intensity values would be less to a certain extent than it would be with a complete set of 36 mm size cubes as dictated by Irwin equations.

Measurements of the flow velocity were made only in the alongwind direction using a single hot-wire anemometer. Although the velocity measurements were mainly concentrated at the centre of the turn-table, where the model will be positioned, measurements were also taken at other locations in order to evaluate the uniformity of the flow, both streamwise and spanwise, in the wind tunnel. The data were analysed to obtain the mean velocity and turbulence intensity profiles, auto-correlation and power spectrum functions and the turbulence length scale, Lux(z) values.

The mean velocity profile obtained at the centre of the turn table is shown in Fig.2. The value of α obtained from the present simulation is in good agreement with Counihan(5) reported values of $0.18 \leq \alpha \leq 0.26$ and with $\alpha = 0.25$ reported by Teunissen (6) for similar sub-urban terrains. Additional comparison of the characteristics of mean velocity profiles are presented in Table 1. Comparison of the turbulence intensity profile obtained from present study with other available data indicated that while the trend of the variation of I(z) with height is in general agreement with the atmospheric data, the intensity of turbulence produced in the experiment is relatively low, especially at lower heights as expected earlier, due to adoption of less rough elements which dominate in the development of the required boundary layer in the lower regions.

By matching the horizontal wind spectrum obtained from this experiment with Karman spectrum, the length scale factor of 1:750 was computed and Fig.3 shows the resonably good agreement of these values of length scales of turbulence with other full-scale atmospheric data (4).

References:

1. Counihan, J. An Improved Method of Simulating an Atmospheric Boundary Layer in a Wind Tunnel. Atmospheric Environment, (3): - 197-214, 1969.

2. Standen N.M. A Spire Array For Generating Thick Turbulent Shear Layers for Natural Wind Simulation in Wind Tunnels. T.R.LTR-LA-94, National Aeronautical Establishment, Ottawa, 1972.

3. Irwin, N.P.A.H. The Design of Spires For Wind Simulation. Jl. Wind Engg. and Industrial Aerodyn., (7): 361-366, 1981.

4. Arunachalam, S. Laboratory Simulation of the Sub-urban Neutral Atmospheric Boundary Layer in a Wind Tunnel. Technical Report, Ruhr University, Bochum, W. Germany,1988.

5. Counihan, J. Adiabatic Atmospheric Boundary Layer : A Review and Analysis of Data from the Period 1880-1972. Atmospheric Environment (5): 299-311, 1975.
6. Teunissen, H.W. Structure of Mean Winds and Turbulence in the Planetary Boundary Layer over Rural Terrain. Boundary Layer Meteorology (9): 187-221, 1990.

Table 1. Characteristic Properties of Mean Velocity Profiles (4)

Experiment	Power law exponent, α	Zo/ δ	Zd/ δ	U*/Uref
Robins (1979) Flow 'C' (14)	0.2 to 0.25	0.0022	0.01	0.065
Present investigation	0.24	0.0027	0.016	0.065

U* = Shear friction velocity

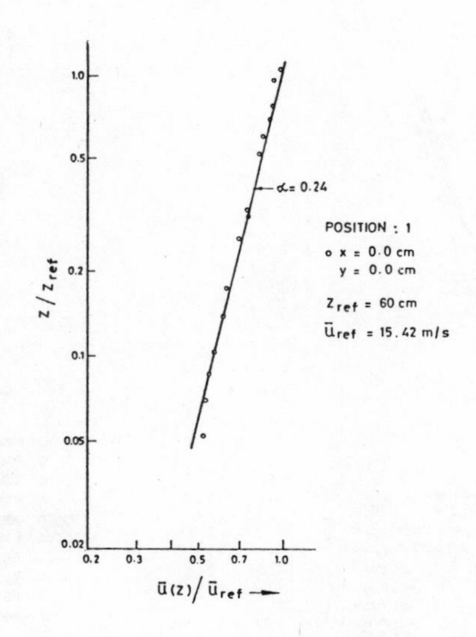

FIG. 1 MEAN VELOCITY PROFILE AT THE CENTRE OF THE TURN-TABLE BASED ON POWER LAW

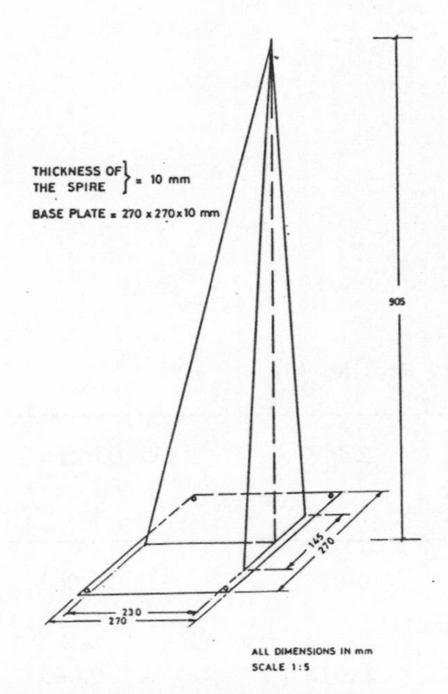

FIG. 1b TRIANGULAR SPIRE WITH TRIANGULAR SPLITTER PLATE

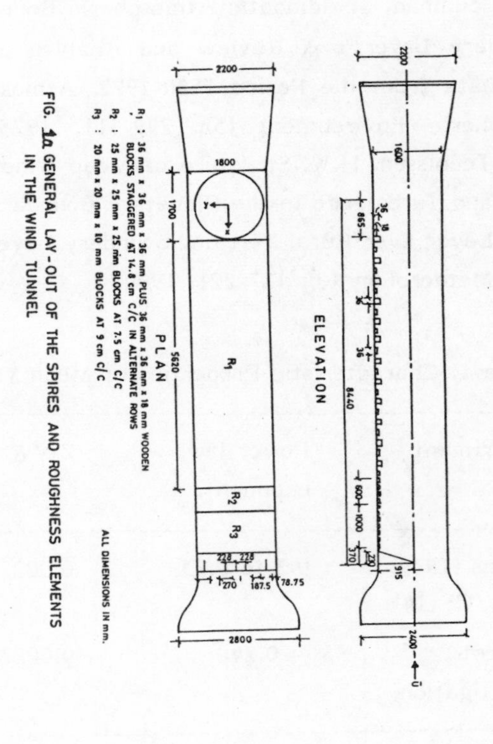

FIG. 1a GENERAL LAY-OUT OF THE SPIRES AND ROUGHNESS ELEMENTS IN THE WIND TUNNEL

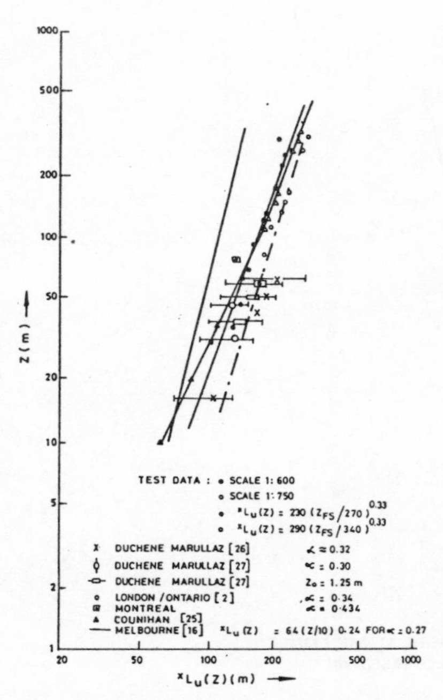

FIG. 3a COMPARISON OF $^xL_u(Z)$ VALUES BETWEEN TEST DATA AND ATMOSPHERIC DATA AVAILABLE IN LITERATURE [1]

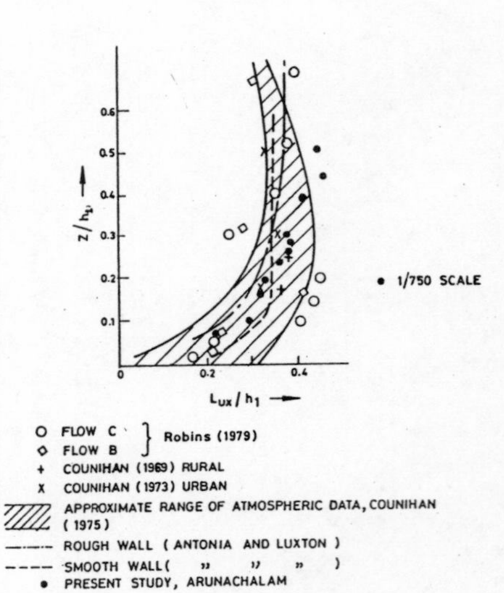

FIG. 3b COMPARISON OF $^xL_u(Z)$ VALUES BETWEEN TEST DATA AND ATMOSPHERIC DATA FROM LITERATURE [1]

Wind Loads on Structures, 1990

SELF EXCITED OSCILLATIONS OF A SQUARE PRISM IN FLUCTUATING FLOW

M. Sathish Ram Kumar *K. Jayaraman* *K. Padmanaban*

Department of Aeronautical Engineering
M.I.T. Campus
Anna University
Madras, INDIA

A quasi steady analysis of the transverse galloping of an infinitely long square prism in a normal steady wind has been carried out by Parkinson & Smith. It involves developing a model for the aerodynamic force, formulating the equation of motion for response of the prism and examining the stability by perturbations about the static equilibrium position. The spring supported model exposed to a steady flow velocity u, spring stiffness k, structural damping r and a mass m per unit length is shown in Fig. 1. The equation of motion for such a system has been written as

$$m\ddot{y} + r\dot{y} + ky = \tfrac{1}{2} C_{F_y} \rho V^2 h l$$

The aerodynamic force coefficient C_{F_y} from measured values on stationary cylinders has been assumed as

$$C_{F_y} = A(\dot{y}/V) - B(\dot{y}/V)^3 + C(\dot{y}/V)^5 - D(\dot{y}/V)^7$$

Figure 2 is a plot of a stationary amplitude ($\bar{y}_s$) against the non-dimensional wind speed (u) exhibiting oscillation hysteresis in region II and below $u < u_1$ (region I) the response is static. At u = 2.82 the system forks with a phase portrait as in Fig. 3. In the region II where $u_1 < u < u_2$ three solutions exist in which the middle one will be the unstable solution. The unstable limit cycle in the range $u_1 < u < u_2$ discriminates between the stationary oscillations on the basis of the initial amplitude. The lower stationary amplitude corresponds to low initial conditions. In the region III (u > 5.26) there is only the asymptotic solution (Fig. 6).

For low initial conditions it is noticed that the static solution exists right through in addition to the stationary oscillations reported by Parkinson & Smith.

In the present study fluctuating flow velocity is considered since atmospheric winds seldom blow steady. A harmonic fluctuating flow of about 2% variation in magnitude with a fluctuating frequency of $\omega = 5$ rad/sec is assumed in velocity leading to

$$u = u_o (1 + 0.02 \sin \omega T)$$

It is found that the system oscillates in this case even in the static solution region I of steady wind but with a frequency ω_n different from the fluctuating frequency (Fig. 7).

In the region II if the initial amplitude is less then $\overline{y}_{s_2}$ the system settles down at $\overline{y}_{s_1}$ with the frequency ω_n as shown in Fig. 8, and for the initial amplitude higher than $\overline{y}_{s_2}$ the stationary amplitude is $\overline{y}_{s_3}$ with the same frequency as shown in Fig. 9. The response frequencies were identified by the FFT of the corresponding time histories.

In the region III the amplitude of oscillations for the system depends upon the initial conditions since the system shows a tendency to settle at a higher amplitude (Fig. 10). It is also noticed that, unlike the steady flow case, in fluctuating flow the system oscillates right from the small wind speeds.

REFERENCE

PARKINSON, G.V. and SMITH, J.D. The Square Prism as an Aeroelastic Non-linear Oscillator, Quart. Journ. Mech. and Applied Math., Vol. XVII, Pt. 2, 1964, pp. 225-239.

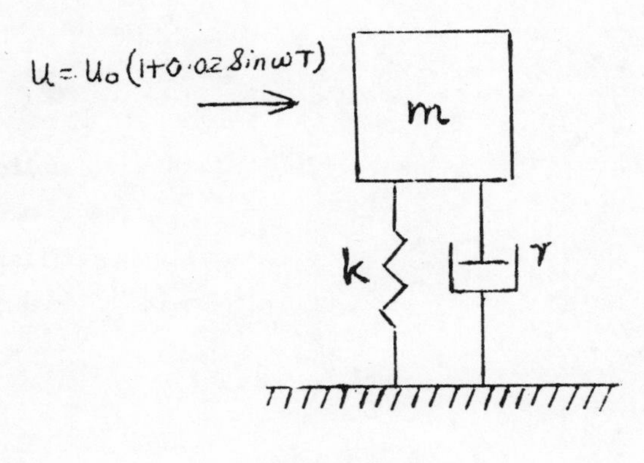

FIG.1

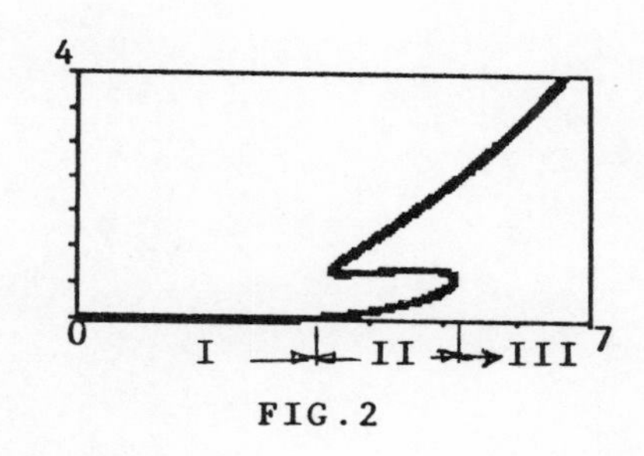

FIG.2

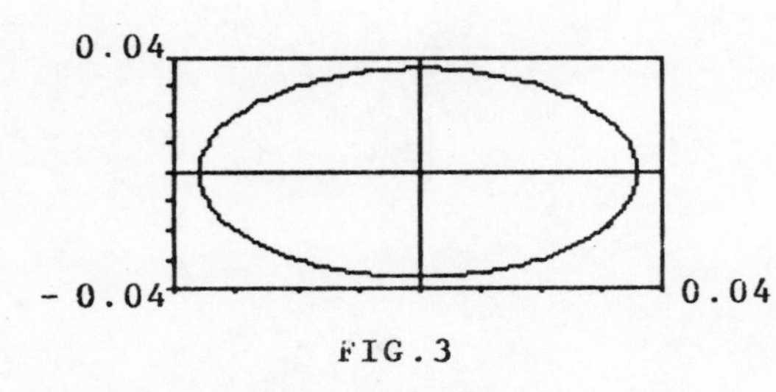

FIG.3

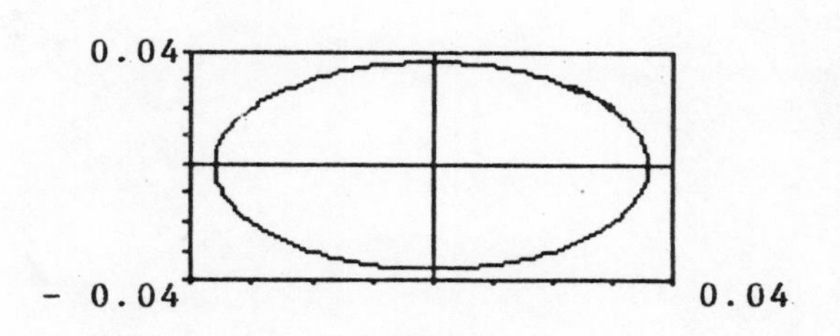

FIG 4

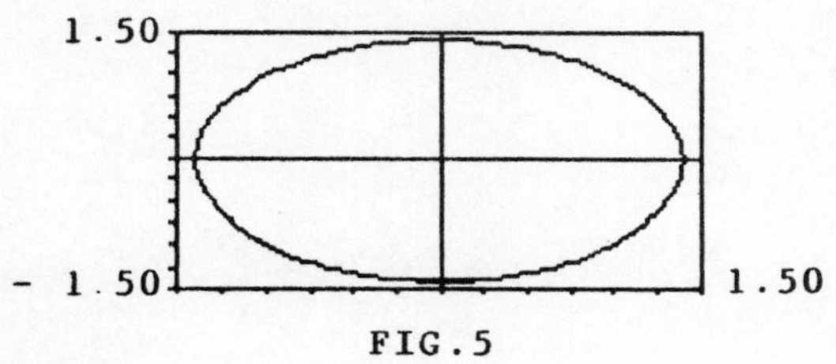

FIG.5

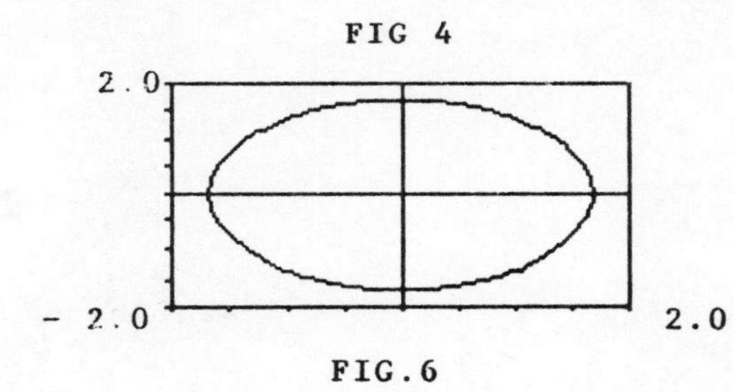

FIG.6

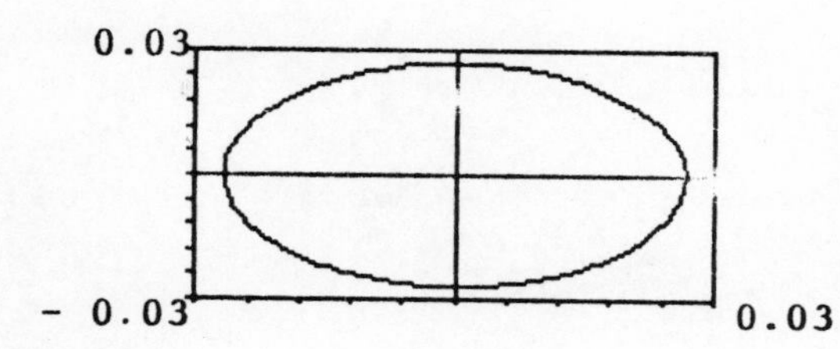

FIG.7

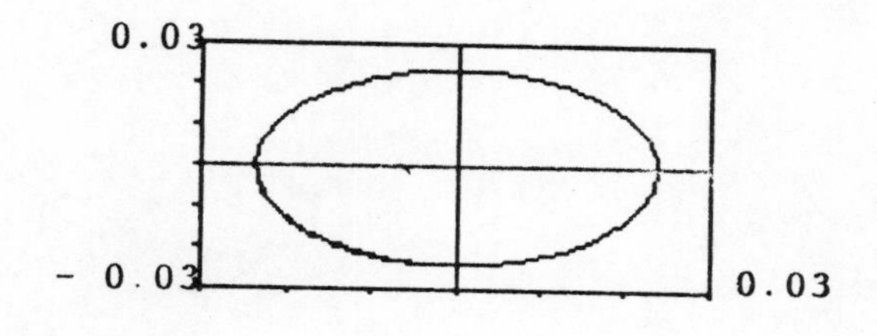

FIG.8

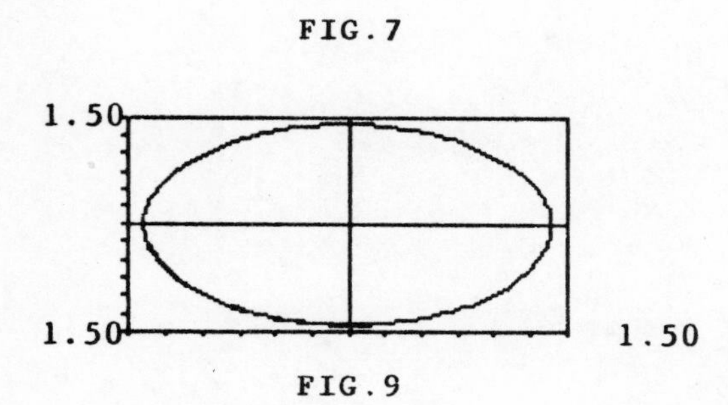

FIG.9

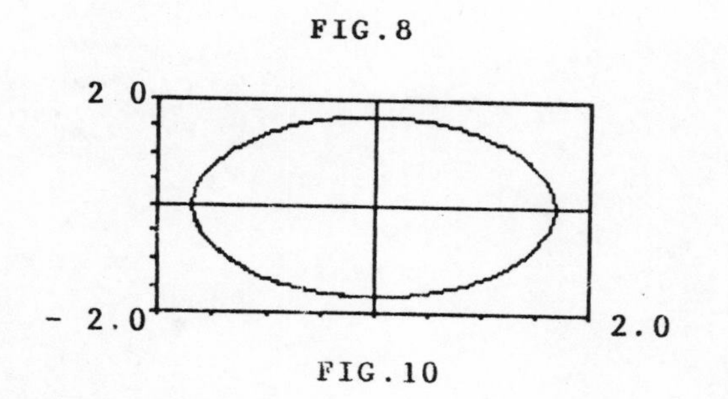

FIG.10

MODEL STUDY ON A 275 m TALL CHIMNEY

N. Nagaraja

Tata Consulting Engineers
Bangalore, INDIA

INTRODUCTION

Experimental analysis of wind effects on tall slender chimneys using aeroelastic models in wind tunnels is an effective tool to determine design wind forces and ascertain interference effect of other structures in close proximity of the chimney.

A 275 m high 7.6 m top diameter single flue RCC chimney forms a part of the 500 MW Sixth Unit of Thermal Power Plant at Trombay, Bombay, put up by Tata Electric Companies (TEC). Tata Consulting Engineers (TCE) were the Consultants. At the request of TEC/TCE, Prof. Melbourne of Australia conducted wind tunnel studies on aeroelastic models of the chimney. The main objective of the studies was to determine experimentally the wind induced moments and shears and compare them with the analytical values. As the proposed chimney was in close proximity of another 152 m high chimney of the earlier unit, as well as of station building and boiler structures, the interference effect of these structures were also studied.

BASIC WIND DATA

From the draft of IS Code available at that time, basic wind speed for the region is 160 km per hr. or 44.4 m/sec. Category two was chosen as most appropriate for the topography of the region. As the region is close to the sea coast, a coastal factor of 1.07 was considered. The reference height was chosen at 229 m corresponding to 5/6 height of the chimney. With these parameters, the mean hourly wind speed at reference height was derived as 49.8 m/sec.

MODEL CHIMNEYS

The aeroelastic models were made for a length scale of 1/150 and a velocity ratio of 1.22. The scales were so chosen that the Reynolds number for the alongwind is greater than 200,000 and that of crosswind greater than 80,000. The models were of welded steel sheet of variable thickness. Mass elements were added to the models to bring up the appropriate mass distribution. Three models of the 275 m chimney were made to correspond with different stages of construction. The first one corresponded to 152 m height equal to the height of the earlier chimney. The second corresponded to such a height that the first mode frequencies of the two chimneys are equal. The third corresponded to the full height of 275 m.

EXPERIMENTAL TECHNIQUE

The models were mounted on a massive support. Damping pads of various thicknesses were clamped at the base to facilitate variation of the structural damping. Analysis was carried out for two values of damping. One of the values was set to be as close to 0.01 as possible.

In the wind tunnel, a shear flow model of the natural wind was developed by using a grid. The test procedure consisted of measuring wind induced moments/shears at the base by means of strain gauge bridges fixed in two perpendicular directions. Mean standard deviation and peak values of both alongwind and crosswind effects were measured. Measurements were done for wind in various directions on individual models as well as on several combinations of the models to study interference effect.

RESULTS OF THE EXPERIMETAL ANALYSIS

The experimental data were corrected for blockage effect to account for the wall constraint of the wind tunnel and damping factor. The mean standard deviation and peak base overturning moments were plotted against reduced velocity $V_r = V/nd$, where, V_r is the mean wind speed at reference height, n is the first mode frequency and d is the outer diameter of chimney at 5/6 height.

CONCLUSIONS

The effect of appurtenances such as caged ladder, lift supporting frame etc. is normally not considered significant and is rarely taken into consideration in design. The results brought out considerable amplification of wind moments due to the effect of the caged ladder. Taking cognition of amplification of the moments, the cage of the ladder was deleted.

No significant interference on the earlier 152 m chimney due to 275 m tall chimney at the intermediate heights of construction were observed. Similarly, the interference effect due to the presence of the 152 m high chimney on the 275 m chimney was also not significant. No significant interference effect due to station building and boiler structures on the 275 m chimney was observed either.

The experimental analysis gave a clear

spectrum of the interference effects thereby dispelling all apprehensions of possible amplifications of wind induced forces due to close proximity of the structures. The analytical analysis was validated by the experimental results.

FULL-SCALE MEASUREMENTS OF WIND SPEEDS IN THE ENVIRONS OF STRUCTURAL ENGINEERING RESEARCH CENTRE, MADRAS

S. Arunachalam *M. Arumugam* *J. Shanmugasundaram*

Structural Engineering Research Centre
Madras, India

Safe and economical design of buildings and structures against wind loading requires assessment of the maximum loads likely to be experienced during the expected lifetime of the structure. Knowledge of different characteristics of the wind at the site such as wind speed and direction, variation of wind speed with height and terrain roughness, and turbulence intensity of the site, is essential for the realistic estimation of wind loads on structures. In the last three decades, many full-scale measurements have been made of the wind and turbulence on structures in the atmospheric boundary layer. Based on these investigations, terrains were classified into different categories and representative values for the power law index, roughness length etc. for these terrains have been recommended. Mathematical models, with empirical or semi-empirical values were also proposed to compute the power spectra of different components of wind speed, turbulence length scales etc. However, in India only limited number of field measurements on collection of wind speed data have been carried out. The need to conduct full-scale studies on tall towers and structures to collect wind data and dynamic response of structures has been emphasised in the recent Code of Practice IS:875 (1987)-Part 3 (1) on wind loading. The above point has also been stressed during the deliberations of many national and international workshops and conferences on Wind Engineering.

Keeping the above objective in view, a 3-cup anemometer was installed on the top of the Fatigue Testing Laboratory of Structural Engineering Research Centre, Madras at an elevation of 19.78 m from ground level to collect wind data and to study the wind characteristics, type of terrain, and power spectrum for the horizontal component of the wind.

The wind speeds were collected in the form of voltage signals using a 3-cup (M/s.Wilh-Lambrecht) anemometer for a period of six months, and only selective data recorded on days when sufficiently strong winds (> 8 m/s) were prevailing, were considered for analysis. The data were acquired through a portable six channel casette recorder. The analog voltage values were converted into digital values using a A/D converter with sampling rates varying between 0.1 sec to 0.4 sec and the total length of record was about 15 minutes. The digitised voltage values were converted into digitised wind speed values using the calibration curve supplied by the manufacturer.

Using the software 'STAR' (STatistical Analysis of Random data) developed at SERC, Madras, the spectral analysis for four different runs of the field data was carried out and the values of mean wind characteristics are given in Table 1.

The above results of our study compare well with field-measured values for an open terrain reported in the literature, in general, and in particular, by Davenport and Teunissen (2,3). It is also seen that the roughness length, Zo value of 3.86 cm reported by the authors compares very well with the value recommended for an open terrain ranging from 1.5 cm to 6.0 cm. Further, it may be noted that the recommended average value of Zo for an open terrain, equal to 2.0 cm, specified implicitely in the recent code IS:875 (Part 3)-1987 compares well with the field experimental value. Besides, the value of K_{10} = 0.0051 obtained from the field data shows an excellent comparison with the value of 0.005 suggested by Davenport. The average power law index, α = 0.136, obtained from this study for the terrain at SERC, Madras, is comparable with α = 0.12 to α = 0.16, recommended by Anna Mani and Mooley (4) for the Meenambakkam Airport Station, Madras, whose terrain conditions are similar to that under study.

The power spectra of horizontal component of wind obtained from the field data for different runs have been compared with spectrum curves suggested by Davenport (5), Harris (5), Simiu (5), Kaimal (3), Teunissen-Kaimal (3) and Davenport-Berman (6). The power spectrum for the data RUN 1 is plotted in Fig.1 and compared with spectra proposed by the above researchers. The comparison indicates that no single spectrum represents the field measured spectra satisfactorily for the whole range of frequencies. However, the spectrum equation given by Simiu is found to predict values closer to the spectrum obtained from the present field data.

References:

1. Indian Standards IS 875 (Part 3)-1987. Indian Standard Code of Practice for Design Loads (other than Earthquake) for Buildings and Structures, Part 3, Wind Loads. Bureau of Indian Standards, New Delhi.
2. Davenport, A.G. Wind Structure and Wind Climate, Safety of Structures under Dynamic Loading. 1, Edited by Ivar Holand, Tapir Publishers, 1978.
3. Teunissen, H.W. Structure of Mean winds and Turbulence in the Planetary Boundary Layer Over Rural Boundary Layer. Boundary Layer Meteorology, 19, 1980.
4. Anna Mani and D.A. Mooley. Wind energy data for India. Allied Publishers, New Delhi, 1983.
5. Simiu E. and R.H.Scanlan. Wind Effects on Structures. John Wiley & Sons, New York, 1978.
6. Duchene-Marullaz, P. Full-Scale Measurements of the Atmospheric Turbulence in a Sub-urban Area. Proc. 4th Int. Conf. on Wind Effects on Structures,UK,1975.

Table 1. Surface and Mean Wind Characteristics of the Terrain under study

Sl. No.	Data	No. of Samples N	Sampling rate Δt (sec)	$T = N.\Delta t$ (min.)	VBAR (m/s)	SIGV (m/s)	$I = \frac{SIGV}{VBAR}$	VFRC (m/s)	Zo (cm)	α	$K_{10} \times 10^3$
1	RUN 1	9550	0.1	15.92	9.3026	1.4362	0.1544	0.5862	3.46	0.1339	4.98
2	RUN 2	4740	0.2	15.8	9.5199	1.4766	0.1551	0.6027	3.56	0.1346	5.04
3	RUN 3	3820	0.25	15.92	8.9348	1.3271	0.1485	0.5417	2.70	0.1288	4.57
4	RUN 4	2380	0.4	15.9	8.8443	1.4828	0.1677	0.6052	5.72	0.1454	6.00
Average					9.1504	1.4307	0.1564	0.584	3.86	0.1357	5.148

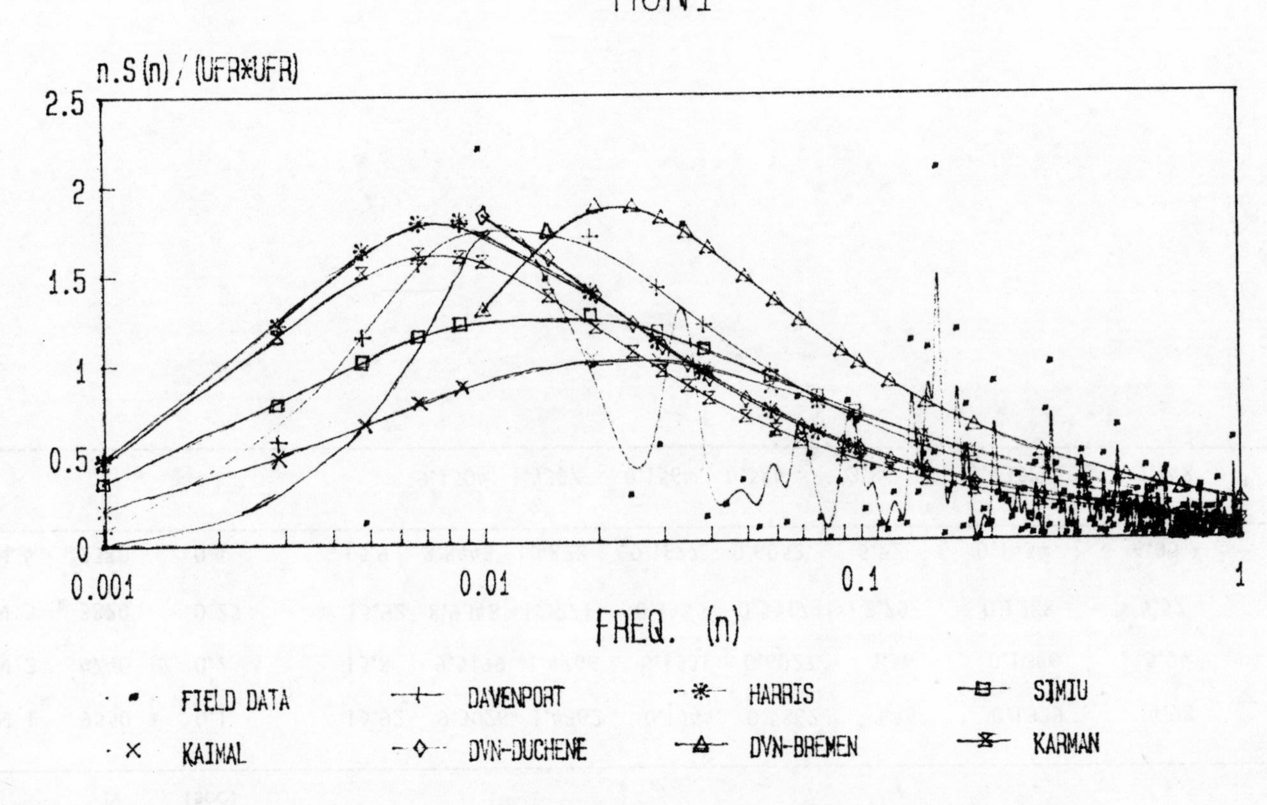

Fig.1. COMPARISON OF ALONGWIND POWER SPECTRUM OF WIND FOR "RUN1" WITH STANDARD SPECTRA IN LITERATURE.

INDEX OF AUTHORS